Oxford Lecture Series in
Mathematics and its Applications 5

Series editors
John Ball   Dominic Welsh

OXFORD LECTURE SERIES IN MATHEMATICS AND ITS APPLICATIONS

1. J. C. Baez (ed.): *Knots and quantum gravity*
2. I. Fonseca and W. Gangbo: *Degree theory in analysis and applications*
3. P. L. Lions: *Mathematical topics in fluid mechanics, Vol. 1: Incompressible models*
4. J. E. Beasley (ed.): *Advances in linear and integer programming*
5. L. W. Beineke and R. J. Wilson (eds): *Graph connections: Relationships between graph theory and other areas of mathematics*

Graph Connections

Relationships between Graph Theory and other Areas of Mathematics

Edited by

Lowell W. Beineke
*Indiana University–Purdue University
Fort Wayne, Indiana, USA*

and

Robin J. Wilson
*The Open University
Milton Keynes, UK*

CLARENDON PRESS · OXFORD

This book has been printed digitally and produced in a standard specification in order to ensure its continuing availability

Great Clarendon Street, Oxford OX2 6DP
Oxford University Press is a department of the University of Oxford.
It furthers the University's objective of excellence in research, scholarship,
and education by publishing worldwide in
Oxford New York
Auckland   Cape Town Dar es Salaam Hong Kong   Karachi
Kuala Lumpur Madrid Melbourne Mexico City   Nairobi
New Delhi Shanghai Taipei Toronto
With offices in
Argentina Austria Brazil Chile Czech Republic France Greece
Guatemala Hungary Italy Japan South Korea Poland Portugal
Singapore Switzerland Thailand Turkey Ukraine Vietnam

Oxford is a registered trade mark of Oxford University Press
in the UK and in certain other countries

Published in the United States
by Oxford University Press Inc., New York

© Oxford University Press, 1997

The moral rights of the author have been asserted

Database right Oxford University Press (maker)

Reprinted 2011

All rights reserved. No part of this publication may be reproduced,
stored in a retrieval system, or transmitted, in any form or by any means,
without the prior permission in writing of Oxford University Press,
or as expressly permitted by law, or under terms agreed with the appropriate
reprographics rights organization. Enquiries concerning reproduction
outside the scope of the above should be sent to the Rights Department,
Oxford University Press, at the address above

You must not circulate this book in any other binding or cover
And you must impose this same condition on any acquirer

ISBN 978-0-19-851497-8

Preface

In a mathematical world of increasing specialization, it is important not to lose sight of how different fields are related to one another, and of how various ideas fit together. Those are the premises on which this book was conceived, as they apply to graph theory. On one level, graphs appear throughout mathematics (indeed, throughout life), since a graph is just a model of a relation. However, that ubiquity may obscure the deeper connections that have been developed between graph theory and other branches of mathematics. The purpose of this book is to present examples of these connections.

The connections are not all of the same kind. Some form a body of material that overlaps two fields, whereas others consist primarily of applications of one area to another. The applications may be to graph theory or from it. Whatever the form of the connections, light is shed on both areas, and that is in itself an excellent reason for examining the connections.

Although we do not claim to cover all of the areas of mathematics which have connections with graph theory, we believe that this collection contains most of the important ones and that there is sufficient diversity in these to illustrate their wide variety.

Wherever feasible, uniform notation and terminology are used throughout the book. Much of this, as well as some other relevant background material, is provided in an introductory chapter. Otherwise, the individual chapters are independent, except for an occasional cross-reference.

The origin of the book was a highly successful one-day conference sponsored by the British Combinatorial Committee and held at the Open University in Milton Keynes in 1994. Several of the chapters are based on talks given there; additional topics were added in order to present a wider range.

That conference was open to all, but was designed primarily for graduate students to learn about 'graph theory across the field of mathematics'. This concept has been carried over to the book; it is a resource for learning about how graph theory interacts with other branches of mathematics. As such, it can function as the basis of a graduate-level seminar, or can be used by individuals or groups interested in particular topics.

Acknowledgements

We are grateful to the authors of the various chapters for their willingness to share their expertise and for their cooperation with our efforts to make the book more than a collection of individual essays.

We are also indebted to the British Combinatorial Committee for supporting the conference that gave rise to the book. Further thanks go to the Department of Mathematical Sciences at Indiana University–Purdue University Fort Wayne, the Mathematical Institute at Oxford University, and the Faculty of Mathematics and Computing at the Open University.

Finally, we want to express our particular thanks to Nicky Kempton, Toni Cokayne, Steve Best and Alison Cadle at the Open University for preparing the manuscript, and to Elizabeth Johnston, Julia Tompson and Keith Mansfield at Oxford University Press for guiding it through the stages of publication.

Fort Wayne, Indiana, USA   L.W.B.
Milton Keynes   R.J.W.
July 1996

Contents

1	**Introduction**	1
	Robin J. Wilson	
1.1	Graphs	1
1.2	Adjacency and incidence	3
1.3	Paths and cycles	4
1.4	New graphs from old	5
1.5	Examples of graphs	7
1.6	Planar graphs	9
1.7	Colouring graphs	11
1.8	The efficiency of algorithms	11
1.9	And finally ...	12
	References	12
2	**Enumeration**	13
	Ronald C. Read	
2.1	Introduction	13
2.2	Labelled graphs and generating functions	13
2.3	Necklaces	15
2.4	Pólya's Enumeration Theorem	16
2.5	Chemical enumeration	19
2.6	The enumeration of graphs	23
2.7	Connected graphs	26
2.8	Trees and rooted trees	28
2.9	Other kinds of graphs	30
2.10	Unsolved problems	31
	References	32
3	**Number Theory**	34
	Roger Cook	
3.1	Introduction	34
3.2	Multiplicative functions	35
3.3	The Möbius function	36
3.4	Euler's function	38
3.5	Pólya's Enumeration Theorem	39
3.6	Eulerian graphs and tournaments	40
3.7	Finite fields and Paley graphs	42
3.8	Quadratic residue tournaments	44
3.9	Hadamard matrices and designs	45
3.10	Ramanujan graphs	46
3.11	Negative Pell equations	48
	References	49

4	**Partial Orders**	52
	Graham Brightwell	
4.1	Introduction	52
4.2	Preliminaries	52
4.3	Dilworth's Theorem	55
4.4	Comparability graphs	58
4.5	Covering graphs and diagrams	60
4.6	Schnyder's Theorem	61
4.7	The incidence order	66
References		67
5	**First-order Logic**	70
	Peter J. Cameron	
5.1	Introduction	70
5.2	First-order logic	70
5.3	First-order properties of graphs	73
5.4	Applications of compactness	75
5.5	The random graph	77
5.6	Homogeneous graphs	78
5.7	$\aleph_0$-categorical graphs	80
5.8	Sparse graphs	81
References		84
6	**Linear Algebra**	86
	Peter Rowlinson	
6.1	Introduction	86
6.2	Graph spectra	87
6.3	Distance-regular graphs	89
6.4	Other algebraic invariants	92
6.5	Eigenvalues and star partitions	95
References		98
7	**Matroids**	100
	James Oxley	
7.1	Introduction	100
7.2	Definitions and examples	101
7.3	Basic matroid operations	103
7.4	Connectivity	106
7.5	Wheels and whirls	108
7.6	Minimally 3-connected graphs and matroids	109
7.7	Excluded minors	111
7.8	Infinite antichains	113
7.9	Conclusion	114
References		114

8	**Codes**	116
	Robert T. Curtis and Tony R. Morris	
8.1	Introduction	116
8.2	Self-dual doubly-even codes	116
8.3	Invariant theory	117
8.4	Constructing binary codes from graphs	119
8.5	How many codes arise like this?	121
8.6	Graphs with the required properties	122
8.7	Counting tetrads	127
References		127
9	**Groups**	128
	Peter J. Cameron	
9.1	Introduction	128
9.2	An example: the Petersen graph	128
9.3	Three kinds of groups	129
9.4	Universal classes	130
9.5	Bounds	132
9.6	Random graphs	133
9.7	Vertex-transitive graphs	134
9.8	Distance-transitive graphs	135
9.9	Local structure	137
9.10	Computational aspects	138
References		139
10	**Geometry**	141
	Edward R. Scheinerman	
10.1	Introduction	141
10.2	Dimension 0: discrete sets	142
10.3	Dimension 1: interval graphs	144
10.4	Dimension 1 and a little: trees and circles	147
10.5	Dimensions 2 and higher	148
10.6	Counting methods via real algebraic geometry	150
References		153
11	**Topology**	155
	Lowell W. Beineke	
11.1	Introduction	155
11.2	Planar graphs	156
11.3	Thickness	158
11.4	Crossing numbers	160
11.5	Orientable surfaces and rotation systems	164
11.6	Genus and chromatic numbers	166
11.7	Non-orientable surfaces	168
11.8	Kuratowski-type theorems	171
References		173

12	**Knots**	176
	Dominic Welsh	
12.1	Introduction	176
12.2	Basic concepts	177
12.3	Tait colourings	179
12.4	Classifying knots	181
12.5	Braids and the Seifert graph	182
12.6	The Jones and Kauffman bracket polynomials	184
12.7	Bivariate polynomials	187
12.8	The Tait conjectures	188
12.9	Two applications	190
	References	192

13	**Probability**	194
	Colin McDiarmid	
13.1	Introduction	194
13.2	Random graphs: usual behaviour	194
13.3	Random graphs: deterministic results	198
13.4	The Lovász Local Lemma	199
13.5	Deterministic graphs: random methods	201
13.6	Concentration of measure and isoperimetric inequalities	204
	References	206

14	**Statistics**	208
	Peter Wild	
14.1	Introduction	208
14.2	Experimental design	209
14.3	A-optimality and closed walks	215
14.4	Bounds	218
14.5	Spanning trees and D-optimality	221
14.6	Line graphs and E-optimality	222
14.7	Row–column designs	224
14.8	Conclusion	225
	References	225

15	**Computing**	227
	Robin Whitty	
15.1	Introduction	227
15.2	Languages	228
15.3	Grammars	228
15.4	Finite automata	231
15.5	Flowgraphs	235
15.6	Program analysis	239
	References	244

16	**Artificial Neural Networks**	247
	Martin Anthony	
16.1	Introduction	247
16.2	Artificial neural networks	248
16.3	Boltzmann machines	248
16.4	Optimization with Boltzmann machines	250
16.5	Feedforward networks	252
16.6	Supervised learning in feedforward networks	253
References		259
17	**International Finance**	261
	Norman Biggs	
17.1	Introduction	261
17.2	Exchange dealing then and now	262
17.3	Exchange rate networks	264
17.4	Some classical theory	266
17.5	The export and import points	268
17.6	Potential theory	271
17.7	Determination of exchange rates by cash flows	273
17.8	Application to a tournament ranking problem	277
17.9	Mechanisms linking exchange rates and trade	277
References		278
Notes on Contributors		280
Index		285

1
Introduction

ROBIN J. WILSON

We present those definitions and theorems in graph theory that are assumed throughout this book. Further explanation of these terms, together with the proofs of stated results, can be found in the standard texts listed below, although not all of the terminology is standardized. Definitions and results not included here are introduced later, as needed.

1.1 Graphs

A *graph* G consists of a finite non-empty set $V(G)$ of elements called *vertices* and a finite set $E(G)$ of distinct unordered pairs of distinct elements of $V(G)$ called *edges* (see Fig. 1.1). We call $V(G)$ the *vertex set* of G and $E(G)$ the *edge set* of G; these are sometimes abbreviated to V and E, respectively. The number n of vertices of G is the *order* of G, and the number of edges of G is denoted by m. The edge $\{v, w\}$ (where v and w are vertices of G) is often denoted by vw.

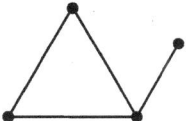

FIG. 1.1

If, in the definition of a graph, we remove the restriction that the edges are distinct, then we obtain a *multigraph* (see Fig. 1.2); two or more edges joining the same pair of vertices are *multiple edges*. If we also remove the restriction that the edges join distinct vertices, thus allowing the existence of *loops*, then the resulting object is a *general graph* (see Fig. 1.3). If loops and multiple edges are excluded, then we use the term *simple graph*.

FIG. 1.2

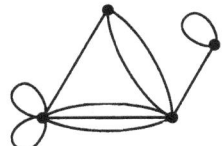

FIG. 1.3

There are many other variations on the concept of a graph. If one vertex is distinguished from the rest, then we have a *rooted graph*; the distinguished vertex is the *root*, indicated by a small square (see Fig. 1.4). A *labelled graph* of order n is a graph whose vertices have been assigned the numbers $1, 2, \ldots, n$ so that no two vertices are assigned the same number (see Fig. 1.5). A *signed graph* is a graph to each edge of which is assigned either $+$ or $-$ (see Fig. 1.6).

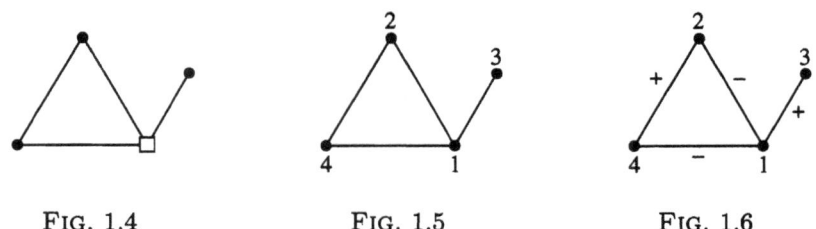

FIG. 1.4   FIG. 1.5   FIG. 1.6

A hypergraph is like a graph, except that the edges consist of any subset of vertices; more formally, a *hypergraph* H consists of a finite non-empty set $V(H)$ of elements called *vertices* and a finite set $E(H)$ of distinct sets of distinct elements of $V(H)$ called *hyperedges* (see Fig. 1.7).

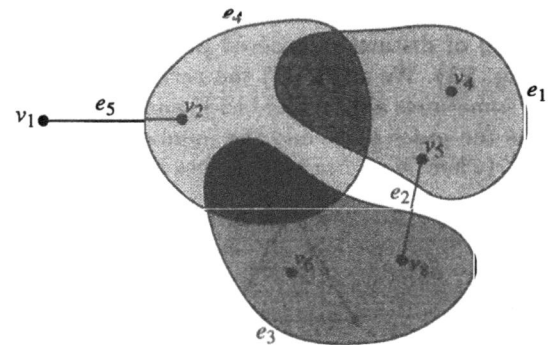

$e_1 = v_3v_4v_5$, $e_2 = v_5v_8$, $e_3 = v_6v_7v_8$, $e_4 = v_2v_3v_7$, $e_5 = v_1v_2$, $e_6 = v_7$

FIG. 1.7

We also define *infinite graphs*, in which we no longer insist that $V(G)$ and $E(G)$ be finite; a *countable graph* is one in which $V(G)$ and $E(G)$ are finite or countably infinite, and a *locally finite graph* is one in which the number of edges incident with each vertex is finite.

Finally, we consider directed graphs, in which each edge is assigned a direction. More formally, a *digraph* D consists of a finite non-empty set $V(D)$ of elements called *vertices* and a finite set $A(D)$ of distinct *ordered* pairs of distinct elements of $V(D)$ called *arcs* (see Fig. 1.8). The arc (v, w) (where v and w are vertices of D) is often denoted by vw. A *simple digraph* is a digraph with no loops vv or multiple arcs. If D is a digraph, then the *underlying graph* of D is the graph or multigraph obtained from D by replacing each arc by an undirected edge joining the same pair of vertices.

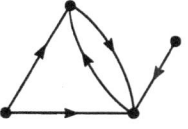

FIG. 1.8

1.2 Adjacency and incidence

If $e = vw$ is an edge of a graph G, then e *joins* the vertices v and w, and these vertices are *adjacent*; in this case, we say that e is *incident* with v and w, and that w is a *neighbour* of v. The *neighbourhood* $N(v)$ of v is the set of all vertices of G adjacent to v. Two edges of G incident with the same vertex are *adjacent edges*.

Two graphs G and H are *isomorphic* (written $G \cong H$) if there is a one-to-one correspondence between their vertex sets that preserves the adjacency of vertices. An *automorphism* of G is a one-to-one mapping ϕ of $V(G)$ onto itself with the property that $\phi(v)$ and $\phi(w)$ are adjacent if and only if v and w are. The automorphisms of G form a group $\Gamma(G)$ under composition, called the *automorphism group* of G; $\Gamma(G)$ is *transitive* if it contains automorphisms mapping each vertex of G to every other vertex, and *edge-transitive* if it contains automorphisms mapping each edge of G to every other edge.

For each vertex v in a graph G, the number of edges incident with v is the *degree* of v, denoted by $\deg(v)$. The maximum degree in G is denoted by Δ. A vertex of degree 0 is an *isolated vertex*, and a vertex of degree 1 is an *end-vertex*. The *degree list* of G is the set of degrees of the vertices of G, often arranged in non-decreasing order; for example, the degree list of the graph in Fig. 1.1 is $(1, 2, 2, 3)$. If all of the vertices of G have the same degree k, then G is *regular of degree k* or *k-regular*. A 3-regular graph is a *cubic graph*.

Analogous concepts can be defined for digraphs. If $e = vw$ is an arc of a digraph D, then v and w are *adjacent*, and e is *incident from* v and *incident to* w. If v is a vertex of a digraph D, then its *out-degree* $\text{outdeg}(v)$ is the number of arcs in D of the form vw, and its *in-degree* $\text{indeg}(v)$ is the number of arcs in D of the form wv.

An *independent* (or *stable*) *set of vertices* in a graph G is a set of vertices of G no two of which are adjacent, and the size of a largest such set is the *independence* (or *stability*) *number* of G. Similarly, an *independent set of edges* or *matching* is a set of edges of G no two of which are adjacent, and the size of a largest such set is the *edge-independence number* of G. An independent set of edges that includes every vertex of G is a *1-factor* or *perfect matching* in G.

Let G be a graph with vertex set $\{v_1, v_2, \ldots, v_n\}$ and edge set $\{e_1, e_2, \ldots, e_m\}$. The *adjacency matrix* of G is the $n \times n$ matrix $\mathbf{A}(G) = (a_{ij})$, where

$$a_{ij} = \begin{cases} 1, & \text{if } v_i \text{ and } v_j \text{ are adjacent,} \\ 0, & \text{if not,} \end{cases}$$

and the *incidence matrix* of G is the $n \times m$ matrix $\mathbf{B}(G) = (b_{ij})$, where

$$b_{ij} = \begin{cases} 1, & \text{if } v_i \text{ is incident with } e_j, \\ 0, & \text{if not.} \end{cases}$$

Note that the eigenvalues of $\mathbf{A}(G)$ are independent of the way in which the vertices are labelled. We refer to them as the *eigenvalues* of G, and to the characteristic polynomial of $\mathbf{A}(G)$ as the *characteristic polynomial* of G.

1.3 Paths and cycles

A sequence of edges of the form $v_0 v_1, v_1 v_2, \ldots, v_{r-1} v_r$ (sometimes abbreviated to $v_0 v_1 \ldots v_r$) is a *walk of length r* between v_0 and v_r. If these edges are all distinct, then the walk is a *trail*, and if the vertices $v_0, v_1, \ldots, v_r$ are also distinct, then the walk is a *path* (or *open path*). Two paths are *edge-disjoint* if they share no common edges, and are *vertex-disjoint* if they share no common vertices, although one frequently relaxes this condition to allow the end-vertices of the paths to coincide. A walk or trail is *closed* if $v_0 = v_r$, and for $r > 0$ a closed walk in which the vertices $v_0, v_1, \ldots, v_{r-1}$ are all distinct is a *cycle*.

A cycle of length 3 is a *triangle*. The length of a shortest cycle in a graph G is the *girth* of G. If v and w are vertices in G, then the length $d(v, w)$ of any shortest path from v to w is the *distance* between v and w. The largest distance between two vertices in G is the *diameter* of G.

These definitions extend to directed graphs and infinite graphs. Thus, a *trail* in a digraph is a sequence of distinct arcs of the form $v_0 v_1, v_1 v_2, \ldots, v_{r-1} v_r$, a *path* is such a sequence in which the vertices are all distinct, and for $r > 0$ a *cycle* is a sequence of arcs of the form $v_0 v_1, v_1 v_2, \ldots, v_{r-1} v_0$, where $v_0, v_1, \ldots, v_{r-1}$ are distinct. In an infinite graph, a *two-way infinite path* is a sequence of distinct edges of the form

$$\ldots, v_{-r} v_{-r+1}, \ldots, v_{-1} v_0, v_0 v_1, \ldots, v_r v_{r+1}, \ldots.$$

A graph G is *connected* if there is a path joining each pair of vertices of G; a graph that is not connected is called *disconnected*. Every disconnected graph can be split into maximal connected subgraphs, called *components*. There are analogous definitions for digraphs; a digraph D is *strongly connected* if there is a (directed) path in D joining each pair of vertices in each direction, and *connected* if the underlying graph is connected.

1.4 New graphs from old

A *subgraph* of a graph $G = (V(G), E(G))$ is a graph $H = (V(H), E(H))$ such that $V(H) \subseteq V(G)$ and $E(H) \subseteq E(G)$. If $V(H) = V(G)$, then H is a *spanning subgraph* of G. If W is any set of vertices in G, then the *subgraph induced by W* is the subgraph of G obtained by joining those pairs of vertices in W that are joined in G. An *induced subgraph of G* is a subgraph that is induced by some subset W of $V(G)$. Similar definitions can be given for digraphs and multigraphs.

If e is an edge of G, then the *edge-deleted subgraph $G - e$* or $G \backslash e$ is the graph obtained from G by removing the edge e; more generally, $G - \{e_1, \ldots, e_k\}$ is the graph obtained from G by removing the edges $e_1, \ldots, e_k$. Similarly, if v is a vertex of G, then the *vertex-deleted subgraph $G - v$* is the graph obtained from G by removing the vertex v together with all its incident edges; more generally, $G - \{v_1, \ldots, v_k\}$ is the graph obtained from G by removing the vertices $v_1, \ldots, v_k$ and all edges incident with any of them. These concepts are illustrated in Fig. 1.9.

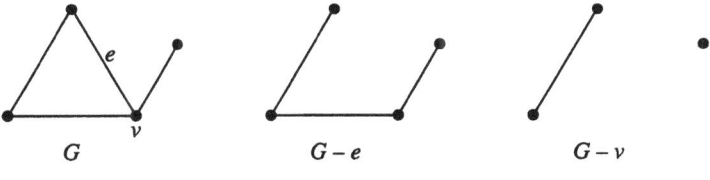

FIG. 1.9

We can also obtain a new graph from G by removing the edge $e = vw$ and identifying v and w so that the resulting vertex is incident to all edges (other than e) that were originally incident with v or w; this is called *contracting the edge e* (see Fig. 1.10), and the resulting graph is denoted by G/e. If the graph H can be obtained from G by a succession of edge contractions such as this, then G is *contractible* to H. A *minor* of G is any graph obtained from G by a succession of edge-deletions and edge-contractions.

FIG. 1.10

If $e = vw$ is an edge of G, then we obtain a new graph by replacing e by two new edges vz and zw, where z is a new vertex; this is called *subdividing the edge* (see Fig. 1.11). Two graphs that can be obtained from the same graph by subdividing its edges are *homeomorphic*.

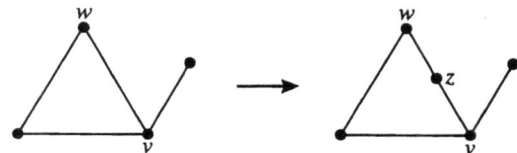

Fig. 1.11

If G and G' are graphs with the same vertex set, then their *intersection* $G \cap G'$ is the graph with edge set $E(G) \cap E(G')$, and their *union* $G \cup G'$ is the graph with edge set $E(G) \cup E(G')$. If G and G' are disjoint graphs, then their *disjoint union* $G \cup G'$ is the graph with vertex set $V(G) \cup V(G')$ and edge set $E(G) \cup E(G')$; the disjoint union of k copies of G is written kG. The *join* $G * G'$ is obtained from the disjoint union of G and G' by adding an edge between each vertex of G and each vertex of G'. The *Cartesian product* $G \times G'$ is the graph with vertex set $V(G) \times V(G')$ in which the vertex (v, w) is adjacent to the vertex (v', w') whenever $v = v'$ and w is adjacent to w', or $w = w'$ and v is adjacent to v'.

The *complement* $\overline{G}$ of G is the graph with the same vertex set as G, but where two vertices are adjacent whenever they are *not* adjacent in G; a graph is *self-complementary* if it is isomorphic to its complement. The *line graph* $L(G)$ of G is the graph whose vertices correspond to the edges of G, and where two vertices are joined whenever the corresponding edges of G are adjacent.

If G is a connected graph, and if the graph $G - e$ is disconnected for some edge e, then e is a *bridge* (or *cut-edge* or *isthmus*) of G. More generally, a *cutset* (or *edge-cut*) in G is a set of edges whose removal disconnects G. A graph G is *k-edge-connected* if every two vertices v and w are connected by at least k edge-disjoint paths, and the *edge-connectivity* $\lambda(G)$ of G is the largest value of k for which G is k-edge-connected.

If G is a connected graph, and if the graph $G - v$ is disconnected for some vertex v, then v is a *cut-vertex* of G. More generally, a *separating set of vertices* in G is a set of vertices whose removal disconnects G. A graph G with at least $k + 1$ vertices is *k-connected* if every two vertices v and w are connected by at least k paths that are pairwise disjoint except for the vertices v and w; a 2-connected graph is a *block* or a *non-separable graph*. The *connectivity* $\kappa(G)$ of G is the largest value of k for which G is k-connected; note that $\kappa(G) \leq \lambda(G)$.

The most important result relating these concepts is *Menger's Theorem*; it takes several forms, among which are the following.

Theorem 1.1 (Menger's Theorem) *Let G be a connected graph with at least $k + 1$ vertices. Then*

(a) *G is k-connected if and only if G cannot be disconnected by the removal of $k - 1$ or fewer vertices;*

(b) *G is k-edge-connected if and only if G cannot be disconnected by the removal of $k - 1$ or fewer edges.*

1.5 Examples of graphs

A graph in which every two vertices are adjacent is a *complete graph*; the complete graph with n vertices and $n(n-1)/2$ edges is denoted by K_n. The *cycle graph* C_n of order n consists of the vertices and edges of an n-gon. The *wheel* W_n is the graph $C_{n-1} * K_1$, and the *path graph* P_n is obtained by removing an edge from C_n. The *null graph* N_n of order n is the graph with n vertices and no edges. The graphs K_5, C_5, W_5, P_5 and N_5 are shown in Fig. 1.12. It is also occasionally useful to introduce the *empty graph* (not really a graph at all), consisting of no vertices or edges.

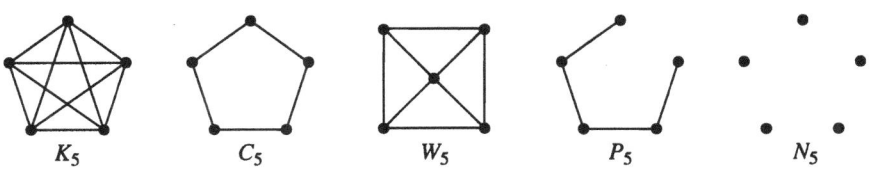

FIG. 1.12

A *clique* in a graph G is a complete subgraph of G, and a *maximum clique* is a clique of maximum order in G. The *clique number* $\omega(G)$ of G is the order of a maximum clique. A *tournament* is an 'oriented complete graph'—that is, a digraph in which every two vertices are joined by exactly one arc.

A *bipartite graph* is a graph whose vertex set can be partitioned into two sets so that each edge joins a vertex of the first set and a vertex of the second set. A *complete bipartite graph* is a bipartite graph in which each vertex in the first set is adjacent to every vertex in the second set; if the two sets contain r and s vertices, then the complete bipartite graph is denoted by $K_{r,s}$. A *complete k-partite graph* is obtained by partitioning the vertex set into k sets, and joining two vertices whenever they lie in different sets; if all of these sets have size r, then the resulting graph is the complement of rK_k, and is denoted by $K_{r,...,r}$ or $K_{k(r)}$. The graphs $K_{3,3}$ and $K_{3,3,3}$ are shown in Fig. 1.13.

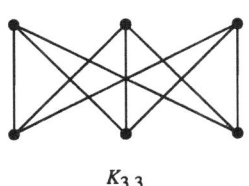 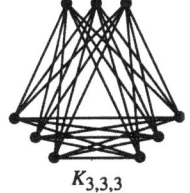 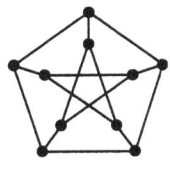

FIG. 1.13 FIG. 1.14

The *Petersen graph* is the graph shown in Fig. 1.14; it is the complement of the line graph of K_5. The *Platonic graphs* are the graphs corresponding to the vertices and edges of the five regular solids—the tetrahedron, cube, octahedron, dodecahedron and icosahedron (see Fig. 1.15). The *k-cube* Q_k is the graph whose vertices correspond to the sequences $(a_1, a_2, \ldots, a_k)$, where each $a_i = 0$ or 1, and whose edges join those pairs of vertices that correspond to sequences differing in just one place; thus Q_3 is the graph of the cube. The *k-dimensional octahedron* is the complement $K_{2,\ldots,2}$ of the graph kK_2.

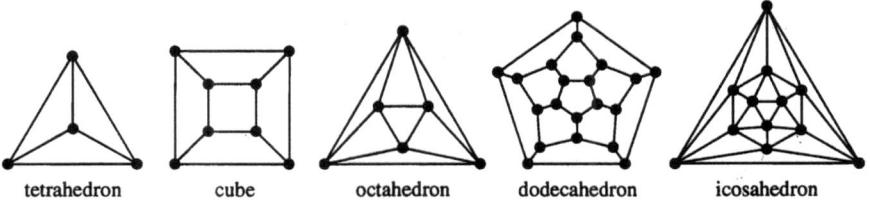

tetrahedron   cube   octahedron   dodecahedron   icosahedron

FIG. 1.15

A connected graph that contains no cycles is a *tree*, and a graph whose components are trees is a *forest* or *acyclic graph*. The trees of order 5 are shown in Fig. 1.16. The *arboricity* of G is the minimum number of forests whose union is G.

FIG. 1.16

Some properties of trees are summarized in the following theorem.

Theorem 1.2   *If T is a tree of order n, then*

(a) *T has $n - 1$ edges;*
(b) *each edge of T is a bridge;*
(c) *each vertex is an end-vertex or a cut-vertex;*
(d) *if $n > 1$, T has at least two end-vertices;*
(e) *there is exactly one path between each pair of vertices.*

If T is any tree, then we obtain another tree by removing all the end-vertices from T. Repeating this procedure as often as necessary, we eventually obtain either a single vertex (the *centre* of T) or two adjacent vertices (the *bicentre* of T); T is *central* or *bicentral* according as T has a centre or bicentre.

If G is a connected graph, then a *spanning tree* in G is a connected spanning subgraph containing no cycles. The number of spanning trees in the complete graph K_n is given by *Cayley's Theorem*.

Theorem 1.3 (Cayley's Theorem)   *The number of distinct spanning trees in K_n is n^{n-2}.*

More generally, one can prove the following result.

Theorem 1.4 (Matrix-Tree Theorem) *Let G be a connected labelled graph with adjacency matrix $\mathbf{A}$, and let $\mathbf{M}$ be the matrix obtained from $-\mathbf{A}$ by replacing each diagonal entry by the degree of the corresponding vertex. Then all cofactors of $\mathbf{M}$ are equal, and their common value is the number of spanning trees in G.*

A connected graph G is *Eulerian* if it has a closed trail that includes every edge of G; such a trail is an *Eulerian trail*. Similarly, a connected digraph D is *Eulerian* if it has a closed directed trail that includes every arc of D. Necessary and sufficient conditions for a graph or digraph to be Eulerian are given in the following theorem.

Theorem 1.5

(a) A connected graph G is Eulerian if and only if each vertex of G has even degree.

(b) A connected digraph D is Eulerian if and only if the in-degree and out-degree of each vertex are equal.

A graph G is *Hamiltonian* if it has a cycle that includes every vertex of $V(G)$; such a cycle is a *Hamiltonian cycle*. More generally, a graph G is *traceable* if it has a path that includes every vertex of $V(G)$; such a path is a *Hamiltonian path*. Analogous definitions can be given for digraphs.

A sufficient condition for a graph to be Hamiltonian is given in *Ore's Theorem*.

Theorem 1.6 (Ore's Theorem) *If G is a simple graph with n (≥ 3) vertices, and if*
$$\deg(v) + \deg(w) \geq n$$
for each pair of non-adjacent vertices v and w, then G is Hamiltonian.

A necessary condition is as follows.

Theorem 1.7 *If G is Hamiltonian, then for each subset S of $V(G)$, the graph $G - S$ has at most S components.*

1.6 Planar graphs

A *planar graph* is a graph that can be embedded in the plane so that no two edges intersect geometrically except at a vertex to which both are incident; a graph so embedded is sometimes called a *plane graph*. The points of the plane not on G are then partitioned into open sets called *regions* or *faces*, one of which, the *infinite region*, is infinite in extent (see Fig. 1.17).

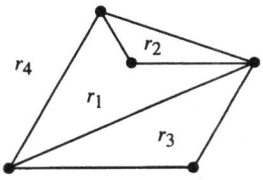

FIG. 1.17

The number of regions is given by *Euler's Polyhedron Formula.*

Theorem 1.8 (Euler's Polyhedron Formula) *Let G be a connected plane graph with n vertices, m edges and r regions. Then*

$$n - m + r = 2.$$

If G has no loops or multiple edges and $n \geq 3$, then each region is bounded by at least three edges; it follows that $2m \leq 3r$, and hence that $m \leq 3n - 6$. Equality holds when each face is bounded by a triangle, and such a graph is a *maximal planar graph* or *triangulation*. If G has no triangles, then $2m \leq 4r$, and so $m \leq 2n - 4$. It is straightforward to check that every simple planar graph has a vertex of degree 5 or less. A 3-connected plane graph is a *map*; and a graph that can be embedded in the plane so that all vertices lie on the boundary of the same region is an *outerplanar graph*.

A necessary and sufficient condition for a graph to be planar has been given by Kuratowski; we present two forms of his result.

Theorem 1.9 (Kuratowski's Theorem)

(a) *A graph G is planar if and only if G has no subgraph homeomorphic to K_5 or $K_{3,3}$.*

(b) *A graph G is planar if and only if G has no subgraph contractible to K_5 or $K_{3,3}$.*

If G is a connected plane graph, then its *dual graph* G^* is the general graph obtained by the following procedure:

(a) place a point inside each region of G—these points are the vertices of G^*;
(b) for each edge e of G, draw a line or simple curve joining the vertices in the two regions bounded by e—these lines are the edges of G^* (see Fig. 1.18).

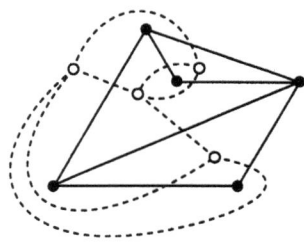

FIG. 1.18

If G^* is a plane graph whose dual graph is isomorphic to G, and if G has n vertices, m edges and r regions, then G^* has r vertices, m edges and n regions.

We can also consider the embedding of graphs on other surfaces. In particular, we can derive an analogue of Euler's Polyhedron Formula for any orientable or non-orientable surface.

1.7 Colouring graphs

The *chromatic number* $\chi(G)$ of a graph G is the minimum number of colours needed to colour the vertices of G so that no two adjacent vertices are assigned the same colour. If $\chi(G) = k$, then G is *k-chromatic*, and if $\chi(G) \leq k$, then G is *k-colourable*. For example, the complete graph K_n is n-chromatic, and the cycle graph C_n is 2-chromatic or 3-chromatic according as n is even or odd. Note that if G is a bipartite graph, then G is 2-colourable. A *colour class* is the set of all vertices of the same colour. A graph G is *perfect* if, for each induced subgraph H of G, the chromatic number $\chi(H)$ is equal to the clique number $\omega(H)$.

An upper bound for the chromatic number of a graph G is given by *Brooks' Theorem*; it involves the maximum degree of G.

Theorem 1.10 (Brooks' Theorem) *Let G be a connected graph which is not a complete graph or a cycle of odd length, and let Δ be the largest degree in G. Then G is Δ-colourable.*

For each graph G, let $P_G(k)$ be the number of ways of colouring the vertices of G so that no two adjacent vertices are assigned the same colour. For example, if $G = K_n$, then $P_G(k) = k(k-1)\ldots(k-n+1)$, and if $G = P_n$, then $P_G(k) = k(k-1)^{n-1}$. It is simple to show that if G has n vertices and m edges, then $P_G(k)$ is a monic polynomial in k of degree n, in which the coefficients alternate in sign, the constant coefficient is 0, and the coefficient of k^{n-1} is $-m$; $P_G(k)$ is the *chromatic polynomial* of G. When working with chromatic polynomials, the following *deletion-contraction formula* is often useful; it is proved by counting the number of colourings of G in which the vertices incident with e have, and have not, different colours.

Theorem 1.11 *Let G be a graph containing an edge e, and let $G\backslash e$ and G/e be the graphs obtained from G by deleting and contracting e. Then*

$$P_G(k) = P_{G\backslash e}(k) - P_{G/e}(k).$$

We end this section with the famous *Four Colour Theorem* for planar graphs, proved by K. Appel and W. Haken in 1976.

Theorem 1.12 (Four Colour Theorem) *Every planar graph is 4-colourable.*

1.8 The efficiency of algorithms

In several chapters of this book we consider the efficiency of graph algorithms. For each graph G under consideration, we associate a number n called its *size*— n is usually the order or the number of edges of G.

The most efficient algorithms are *polynomial-time algorithms*, which can be solved in time cn^k, for some constants c and k. The class of problems that can be solved in polynomial time is denoted by P. A problem is an *NP-problem* if its solution, when given, can be checked in polynomial time, even if it took exponential time to find the solution originally. It is clear that P $\subseteq$ NP, where NP denotes the class of all NP-problems. It is unknown whether P = NP, although it is generally believed that this is *not* the case.

Finally, there is a class of important problems known as *NP-complete problems*. These problems, which include the travelling salesman problem, the graph isomorphism problem, the Hamiltonian cycle problem (is G Hamiltonian?) and the k-colourability problem (is G k-colourable?), have the property that if any one of them is in P, then so are all of them. It is generally believed that *none* of them is in P. Further information on the topic is given in [5].

1.9 And finally ...

If S is a finite set, we denote the number of elements in S by $|S|$; the empty set is denoted by $\emptyset$. We use $\lfloor x \rfloor$ for the largest integer not greater than x, and $\lceil x \rceil$ for the smallest integer not smaller than x (so that, for example, $\lfloor \pi \rfloor = 3$ and $\lceil \pi \rceil = 4$). The sets of real numbers, rational numbers, integers and positive integers are denoted, respectively, by **R**, **Q**, **Z** and **N**. The end of a proof is denoted by $\square$.

In the references for each chapter, Mathematical Reviews numbers are indicated by *MR* **15**–234 (page 234 of Volume **15**), *MR* **35**#234 (review number 234 of Volume **35**), or *MR* **80g**: 05123 (review number 123 of Section 05 in issue **g** of 1980).

References

References [2], [3], [4], [7] and [8] below are standard texts in graph theory, and references [1] and [6] are standard texts in combinatorial mathematics.

1. I. Anderson, *A First Course in Combinatorial Mathematics*, 2nd edn, Oxford University Press, 1989; *MR* **49**#2402.
2. C. Berge, *Graphs*, North-Holland, 1985; *MR* **50**#9640.
3. J. A. Bondy and U. S. R. Murty, *Graph Theory with Applications*, American Elsevier, 1979; *MR* **54**#117.
4. G. Chartrand and L. Lesniak, *Graphs and Digraphs*, 2nd edn, Wadsworth & Brooks Cole, 1986; *MR* **55**#5449.
5. M. Garey and D. S. Johnson, *Computers and Intractability. A Guide to the Theory of NP-Completeness*, W. H. Freeman, 1979; *MR* **80g**: 68056.
6. M. Hall, Jr, *Combinatorial Theory*, Blaisdell, 1967; *MR* **37**#80.
7. F. Harary, *Graph Theory*, Addison-Wesley, 1969; *MR* **41**#1566.
8. R. J. Wilson, *Introduction to Graph Theory*, 4th edn, Addison Wesley Longman, 1996; *MR* **50**#9643.

2
Enumeration

RONALD C. READ

In this chapter we present the basic problem of graphical enumeration—the enumeration of unlabelled graphs. In particular, we look at Pólya's Enumeration Theorem and some problems to which it can be applied. We then consider the enumeration of some other kinds of graphs. The chapter concludes with some remarks about unsolved problems of graphical enumeration.

2.1 Introduction

How many graphs are there on four vertices? This question can easily be resolved by a process of trial and error, but if we ask how many graphs there are on ten vertices, the answer can be reliably found only by theoretical means, while the similar question for an indefinite number of vertices clearly requires a formula of some kind. In this chapter we shall be largely concerned with deriving such a formula, and some similar results.

2.2 Labelled graphs and generating functions

Let us first dispose of the easy case, that of labelled graphs. If the vertices of the graph are labelled, then the problem is almost trivial. If there are n vertices, then we have $n(n-1)/2$ pairs of vertices which may or may not correspond to edges. The edges can be chosen or not chosen to be in the graph independently of each other. Hence, with two choices (edge or non-edge) for each pair of vertices, we have a total of 2^N possible graphs, where, for convenience, we write $N = \binom{n}{2} = n(n-1)/2$.

We can do better than this and break this number down according to the number of edges. To form a labelled graph on n vertices and e edges, we must choose e of the N vertex-pairs. This can be done in

$$\binom{N}{e} = \frac{N!}{e!(N-e)!} \qquad (2.1)$$

ways.

That more or less wraps up the basic problem of enumerating labelled graphs, but before we leave this topic let us introduce the concept of a generating function—a concept that will be of great importance later on.

Suppose that we have a sequence (finite or infinite) of numbers. It is often convenient to handle such a sequence by making its elements the coefficients in a power series. Thus, for a sequence $\{a_n\}$, we can define a power series

$$a_0 + a_1 x + a_2 x^2 + a_3 x^3 + \cdots .$$

Such a power series is called a *generating function* for the sequence. (Strictly speaking, the generating function is not the power series, but rather the function to which the series converges, if it does; in practice, the term 'generating function' is also used for the series, even if it does not converge!)

Consider the labelled graphs with five vertices. By formula (2.1), the numbers of such graphs, listed according to the numbers of edges, are given by the finite sequence

$$1, 10, 45, 120, 210, 252, 210, 120, 45, 10, 1.$$

Thus the generating function for this sequence is

$$1 + 10x + 45x^2 + 120x^3 + 210x^4 + 252x^5 + 210x^6 + 120x^7 + 45x^8 + 10x^9 + x^{10},$$

which is $(1+x)^{10}$. In general, the generating function for labelled graphs on n vertices is $(1+x)^N$.

The problem of enumerating unlabelled graphs is much more difficult. By an *unlabelled graph* we mean an equivalence class of graphs under isomorphism. Now the number of labelled graphs corresponding to a given unlabelled graph G—or, to put it another way, the number of ways of labelling G—depends on how much symmetry G has (that is, on the number of automorphisms of G).

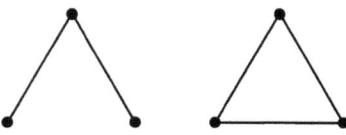

FIG. 2.1

Consider the two graphs in Fig. 2.1. The first graph has two automorphisms: the identity and the automorphism that interchanges the two vertices of degree 1. The second graph has six automorphisms, corresponding to the six permutations of the three vertices. This example illustrates a general result, that if a graph has n vertices, m automorphisms and l labellings, then $ml = n!$. We shall not need this result except to demonstrate that we cannot expect to find any easy correspondence between labelled and unlabelled graphs. Any enumeration of unlabelled graphs will involve the numbers of automorphisms of the graphs in some essential way. We shall eventually see how this is done, but to do so we approach the matter indirectly and study first an apparently completely different problem.

2.3 Necklaces

Suppose that we want to make a necklace out of four white beads and two red beads. In how many ways can we do this? (We assume that there is no clasp or any other such way of distinguishing one part of the necklace from another.)

If this were just a question of choosing two of the six possible positions in which a red bead could be placed, this would be a very simple problem; but we must take into account the fact that rotating a necklace, or turning it over, gives essentially the same necklace. Thus, for example, in Fig. 2.2, diagram (b) is a rotation of diagram (a), while diagram (c) is a *flipped* version of diagram (a) or (b). These three diagrams all correspond to the same necklace, yet they correspond to different choices of the six positions at which the beads can be placed, if we regard these positions as being fixed.

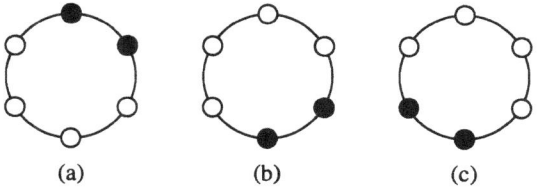

(a)   (b)   (c)

FIG. 2.2

We have here a common type of problem, which can be expressed in general terms as follows. We have a certain number of what we call *sites*. At each site we can place exactly one of a certain collection of objects, which we call *figures*. The result of placing one figure at each site is called a *configuration*. Thus, for the necklace problem above, the sites are the six positions around the string on which the beads are to be threaded, and the collection of figures has just two elements—a white bead and a red bead. Note that, when choosing figures, we allow repetitions; thus we could choose all white beads, or all red beads, or some of each. The matter of controlling how many of each we choose will be addressed shortly.

We now introduce a group acting on the set of sites. If A is a group of permutations of the sites, we say that two configurations are *equivalent* if one can be obtained from the other by permuting the sites (together with the figures placed there) by some element of A. For the necklace problem, the group is generated by a rotation of the necklace through $60°$, and a turning over of the necklace. This group is the dihedral group of degree 6 and order 12, usually denoted by D_6.

There is one more thing. With each figure we associate a non-negative integer, called the *content* of the figure, and we define the *content* of the configuration to be the sum of the contents of the figures chosen for this configuration. The general problem for which we seek an answer is then *to find the number of configurations with a given content*.

For the necklace problem, we want the number of six-bead necklaces with just two red beads. If we define the content of a white bead to be 0 and that of a red bead to be 1, then the content of a configuration (necklace) is precisely the number of red beads. Thus this necklace problem, and similar problems with different numbers of beads, appear as special cases of the very general problem outlined above. If we had a theorem giving the solution to the general problem, then we could use it for any necklace problem and much more besides. Fortunately, such a theorem exists, and we discuss it in the next section.

2.4 Pólya's Enumeration Theorem

The theorem which provides the solution of all problems of the kind considered above is one that Georg Pólya presented in 1937 [7, 8]. It is universally known as *Pólya's Theorem*, even though it was anticipated to some extent by J. H. Redfield [12]. To see how it works, let us set out the essentials of the problem with which it deals.

We have a set of (undefined) figures, each figure having a *content*. We have a set of *sites*, and a group A of permutations of these sites. A *configuration* is obtained by placing one figure at each site, and the *content of the configuration* is defined as the sum of the contents of the figures that make up the configuration. Two configurations are *equivalent* if either can be obtained from the other by a permutation belonging to A. The problem then is: *given the relevant information about the figures and the group A, compute the number of inequivalent configurations with given content.*

Pólya's Theorem handles this problem by using generating functions. To summarize the relevant information about the figures, we construct a generating function

$$f(x) = a_0 + a_1 x + a_2 x^2 + a_3 x^3 + \cdots ,$$

where a_i is the number of figures with content i. This is called the *figure generating function*. The answer to the problem then appears as a generating function

$$F(x) = A_0 + A_1 x + A_2 x^2 + A_3 x^3 + \cdots ,$$

where A_i is the number of configurations with content i; this is known as the *configuration generating function*. Clearly, if we know what $F(x)$ is, we can answer any question about the number of configurations with a specific content. The link by which we go from the figure generating function to the configuration generating function depends on the cycle structure of the group A.

Enumeration

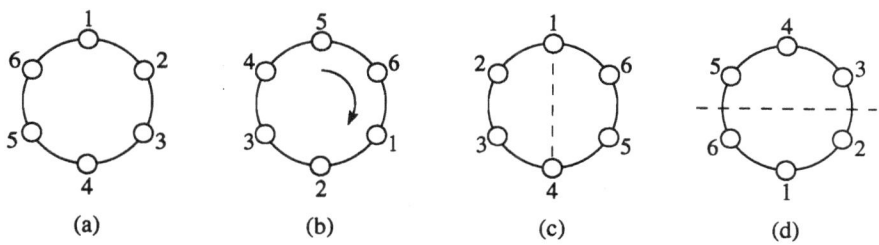

FIG. 2.3

Suppose that we number the six sites around the necklace with the numbers 1, 2, 3, 4, 5, 6 (see Fig. 2.3(a)). Then the permutation corresponding to a clockwise rotation through 120° (Fig. 2.3(b)) is

$$\begin{pmatrix} 1 & 2 & 3 & 4 & 5 & 6 \\ 3 & 4 & 5 & 6 & 1 & 2 \end{pmatrix}.$$

In this permutation 1 maps onto 3, 3 onto 5, and 5 onto 1, completing a cycle of length 3, which we write as (135). Similarly, the remaining elements form another cycle of length 3, namely (246). Thus

$$\begin{pmatrix} 1 & 2 & 3 & 4 & 5 & 6 \\ 3 & 4 & 5 & 6 & 1 & 2 \end{pmatrix} = (135)(246),$$

and this permutation splits into two cycles of length 3. In general, any permutation can be expressed uniquely (except for order) as a product of disjoint cycles.

With each permutation we associate a monomial

$$s_1^{j_1} s_2^{j_2} s_3^{j_3} \ldots,$$

where the s_i are indeterminates and j_i denotes the number of cycles of length i. We call this monomial the *cycle type* of the permutation. For the permutation just considered, with two cycles of length 3, the cycle type is s_3^2. For the permutation obtained by flipping the necklace about a line joining sites 1 and 4 (Fig. 2.3(c)), we have the cycles

$$(1)(26)(35)(4),$$

and the corresponding cycle type is $s_1^2 s_2^2$.

Pólya defined the *cycle index* of a group A to be the 'average' of all the cycle types for the elements of A. In other words, to find the cycle index of A we calculate the cycle type of each element of A, add these together and divide by the number of elements. The cycle index of A is denoted by $Z(A; s_1, s_2, s_3, \ldots)$.

We are now in a position to state Pólya's Theorem.

Theorem 2.1 (Pólya's Enumeration Theorem) *The configuration generating function is obtained by substituting the figure generating function into the cycle index of A.*

We still need to state what is meant by *substituting* the figure generating function into the cycle index. It means that in the cycle index we replace every occurrence of s_i by $f(x^i)$. Thus Pólya's Theorem says that, in the notation we used above,

$$F(x) = Z(A; f(x), f(x^2), f(x^3), \ldots).$$

At last we can solve our necklace problem. As before, we define the content of a white bead as 0 and the content of a red bead as 1, thus making the content of the configuration the number of red beads. Then the figure generating function is just $x^0 + x^1$, or $1 + x$.

Now we must find the cycle index of D_6. Two of the cycle types have already been found, and the remainder can be found similarly. For the rotations we have:

> the identity rotation: 6 cycles of length 1, cycle type s_1^6;
> rotations through $\pm 60°$:   1 cycle of length 6, cycle type s_6;
> rotations through $\pm 120°$:   2 cycles of length 3, cycle type s_3^2;
> a rotation through $180°$:   3 cycles of length 2, cycle type s_2^3.

A flip of the necklace can be performed about an axis of symmetry passing through two opposite beads (Fig. 2.3(c)), or about one of three other axes (Fig. 2.3(d)). The first three have cycle types of the form $s_1^2 s_2^2$, as we have already seen; the others have cycle types of the form s_2^3.

Putting this all together, we see that the cycle index of D_6 is

$$Z(D_6) = \tfrac{1}{12}(s_1^6 + 2s_6 + 2s_3^2 + 4s_2^3 + 3s_1^2 s_2^2).$$

To apply Pólya's Theorem, we substitute $f(x) = 1 + x$ into $Z(D_6)$. This gives

$$\tfrac{1}{12}\left[(1+x)^6 + 2(1+x^6) + 2(1+x^3)^2 + 4(1+x^2)^3 + 3(1+x)^2(1+x^2)^2\right].$$

Expanding this expression in powers of x, we obtain the configuration generating function

$$1 + x + 3x^2 + 3x^3 + 3x^4 + x^5 + x^6.$$

Thus the number of necklaces with two red beads and four white beads is 3, the coefficient of x^2.

Well, the mountain has laboured and brought forth its tiny mouse. We certainly did not need all the power of Pólya's Theorem to tell us that! However, the necklace problem was introduced simply to illustrate the method. In the next section we shall look at some more significant applications.

2.5 Chemical enumeration

The structural formula for a chemical compound is essentially a graph in which the vertices are atoms (which can be of different kinds) and the edges represent the chemical bonds between them. Note that in chemical compounds there can be double or even triple bonds, but we shall not consider this possibility.

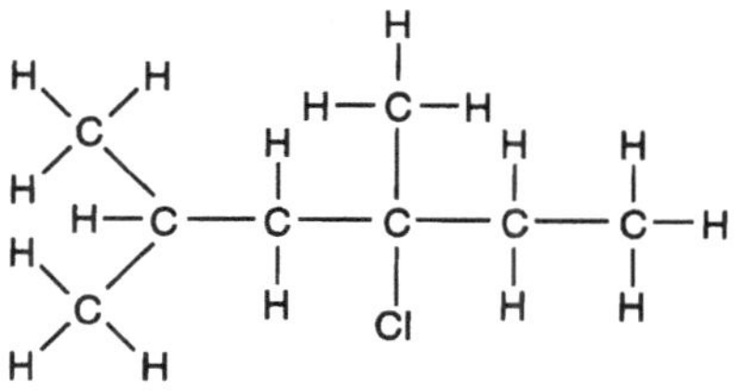

FIG. 2.4

Such a formula is shown in Fig. 2.4. This particular formula is an example of what are known as *monosubstituted alkanes*, defined as follows. There are three kinds of atom: carbon atoms (symbol C) with degree (valency) 4, hydrogen atoms (symbol H) with degree 1, and one other atom (symbol X) of degree 1; in Fig. 2.4 this is taken to be a chlorine atom (symbol Cl). There are no double or triple bonds.

These conditions imply that if there are n carbon atoms, then there must be $2n + 1$ hydrogen atoms, and the simple formula for a monosubstituted alkane is therefore $C_n H_{2n+1} X$. We ask how many different alkanes there are for any given value of n. If $A(x)$ denotes the corresponding generating function, then the number we seek is the coefficient of x^n in $A(x)$.

The presence of the atom X where there might have been another hydrogen atom makes the situation simpler than it would otherwise be, since it gives us a starting point for tackling the problem. (The problem of enumerating alkanes, in which the X is just another hydrogen atom, is more difficult and we shall not attempt it here.) Now we have one carbon atom adjacent to the atom X, which is thereby distinguished from the other carbon atoms, and at this atom we have three bonds at the ends of which are three portions of the molecule (see Fig. 2.5).

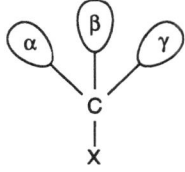

FIG. 2.5

Each of these portions of the molecule is itself a monosubstituted alkane. We can now set up a Pólya-type problem. We have three sites α, β and γ, at each of which we place what is essentially a monosubstituted alkane, with the 'root' carbon atom (circled in Fig. 2.5) playing the role of X. Note that there might be a hydrogen atom adjacent to the root, and we take care of this by including the case $n = 0$ in the definition of $A(x)$.

How are we to permute the sites? The four bonds at the root carbon atom can be thought of as being in fixed positions relative to it. Therefore, since the bond going to X is different from the others, the only permutations of the sites that can give an equivalent graph are cyclic permutations of α, β and γ, forming the cyclic group $\mathbf{Z}_3$.

It is easy to find the cycle index of $\mathbf{Z}_3$. Apart from the identity, there are the cyclic rotations
$$\begin{pmatrix} 1 & 2 & 3 \\ 2 & 3 & 1 \end{pmatrix} \quad \text{and} \quad \begin{pmatrix} 1 & 2 & 3 \\ 3 & 1 & 2 \end{pmatrix},$$
which are of cycle type s_3. Hence
$$Z(\mathbf{Z}_3) = \tfrac{1}{3}(s_1^3 + 2s_3).$$

Applying Pólya's Theorem, we see that the number of different configurations for the part of the molecule containing the portions α, β and γ is given by the generating function
$$Z(\mathbf{Z}_3; A(x), A(x^2), A(x^3), \ldots) = \tfrac{1}{3}[A^3(x) + 2A(x^3)], \tag{2.2}$$

and the coefficient of x^n in this expression is the number of configurations in which the portions α, β and γ have n carbon atoms between them. But we want just $n - 1$ carbon atoms so that, with the root atom, we get n carbon atoms altogether. We take care of this by multiplying by x in the generating function, to obtain
$$\tfrac{1}{3}x[A^3(x) + 2A(x^3)]. \tag{2.3}$$

What is this generating function? It is clearly the generating function for monosubstituted alkanes, whose generating function is $A(x)$. Well, almost. One thing is not quite right. The generating function (2.3) has no constant term, whereas $A(x)$ has constant term 1, for the case $n = 0$. This discrepancy arises because the case $n = 0$ corresponds to the molecule X—H, which has no carbon atom and therefore cannot arise from the process that yielded the generating function (2.2). This is the only exception, however, so we just add the required constant term to the expression (2.3), and we then do indeed obtain the generating function for monosubstituted alkanes. We are thus led to the equation
$$A(x) = 1 + \tfrac{1}{3}x[A^3(x) + 2A(x^3)]. \tag{2.4}$$

Now it would be nice if we could manipulate equation (2.4) so as to get an explicit expression for $A(x)$, but this does not appear to be possible. However, this equation can be used to compute the coefficient in $A(x)$ as far as we like, using a recursive procedure. For, if we have computed the coefficients up to (say) x^{10} and we substitute them into the right-hand side of equation (2.4), then the result contains the term in x^{11}, because of the multiplier x. Hence each substitution yields one further coefficient.

In this way we find that the first few coefficients are

$$1, 1, 1, 2, 5, 11, 28, 74, 199, 551, 1553, 4436, 12832,$$

and that, for example, the coefficient of x^{30} is 6641338630714.

How realistic are these answers? For small values of n, the results are perfectly valid and the many different molecules (*isomers*) with the same general formula $C_n H_{2n+1} X$ have been investigated by chemists. But suppose that we construct a monosubstituted alkane in which the root is attached to three carbon atoms, each of which is attached to three more carbon atoms, and so on, until we get to a large number of carbon atoms all at distance k (say) from the root, and we then finish the molecule by adding hydrogen atoms to these carbons (see Fig. 2.6 for small values of k). Then it might seem that, for large values of k, there would be overcrowding near the periphery of the molecule.

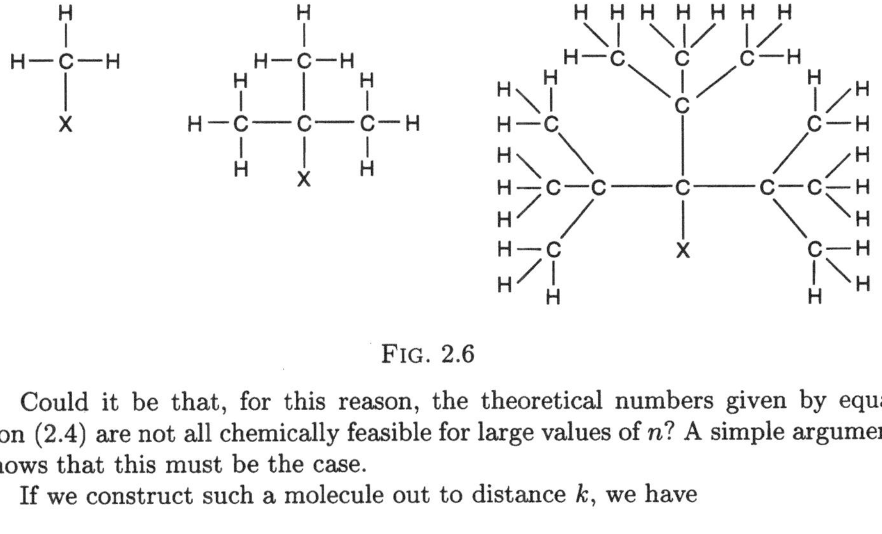

FIG. 2.6

Could it be that, for this reason, the theoretical numbers given by equation (2.4) are not all chemically feasible for large values of n? A simple argument shows that this must be the case.

If we construct such a molecule out to distance k, we have

$$1 + 3 + 3^2 + 3^3 + \cdots + 3^k = \tfrac{1}{2}(3^{k+1} - 1)$$

carbon atoms. These atoms are contained in a sphere centred at the root and with radius k. Since the bonds do not all go directly outwards from the root, these atoms are in fact contained in a sphere of radius less then k, but no matter—a sphere of radius k certainly contains at least $\tfrac{1}{2}(3^{k+1} - 1)$ atoms. If the radius of a carbon atom is r, we have to pack carbon atoms with a total volume of $\tfrac{4}{3}\pi r^3 \cdot \tfrac{1}{2}(3^{k+1} - 1)$, or approximately $2\pi r^3 \cdot 3^k$, into a sphere of volume $\tfrac{4}{3}\pi k^3$.

This is clearly impossible for k sufficiently large, since 3^k increases much faster than k^3.

Hence there are limits to the extent to which equation (2.4) can be regarded as applicable to real-life chemistry. Nevertheless, it has provided chemists with much useful information.

Another application of Pólya's Theorem concerns *alkyl-substituted benzenes*. For our purposes, a benzene molecule is a hexagon, at the corners of which we can attach atoms or more complicated structures. It has long been known to chemists that there are three essentially different ways of attaching two atoms to such a 'benzene ring'; they are shown in Fig. 2.7, together with the prefixes which are customarily used to distinguish them. We see at once that this is precisely the result for the necklace problem in Section 2.2.

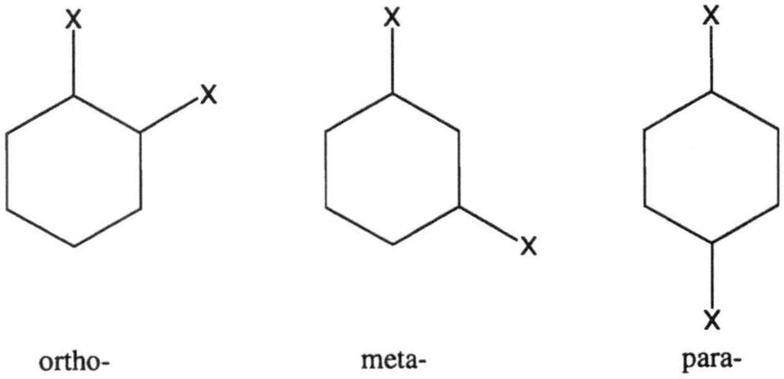

FIG. 2.7

An alkyl radical can be thought of as a monosubstituted alkane with the X removed. This leaves a 'free' bond which can be used to attach the radical to something else. If one or more such alkyl radicals are attached at the vertices of a benzene ring, we get an alkyl-substituted benzene (see Fig. 2.8).

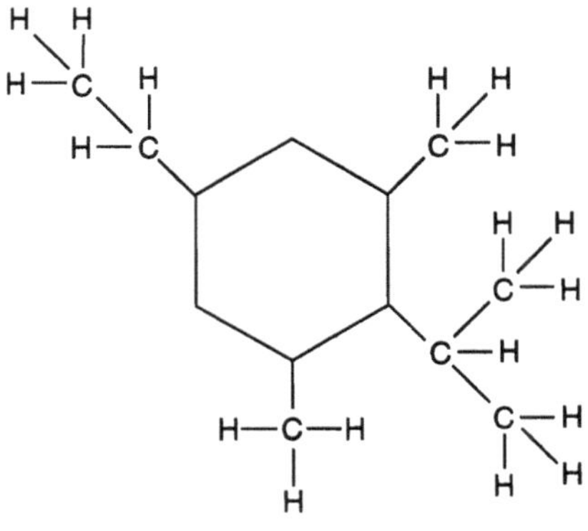

FIG. 2.8

How many such compounds are there with a given number of carbon atoms, excluding the six carbon atoms that make up the benzene ring? This is a straightforward Pólya-type problem, and we have all the information necessary to solve it. We have six sites, permuted by the dihedral group D_6, exactly as in the necklace problem. But now the figure generating function is $A(x)$, instead of just $1 + x$. Thus the solution is given by the generating function

$$\tfrac{1}{12}[A^6(x) + 2A(x^6) + 2A^2(x^3) + 4A^3(x^2) + 3A^2(x)A^2(x^2)].$$

Knowing the expansion of $A(x)$, we can compute the coefficients in this generating function as far as we wish. The first few terms in the expansion are

$$1 + x + 4x^2 + 8x^3 + 23x^4 + 57x^5 + 165x^6 + 452x^7 + 1311x^8 + 3773x^9 + 11091x^{10} + \cdots.$$

2.6 The enumeration of graphs

We now return to the main aim of this chapter, the problem of enumerating unlabelled graphs, since we now have the necessary apparatus to solve it. Consider graphs with n vertices. A pair of vertices may correspond to an edge or it may not. Thus we have $N = n(n-1)/2$ possible pairs of vertices, which will be the 'sites' for our problem, and at each site we can place one of two figures, 'edge' or 'non-edge'. If we give these figures content 1 and 0, respectively, then the content of the configuration (which is a graph) is the number of edges.

Since the vertices are unlabelled, we can permute them by any permutation of the symmetric group S_n of all permutations. Any permutation of vertices induces a permutation of the pairs of vertices (the sites). It is this set of permutations of the sites, denoted by $S_n^{(2)}$, that is relevant to this problem. We must therefore find its cycle index.

The case $n = 4$ provides some insight into what is involved. In this case we have six sites for possible edges. Permutations of the vertices having the same cycle type induce permutations of the sites with the same cycle type, although the converse is not true. Hence we need look at only one example of each cycle type. One such is the identity, which leaves all the sites unmoved; so, corresponding to the cycle type s_1^4 in $Z(S_4)$, we have s_1^6 in $Z(S_4^{(2)})$. Another type of permutation is one that interchanges just two vertices, say 1 and 2, leaving the others fixed (as in Fig. 2.9(a)). This maps the pairs (1,2) and (3,4) onto themselves, and it is easily verified that the other pairs interchange in twos. Hence the cycle type $s_1^2 s_2$ in $Z(S_4)$ gives rise to a term $s_1^2 s_2^2$ in $Z(S_4^{(2)})$.

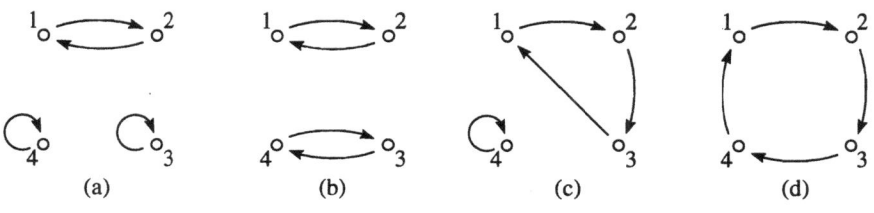

Fig. 2.9

Proceeding similarly with the three other types of permutations of four objects, illustrated diagrammatically in Fig. 2.9, we find the following relations.

in $Z(S_4)$	in $Z(S_4^{(2)})$	number
s_1^4	s_1^6	1
$s_1^2 s_2$	$s_1^2 s_2^2$	6
$s_1 s_3$	s_3^2	8
s_2^2	$s_1^2 s_2^2$	3
s_4	$s_2 s_4$	6

A standard result, which we quote without proof, is that the number of permutations in S_n with cycle type $s_1^{j_1} s_2^{j_2} \ldots$, where $j_1 + 2j_2 + 3j_3 + \cdots + nj_n = n$, is

$$\frac{n!}{1^{j_1}\, j_1!\; 2^{j_2}\, j_2! \ldots n^{j_n}\, j_n!}. \tag{2.5}$$

Computing these numbers in the case $n = 4$, we obtain the third column in the table above. From this, we deduce that

$$Z(S_4^{(2)}) = \tfrac{1}{24}\left[s_1^6 + 9s_1^2 s_2^2 + 8s_3^2 + 6s_2 s_4\right].$$

We now substitute the figure generating function $1 + x$ into this cycle index to obtain the generating function for these graphs. After simplification, it becomes

$$1 + x + 2x^2 + 3x^3 + 2x^4 + x^5 + x^6.$$

Here again we might be tempted to say that we have gone a long way round to obtain a result that could have been found easily by constructing the graphs themselves by a process of trial and error, taking due care to avoid duplicates or omissions. This is true; but the above method can be applied to the general case of n vertices, and a general formula can be obtained. We shall not attempt to give a full derivation of this formula, but here are some indications of what has to be done.

For the general cycle type, for which formula (2.5) gives the number of permutations, we have to consider what happens to a pair of vertices when the vertices are permuted by a permutation with this cycle type. In doing this we must consider separately

(a) pairs of vertices from the same cycle;
(b) pairs of vertices from different cycles of the same length;
(c) pairs of vertices from cycles of different lengths.

In (b), if the cycle length is h, then there are $\binom{h}{2}$ ways in which the two cycles can be chosen. In (c), if the cycles are of lengths h and k, then there are $j_h j_k$ choices; moreover, as the two vertices move around their respective cycles under repeated applications of the permutation, the number of steps before they return to their original positions, thus completing a cycle, is the lowest common multiple of h and k. These observations provide a few points of reference in the following general formula:

$$Z(S_n^{(2)}) = \frac{1}{n!} \sum \left\{ \frac{n!}{\Pi i^{j_i} \cdot j_i!} \prod_i s_{2i+1}^{ij_{2i+1}} \cdot \prod_i (s_i s_{2i}^{i-1})^{j_{2i}} s_i^{i\binom{j_i}{2}} \cdot \prod_{h<k} s_{[h,k]}^{(h,k)j_h j_k} \right\},$$

where $[h, k]$ is the lowest common multiple of h and k, (h, k) is their highest common factor, and the summation is over all cycle types. Substitution of $1 + x$ in this cycle index gives the numbers of graphs with n vertices, counted by numbers of edges.

Formidable though this formula may appear, it can be used to compute actual values. This has been done in many different places; an early list of these numbers appears in Stein and Stein [13], which gives all the numbers for graphs with up to 18 vertices.

Note that if we put $x = 1$ in the generating function for graphs on n vertices, we obtain the total number of graphs, irrespective of the number of edges. If the total number is all that we want to know, it is easier to put $x = 1$ *before* the substitution of $1 + x^i$ for s_i in the cycle index; that is, we replace each s_i in the cycle index by 2. This gives a much simpler result—namely,

$$\sum \frac{2^X}{\Pi i^{j_i} \cdot j_i!},$$

where

$$X = \sum_i ij_{2i+1} + \sum_i ij_{2i} + \sum_i i\binom{j_i}{2} + \sum_{h<k}(h,k)j_h j_k,$$

and the main summation is over all partitions of n. Total numbers of graphs were computed many years ago, up to $n = 24$, by King and Palmer [5]; their results are given in Table 2.1.

Table 2.1

n	graphs
1	1
2	2
3	4
4	11
5	34
6	156
7	1 044
8	12 346
9	274 668
10	12 005 168
11	1 018 997 864
12	165 091 172 592
13	50 502 031 367 952
14	29 054 155 657 235 488
15	31 426 485 969 804 308 768
16	64 001 015 704 527 557 894 928
17	245 935 864 153 532 932 683 719 776
18	1 787 577 725 145 611 700 547 878 190 848
19	24 637 809 253 125 004 524 383 007 491 432 768
20	645 490 122 795 799 841 856 164 638 490 742 749 440
21	32 220 272 899 808 983 433 502 244 253 755 283 616 097 664
22	3 070 846 483 094 144 300 637 568 517 187 105 410 586 657 814 272
23	559 946 939 699 792 080 597 976 380 819 462 179 812 276 348 458 981 632
24	195 704 906 302 078 447 922 174 862 416 726 256 004 122 075 267 063 365 754 368

2.7 Connected graphs

For many purposes, we are interested in the numbers of connected graphs rather than of all graphs. After all, a graph that is not connected is simply a collection of components, each of which is a connected graph. Thus we can, so to speak, build up any graph by suitably choosing a set of connected graphs. We shall use this fact to establish a link between the numbers of all graphs (which we now know) and the numbers of connected graphs.

Imagine that we have solved the enumeration problem for connected graphs, and that we have the answer in the form of a generating function of the form

$$c(x) = c_1 x + c_2 x^2 + c_3 x^3 + c_4 x^4 + \cdots + c_n x^n + \cdots,$$

where c_n is the total number of connected graphs with n vertices.

From the information contained in this equation, we shall now build up the generating function $g(x)$ for all graphs, connected or not. Any graph G can be specified uniquely by listing how many times each connected graph occurs as a component of G. Consider one particular connected graph with n vertices. If it occurs r times as a component of G, then these components contribute rn to the number of vertices of G. The possible contributions, for different values of r, are summarized by the generating function

$$1 + x^n + x^{2n} + x^{3n} + \cdots = (1 - x^n)^{-1}.$$

Now there are c_n connected graphs with n vertices, and each contributes a factor of $(1 - x^n)^{-1}$ to the generating function $g(x)$. Hence, together, they contribute $(1 - x^n)^{-c_n}$. Repeating this argument for all values of n, we see that

$$g(x) = \prod_{n \geq 1} (1 - x^n)^{-c_n}. \tag{2.6}$$

Now the coefficients in $g(x)$ are the numbers g_n given by the above table. Thus, although these numbers are not easy to compute, we can take them to be known quantities, and we therefore have the task of using equation (2.6) in reverse; so to speak, to compute the as yet unknown values c_n. This is not difficult. Comparing coefficients of $x, x^2, x^3, \ldots$ on both sides, we can determine the coefficients $c_1, c_2, c_3, \ldots$ one by one. The numbers of connected graphs with up to 20 vertices are given in Table 2.2.

Table 2.2

n	connected graphs
1	1
2	1
3	2
4	6
5	21
6	112
7	853
8	11 117
9	261 080
10	11 716 571
11	1 006 700 565
12	164 059 830 476
13	50 335 907 869 219
14	29 003 487 462 848 061
15	31 397 381 142 761 241 960
16	63 969 560 113 225 176 176 277
17	245 871 831 682 084 026 519 528 568
18	1 787 331 725 248 899 088 890 200 576 580
19	24 636 021 429 399 867 655 322 650 759 681 644
20	645 465 483 198 722 799 426 731 128 794 502 283 004

The numbers of connected graphs, broken down by the number of edges as well as of vertices, can be obtained from the corresponding numbers of all graphs in much the same way. It is more complicated, however, since the generating functions have to be functions of two variables, an extra variable being needed to keep track of the number of edges. We shall not discuss this problem here.

Note that the generating function $g(x)$ has the constant term 1. We can, if we wish, associate this term with a hypothetical 'null graph' with no vertices and no edges. This is quite logical, but leads to the strange conclusion that this null graph is a *disconnected* graph. For, if it were connected, we would have $c_0 = 1$ and we would have to allow $n = 0$ in the right-hand side of equation (2.6), which would then vanish altogether! Many graph theorists prefer not to interpret the constant in $g(x)$ in this way; a discussion of the pros and cons of this matter can be found in [4].

2.8 Trees and rooted trees

A *tree* is a connected graph without cycles. An enumeration problem of considerable interest is that of determining how many trees there are with n vertices. The generating function that answers this question will be quoted later, but it would take us too much out of our way to derive it rigorously. We can, however, with very little trouble, derive the generating function for rooted trees, a function which will be needed to express the quoted result for unrooted trees.

A *rooted tree* is a tree in which one vertex, called the *root*, has been distinguished in some way from the others. The monosubstituted alkanes that we met in Section 2.5 were basically rooted trees. A *rooted forest* is a graph in which each component is a rooted tree. There is an immediate connection between the numbers of rooted trees and the numbers of rooted forests; for, if we delete the root of a rooted tree, we obtain a forest, as shown in Fig. 2.10.

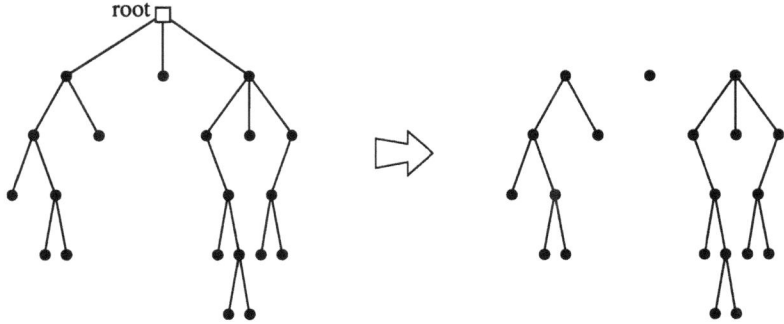

FIG. 2.10

Moreover, this forest is a rooted forest, since in each component the vertex that was adjacent to the original root is, for this very reason, distinguished from the other vertices in the component.

This correspondence between rooted trees and rooted forests is easily seen to be one-to-one, and we deduce that the number of rooted trees with n vertices is equal to the number of rooted forests with $n-1$ vertices.

Let
$$T(x) = T_1 x + T_2 x^2 + T_3 x^3 + \cdots$$
be the generating function for rooted trees, so that T_n is the number of rooted trees with n vertices. We can now use the same argument by which we derived equation (2.6) to write down the generating function for graphs in which every component is a rooted tree—that is, the generating function for rooted forests. It is
$$\prod_{n \geq 1} (1 - x^n)^{-T_n}.$$

It follows that the number of rooted trees with n vertices is the coefficient of x^{n-1} in this formula, or, what amounts to the same thing, the coefficient of x^n in
$$x \prod_{n > 1} (1 - x^n)^{-T_n}.$$

This formula is the generating function for rooted trees, which is what we have already called $T(x)$. Thus we are led to the equation
$$T_1 x + T_2 x^2 + T_3 x^3 + T_4 x^4 + \cdots$$
$$= x(1-x)^{-T_1}(1-x^2)^{-T_2}(1-x^3)^{-T_3}(1-x^4)^{-T_4}\ldots, \qquad (2.7)$$
a result first discovered by Cayley [1].

This equation does not yield an explicit formula for $T(x)$, but the coefficients can be computed one by one from it. For if we know $T_1, T_2, \ldots, T_n$ then, because of the multiplier x, the right-hand side of equation (2.7) gives us the value of T_{n+1}.

By an ingenious argument, too lengthy to be given here, it can be shown that the generating function $t(x)$ for *unrooted* trees (that is, trees in the usual sense) depends on $T(x)$ according to the equation
$$t(x) = T(x) - \tfrac{1}{2}[T^2(x) - T(x^2)].$$

This is known as *Otter's formula*, after its discoverer [6]. A more recent derivation of it can be found in [3].

The numbers of rooted and unrooted trees with up to 20 vertices are given in Table 2.3.

Table 2.3

n	rooted trees	unrooted trees
1	1	1
2	1	1
3	2	1
4	4	2
5	9	3
6	20	6
7	48	11
8	115	23
9	286	47
10	719	106
11	1 842	235
12	4 766	551
13	12 486	1 301
14	32 973	3 159
15	87 811	7 741
16	235 381	19 320
17	634 847	48 629
18	1 721 159	123 867
19	4 688 676	317 955
20	12 826 228	823 065

2.9 Other kinds of graphs

The enumeration presented in Section 2.6 was first set out by Frank Harary in 1955. His paper [2], still a useful reference for this type of problem, also dealt with a number of other graphical enumeration problems. We shall not deal with these in detail here, but look at three problems and see how they fall within the general ambit of Pólya's Theorem.

Directed graphs For directed graphs, the sites are *ordered* pairs of vertices. The enumeration goes through as before, using Pólya's Theorem, except that the group is now the group $S_n^{[2]}$ of permutations of these ordered pairs. Using the cycle index of this group, we find that the enumeration is given by the formula

$$Z(S_n^{[2]}; 1+x, 1+x^2, 1+x^3, \ldots).$$

The formula for $Z(S_n^{[2]})$ is similar to that for $Z(S_n^{(2)})$, and can be found in [2] or [3]. Note, for example, that

$$Z(S_4^{[2]}) = \tfrac{1}{24}(s_1^{12} + 6s_1^2 s_2^5 + 3s_2^6 + 8s_3^4 + 6s_4^3),$$

and that substitution of $1+x$ in this equation gives

$$1 + x + 5x^2 + 13x^3 + 27x^4 + 38x^5 + 48x^6 + 38x^7 + 27x^8 + 13x^9 + 5x^{10} + x^{11} + x^{12},$$

yielding a total of 218 digraphs with four vertices.

Multigraphs In a multigraph there can be many edges between the same pair of vertices. Hence the enumeration for multigraphs is much the same as for graphs, except that the set of figures for the sites is different. Instead of having just two figures, 'edge' and 'non-edge', we now have 'r-fold edge', where r can be any non-negative integer. We want the content of a configuration (multigraph), and hence that of a figure, to be the number of edges. Since we have exactly one figure of each content from 0 upwards, the figure generating function is

$$1 + x + x^2 + x^3 + \cdots = (1-x)^{-1}. \tag{2.8}$$

On substituting this into the cycle index of $S_n^{(2)}$, we obtain the generating function for multigraphs.

Multigraphs of strength s These are multigraphs in which the number of edges between two vertices does not exceed a given integer s. This is an easy variation on the preceding problem; we simply truncate the figure generating function (2.8) at x^s.

For further graph enumeration problems, see the previously mentioned paper by Harary or, for a wider survey, the book by Harary and Palmer [3].

2.10 Unsolved problems

Although Pólya's Theorem is a powerful weapon for enumerating graphs and many other combinatorial configurations, it is by no means all-powerful. Other theorems have been used for problems that Pólya's Theorem is unable to cope with. A theorem, commonly known as *Burnside's Lemma*, lies behind Pólya's Theorem and can be used to enumerate various kinds of oriented graphs. A theorem of De Bruijn, later generalized by Harary and Palmer as the *Power Group Enumeration Theorem* (see [3]), enables self-complementary graphs and digraphs and some other configurations to be counted. The *Superposition Theorem*, which I introduced in 1958, can be used to count some kinds of graphs with given degree sequence (see [9, 11]). Other techniques have been brought into play to enumerate yet more kinds of graphs.

For all that, there are many types of graphs for which no method of enumeration is known. To conclude this chapter, we list some outstanding unsolved problems in this area.

Identity graphs These are graphs with no automorphisms except the identity. The problem of enumerating these seems to be completely intractible.

Labelled self-complementary graphs Problems about labelled graphs are usually much easier than the corresponding problems for unlabelled graphs—recall the situation for ordinary graphs, as given above. This is not so with self-complementary graphs. Strictly speaking, there is no such thing as a labelled self-complementary graph on more than one vertex. For if the vertices are labelled, then the complement of a graph *has* to be different from the original graph. The enumeration problem in question should really be re-stated as: for a given number n of vertices, what total do we get if we count every self-complementary graph on n vertices as many times as it has labellings? Rephrased in this way, the problem is seen to be a somewhat artificial one—a fact which no doubt explains its difficulty.

Graphs with given degree sequence If multiple edges and loops are allowed, then there is an explicit (although not entirely practical) solution to this problem using the Superposition Theorem (see [11]). But for graphs in the usual sense, there seems little prospect of a solution.

Triangle-free graphs These are graphs with no cycles of length 3. Labelled triangle-free cubic graphs have been enumerated (see [14]), but the unlabelled problem remains open.

Hamiltonian graphs The enumeration of graphs with a Hamiltonian cycle is still open.

Planar graphs Although some special classes of planar graphs have been enumerated, the general problem remains open.

Self-complementary and self-converse digraphs The *complement* of a digraph is obtained by replacing each arc by a non-arc, and vice versa; the *converse* of a digraph is obtained by reversing the directions of all the arcs. This is an example of an enumeration problem in which the digraphs are required to have two properties simultaneously; such problems tend to be very difficult.

These are just a few of many unsolved problems in graphical enumeration. Many more can be found in Harary and Palmer's book [3]. This book is a good introduction and starting point for those who wish to pursue further studies of the subject, and join the ranks of the 'graph theorists who count'.

References

1. A. Cayley, On the theory of the analytical forms called trees, *Phil. Mag.* **13** (1857), 172–176 = *Math. Papers* **3** (1890), 242–246.
2. F. Harary, The number of linear, directed, rooted and connected graphs, *Trans. Amer. Math. Soc.* **78** (1955), 445–463; *MR* **16**–844.
3. F. Harary and E. M. Palmer, *Graphical Enumeration*, Academic Press, 1973; *MR* **50**#9682.
4. F. Harary and R. C. Read, Is the null-graph a pointless concept?, *Graphs and Combinatorics* (ed. R. A. Bari and F. Harary), Lecture Notes in Math. **406**, Springer-Verlag, 1974, pp. 37–44; *MR* **50**#12819.
5. C. King and E. M. Palmer, Calculation of the number of graphs of order $p = 1(1)24$, unpublished report.
6. R. Otter, The number of trees, *Ann. of Math. (2)* **49** (1948), 583–599; *MR* **10**–53c.
7. G. Pólya, Kombinatorische Anzahlbestimmungen für Gruppen, Graphen und chemische Verbindungen, *Acta Math.* **68** (1937), 145–254.
8. G. Pólya and R. C. Read, *Combinatorial Enumeration of Groups, Graphs, and Chemical Compounds*, Springer-Verlag, 1987; *MR* **89f**: 05013.
9. R. C. Read, The enumeration of locally restricted graphs, I, II, *J. London Math. Soc.* **34** (1959), 417–436, and **35** (1960), 344–451; *MR* **21**#7162 and **25**#3863.

10. R. C. Read, On the number of self-complementary graphs and digraphs, *J. London Math. Soc.* **38** (1963), 99–104; *MR* **26**#4339.
11. R. C. Read, The use of S-functions in combinatorial analysis, *Canad. J. Math.* **20** (1968), 808–841; *MR* **37**#5108.
12. J. H. Redfield, The theory of group-reduced distributions, *Amer. J. Math.* **49** (1927), 433–455.
13. M. L. Stein and P. R. Stein, *Enumeration of linear graphs and connected linear graphs up to $p = 18$ points*, Report LA–3775, Los Alamos Scientific Lab. of the Univ. of California, 1967.
14. N. C. Wormald, The number of labelled cubic graphs with no triangles, *Congr. Numer.* **33** (1981), 359–378; *MR* **84d**: 05097.

3
Number Theory

ROGER COOK

Number theory is concerned with the properties of the positive integers. It includes topics such as primality and factorization, multiplicative functions, congruences and quadratic residues, and Diophantine equations. Examples in this chapter show that each of these topics has relevance to graph theory.

3.1 Introduction

Number theory is a collection of various results and topics, arising from the properties of the set **N** of positive integers. The central topics that appear in most undergraduate courses in number theory (see, for example, Hardy and Wright [20] or Schroeder [38]) are the following.

Prime numbers and factorization A positive integer $n > 1$ is *prime* if it has no positive integer divisors other than 1 and itself. The positive integers have the Unique Factorization Property: any integer $n > 1$ can be expressed as the product of prime powers

$$n = p_1^{a_1} p_2^{a_2} \ldots p_r^{a_r},$$

where the p_i are distinct primes, and this representation is unique up to the ordering of the primes.

Multiplicative functions A function $f : \mathbf{N} \to \mathbf{C}$ is *multiplicative* if

$$f(mn) = f(m)f(n),$$

whenever m and n are relatively prime—that is, they have greatest common divisor 1. If n has the above factorization, then

$$f(n) = f(p_1^{a_1}) \ldots f(p_r^{a_r}).$$

Note that multiplicative functions are determined by their values at the prime powers. Two important multiplicative functions that have several applications in graph theory are the *Möbius function* μ and *Euler's function* ϕ. The Möbius function is defined by

$$\mu(n) = \begin{cases} 1, & \text{if } n = 1, \\ (-1)^r, & \text{if } n \text{ is a product of } r \text{ distinct primes}, \\ 0, & \text{if } n \text{ has a repeated prime factor}. \end{cases}$$

Euler's function $\phi(n)$ is the number of integers k with $1 \leq k \leq n$ and k relatively prime to n.

Congruences and quadratic residues The idea of congruence is commonly introduced as an example of an equivalence relation. The residue classes modulo m form a ring $\mathbf{Z}_m$, and for a prime p the residue classes modulo p form a field $\mathbf{Z}_p$. If p does not divide a, then a is a *quadratic residue* $(\bmod\, p)$ if there are solutions x to the congruence
$$x^2 \equiv a \,(\bmod\, p).$$
Then $(p-1)/2$ of the non-zero residues $(\bmod\, p)$ are quadratic residues and the remaining $(p-1)/2$ are quadratic non-residues. The *Legendre symbol*
$$\left(\frac{a}{p}\right) = \begin{cases} 1, & \text{if } a \text{ is a quadratic residue,} \\ 0, & \text{if } p \text{ divides } a, \\ -1, & \text{if } a \text{ is a quadratic non-residue,} \end{cases}$$
is an example of a character in a finite field.

In general, a finite field F has a prime power $q = p^k$ elements. A *character* χ of F is a mapping $\chi : F \to \mathbf{C}$ such that $\chi(0) = 0$ and $\chi(mn) = \chi(m)\chi(n)$ for all m and n in F. Estimates for character sums lead to important properties for several classes of graphs.

Diophantine equations If d is square-free, solutions (x, y) to Pell's equation
$$x^2 - dy^2 = 1$$
can be obtained by using continued fractions. For the 'negative' Pell's equation
$$x^2 - dy^2 = -1,$$
the situation is less clear. Recent density results, due to Cremona and Odoni [10], use graph-theoretic enumeration to show that the equation is soluble for a set of values d with positive density.

The examples in this chapter show some of the many links between number theory and graph theory. Hopefully, they may point to other areas where the connections may be developed further.

3.2 Multiplicative functions

Multiplicative functions are determined by their values at the prime powers and, in some sense, they preserve the multiplicative structure of $\mathbf{N}$ in the image space. A more complete account of multiplicative functions and their uses can be found in Apostol [1].

The generating functions most naturally associated with multiplicative functions are the *Dirichlet series*
$$F(s) = \sum_{n=1}^{\infty} f(n) n^{-s},$$

which we treat formally. The best-known example is the *Riemann zeta function*

$$\zeta(s) = \sum_{n=1}^{\infty} n^{-s}.$$

Multiplying two Dirichlet series together and rearranging the terms gives

$$F(s)G(s) = \sum_{m=1}^{\infty} \sum_{n=1}^{\infty} f(m)g(n)(mn)^{-s} = \sum_{k=1}^{\infty} h(k)k^{-s},$$

where

$$h(k) = \sum_{mn=k} f(m)g(n) = \sum_{d|k} f(d)g(k/d).$$

If we write the generating function for the Möbius function as

$$M(s) = \sum_{n=1}^{\infty} \mu(n)n^{-s},$$

then it is easy to see that

$$M(s) = 1/\zeta(s). \tag{3.1}$$

The use of generating functions and Dirichlet series represents an analytic approach to multiplicative functions. A more algebraic approach was developed by Bell (see Apostol [1, p. 29]). The *Dirichlet convolution* $*$ is defined on the class of multiplicative functions by

$$(f * g)(k) = \sum_{mn=k} f(m)g(n).$$

Then $*$ is closed, commutative and associative on the class of multiplicative functions (see Apostol [1]). The identity for $*$ is the function

$$i(n) = \begin{cases} 1, & \text{if } n = 1, \\ 0, & \text{if } n > 1. \end{cases}$$

The Dirichlet series associated with i is the constant 1, and equation (3.1) can be written in the form $\mu * u = i$, where the function u is identically 1, so that its Dirichlet series is the zeta function. This leads to the following result.

Theorem 3.1 (Möbius Inversion Formula) $f = \mu * g$ *if and only if* $g = f * u$.

3.3 The Möbius function

Rota [35] gave a treatment of the Möbius function in a more general combinatorial setting. If P is a partially ordered set, the partial order on P defines a function

$$\zeta(i,j) = \begin{cases} 1, & \text{if } i \leq j, \\ 0, & \text{otherwise;} \end{cases}$$

this is called the *zeta function of the set* P. For any complex-valued function f on P, we can define a second function g on P by

$$g(x) = \sum_{y \leq x} f(y).$$

Note that when P is the set of natural numbers ordered by division, we have $g = f * u$. We can recover the function f by a Möbius Inversion Formula

$$f(x) = \sum_{y \leq x} \mu(y,x) g(y),$$

where μ is the inverse function for ζ. Thus, whenever we can compute the sum $g(x)$, we can recover the function $f(x)$ by Möbius inversion.

Wilf [42] applied this generalized inversion formula to the chromatic polynomial of a graph. Suppose that we arbitrarily assign t colours to the n vertices of a graph G. Some edges will be *bad*, in that both ends are assigned the same colour. The set of bad edges in G that result from a particular colouring is called the *bond* of the colouring. A set S of edges of G which can occur as the bond of a colouring is called *bond-closed*. For each bond-closed set S, let $p(t, S)$ denote the number of ways of assigning t colours to the vertices of G so that the bond of the colouring is precisely S.

Suppose that S has v vertices, m edges and c components. Any colouring of G whose bond T contains S must colour all the vertices in each connected component of S with the same colour. Thus there are t^c ways of colouring the vertices of S. When we form the sum

$$\sum_{T \supseteq S} p(t, T),$$

the $n - v$ vertices of G that are not in S can be coloured in any of the t colours, so that

$$\sum_{T \supseteq S} p(t, T) = t^{n-v+c}.$$

This sum can be inverted, using Möbius inversion, to give $p(t, S)$. By taking the partially ordered set to be the subsets of vertices of G, ordered by set inclusion, we get an explicit expression for $p(t, S)$.

Theorem 3.2 *Let G be a graph, and let S be a bond-closed set of edges of G. Then the number of ways of colouring the vertices of G in t colours, so that the bond of the colouring is precisely S, is*

$$p(t, S) = \sum_{T \supseteq S} \mu(S, T) t^{n-v(T)-c(T)}.$$

3.4 Euler's function

Classifying the integers $1, 2, \ldots, n$ by their greatest common divisor with n, we have
$$n = \sum_{d|n} \sum_{\substack{k=1 \\ (k,n)=d}}^{n} 1 = \sum_{d|n} \phi(n/d).$$

If f is the function with $f(n) = n$ identically, then $f = u*\phi$. By Möbius inversion, $\phi = \mu * f$; that is,
$$\phi(n) = n \sum_{d|n} \frac{\mu(d)}{d}.$$

Since ϕ is the Dirichlet convolution of two multiplicative functions, ϕ is itself multiplicative (see Apostol [1, p. 35]). Euler's function, defined in terms of the integers $1, 2, \ldots, n$, plays an important part in the structure of the cyclic group $\mathbf{Z}_n$, and it therefore occurs in many enumeration problems.

Let A be a permutation group acting on the set $X = \{1, 2, \ldots, n\}$. Any permutation a in A can be written uniquely as a product of disjoint cycles. For $k = 1, 2, \ldots, n$, let $j(a; k)$ denote the number of cycles of length k in this representation of a. The *cycle index* of A, denoted by $Z(A)$ or $Z(A; s_1, \ldots, s_n)$, is defined by
$$Z(A) = |A|^{-1} \sum_{a \in A} \prod_{k=1}^{n} s_k^{j(a;k)}.$$

This polynomial was discovered independently by Redfield [33] in 1927 and Pólya in 1935 (see [31]); it has already been discussed in Chapter 2. Our emphasis here is to give examples that show how number-theoretic functions arise in enumeration problems.

Let $\mathbf{Z}_n$ be the cyclic group of order n, the group acting on the set $\{1, 2, \ldots, n\}$ generated by the cycle $(12 \ldots n)$. Redfield [33] showed that
$$Z(\mathbf{Z}_n) = n^{-1} \sum_{d|n} \phi(d) s_d^{n/d}.$$

Now let B be a finite permutation group acting on a countable set Y with at least two elements. The *power group* B^A acts on the collection Y^X of functions $f : X \to Y$. The permutations in B^A are the ordered pairs (a, b), with a in A and b in B. The image of any function f in Y^X under the permutation (a, b) is defined by
$$((a, b)f)(x) = bf(ax),$$
for each x in X. Then (see Harary [17, p. 184]) the cycle index is given by
$$Z(B^A) = (|A| \cdot |B|)^{-1} \sum_{(a,b)} s_k^{j((a,b);k)},$$

where
$$j((a,b);1) = \prod_{k=1}^{d}(\sum_{s|k} sj(b;s))^{j(a;k)}$$

and, for $k > 1$,
$$j((a,b);k) = k^{-1}\sum_{s|k} j((a,b);1)\mu(k/s),$$

where μ is the usual Möbius function.

3.5 Pólya's Enumeration Theorem

We recall from Chapter 2 that if $c(x)$ denotes the figure generating function and $Z(A;c(x))$ the cycle index

$$Z(A;c(x),c(x^2),c(x^3),\ldots),$$

then the central result in graphical enumeration can be stated in the following form.

Theorem 3.3 (Pólya's Enumeration Theorem) *The configuration generating function $C(x)$ is determined by substituting for each variable s_k in $Z(A)$ the figure generating function $c(x^k)$.*

Symbolically, we have
$$C(x) = Z(A;c(x)).$$

An *r-set* of the object set X is a subset with r elements. Two r-sets S and S' of X are *A-equivalent* when $S' = aS$ for some a in A.

Corollary 3.4 *The coefficient of x^r in $Z(A;1+x)$ is the number of A-equivalence classes (or orbits) of r-sets of X.*

Many results can be obtained by using this corollary rather than the full version of Pólya's Enumeration Theorem. For the cyclic group $\mathbf{Z}_n$,

$$Z(\mathbf{Z}_n;1+x) = n^{-1}\sum_{k|n} \phi(k)(1+x^k)^{n/k}$$

(see Harary and Palmer [18] and Chapter 2).

The enumeration of trees is particularly important, since it has applications to the enumeration of chemical structures. The earliest molecular graph enumeration is due to Cayley [7], and in 1948 Otter [29] gave enumeration formulas for trees. A *rooted tree* has one of its vertices, called the *root*, distinguished from the others. A *plane tree* consists of a tree together with a particular embedding in the plane. An *isomorphism of plane trees* is a graph isomorphism that preserves the cyclic order of the edges at each vertex. A *planted tree* is a rooted tree in which the root has degree 1. Harary, Prins and Tutte [19] used Pólya's Theorem to obtain the counting function $R(x)$ for rooted plane trees in terms of the cycle index sum of the cyclic group and the corresponding function $P(x)$ for planted plane trees.

Theorem 3.5 *If $P(x)$, $Q(x)$ and $R(x)$ are the counting functions for planted, ordinary and rooted plane trees, then*

$$P(x) = \sum_{n=1}^{\infty} n^{-1} \binom{2n-2}{n-1} x^{n+1}, \quad R(x) = x \sum_{n=0}^{\infty} Z(\mathbf{Z}_n; P(x)/x)$$

and

$$Q(x) = R(x) - (1/2x^2)(P^2(x) - P(x^2)).$$

The coefficients of

$$P(x) = 1 + x + 2x^2 + 5x^3 + \cdots$$

are the *Catalan numbers*

$$u_n = \frac{1}{n+1}\binom{2n}{n}.$$

Walkup [40] observed that there is a one-to-one correspondence between the set of rooted plane trees with n edges and a root of degree m ($1 \leq m \leq n$) and the set of equivalence classes of sequences of m non-trivial planted plane trees with a total of n edges. Combining this observation with Pólya's Theorem, he obtained the following result.

Theorem 3.6 *Let $r(n)$ denote the number of non-isomorphic unlabelled rooted plane trees with n edges. Then, for $n > 1$,*

$$r(n) = \frac{1}{2n} \sum_{d|n} \phi(n/d) \binom{2d}{d},$$

where ϕ is Euler's function.

3.6 Eulerian graphs and tournaments

The simplest non-trivial congruence problem is to determine whether an integer is even or odd. Even such simple considerations play a fundamental role in one of the oldest results in graph theory. The problem of the Königsberg bridges is usually regarded as the origin of graph theory (see Biggs, Lloyd and Wilson [5] or Fleischner [14]). Euler showed why it is impossible to walk round the bridges of Königsberg in such a way that each bridge is crossed exactly once.

An *Eulerian trail* in a graph is a trail that contains each edge of the graph exactly once. A graph is *Eulerian* when it has a closed Eulerian trail that passes through all the vertices. Using different terminology, Euler (see [5, p. 3]) gave necessary and sufficient conditions for a graph to be Eulerian, but proved only the necessity. The sufficiency was shown by Hierholzer in 1873 (see [5, p. 11]); apparently he was unaware of Euler's work.

Theorem 3.7 *A connected graph G is Eulerian if and only if each vertex has even degree.*

Closely related to the question of Eulerian trails is the *Chinese postman problem*, first stated in 1960 by Meigu Guan (Mei-Ko Kwan) [23]: *find, in a connected graph G, a shortest closed walk covering each edge at least once.* If G is a connected graph with m edges, then the shortest such walk contains between m and $2m$ (inclusive) edges. It has m edges if and only if G is Eulerian. Edmonds and Johnson [12] considered a more general problem, in which each edge e of the graph G has an associated cost $c(e)$. They showed that finding a postman's walk of minimum cost in G is equivalent to solving the following programming problem: *find non-negative integers $x(e)$ that minimize*

$$\sum_{e \in E(G)} c(e) x(e),$$

subject to the constraint that for each vertex v of G,

$$\sum (1 + x(e)) \equiv 0 \,(\mathrm{mod}\, 2),$$

where the summation extends over all edges e incident with v.

A round-robin sporting tournament, in which each of n teams plays each other team once, can be identified with a complete graph on n vertices where each edge ij is given the orientation $i > j$ to signify that team i has beaten team j. More formally, a *tournament* is a directed graph in which each pair of vertices is joined by exactly one arc. The problem of scheduling tournaments can be solved using congruences, with an algorithm that has been rediscovered many times (see Moon [27]).

If n is odd, then not all teams can be scheduled in each round, and so we introduce a dummy team. If a team is paired with the dummy team, then it has a bye in that round. We can thus suppose that there are an even number of teams, labelled $1, 2, \ldots, n$. Teams i ($i \neq n$) and j ($j \neq n$) with $i \neq j$ are scheduled to meet in round k, where

$$i + j \equiv k \,(\mathrm{mod}\,(n-1)).$$

This schedules all the games in round k, apart from team n and the unique team i with

$$2i \equiv k \,(\mathrm{mod}\,(n-1)),$$

who are thus scheduled to meet. Since n is even, $(2, n-1) = 1$ and this congruence has a unique solution. It is then easy to check that each team plays each of the others exactly once. In Fig. 3.1, we illustrate this construction for the case $n = 6$.

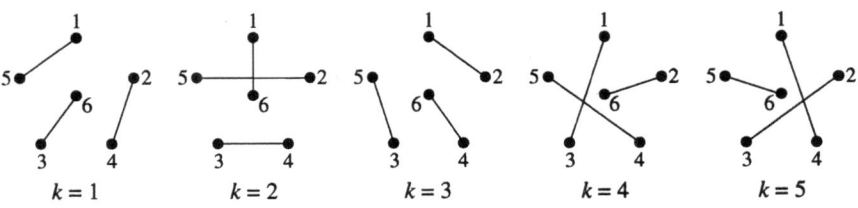

FIG. 3.1

3.7 Finite fields and Paley graphs

If p is prime, the residue classes modulo p form the finite field F_p. Recall that the order of every finite field $F = F_q$ is a prime power, say $q = p^k$. The elements of F_q can be identified with polynomials of degree at most $q - 1$ over the field F_p, subject to the relation $x^q = x$. A non-zero element a in F is a quadratic residue when there is a solution y in F to the equation $y^2 = a$. Then -1 is a quadratic residue in F_q when $p \equiv 1 \pmod{4}$, and it is not a quadratic residue when $p \equiv 3 \pmod{4}$.

The *Paley graph*, or *quadratic residue graph*, P_q can be constructed whenever $p \equiv 1 \pmod{4}$. It has q vertices, identified with the elements of F_q. Two vertices x and y are adjacent if and only if $x - y$ is a quadratic residue. Because -1 is a quadratic residue, this properly defines an undirected graph. The Paley graph P_{13} is shown in Fig. 3.2.

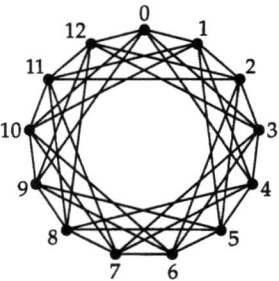

FIG. 3.2

The non-zero elements of F_q form a cyclic group F^* under multiplication. If g is a generator of this group, then the quadratic residues are precisely the even powers of g, so there are $(q-1)/2$ quadratic residues. Thus the Paley graph P_q is a subgraph of the complete graph K_q in which each vertex has degree $(q-1)/2$. Its behaviour is similar to a random subgraph of K_q, where each edge occurs with probability $\frac{1}{2}$ (see Bollobás [6]).

Theorem 3.8 *The graph P_q is regular of degree $(q-1)/2$, any two adjacent vertices have $(q-5)/4$ common neighbours, and any two non-adjacent vertices have $(q-1)/4$ common neighbours.*

Further refinements of this result can be obtained by using deep estimates for character sums. The characters χ of F are obtained by taking a generator g of F^* and defining χ on F^* by taking $\chi(g)$ to be a $(q-1)$th root of unity and

$$\chi(g^r) = \chi(g)^r.$$

The character is then extended to F by taking $\chi(0) = 0$. The principal character χ_0 is given by taking $\chi_0(g) = 1$, so that χ_0 is identically 1 on F^*. The quadratic residue character is obtained by taking $\chi(g) = -1$, so that χ takes the value 1 at quadratic residues and -1 at quadratic non-residues. The characters form a group, the *dual* of F^*, and the order of a character χ is the least positive integer d such that χ^d is the principal character χ_0. If χ is a quadratic character and f is a polynomial with coefficients in the field F_q, then

$$\sum_{x \in F} \chi(f(x)) = S - S',$$

where S and S' are the numbers of values x for which f is a quadratic residue and non-residue, respectively. This observation can be used to give bounds for character sums, but similar results hold for arbitrary characters. The general result is due to Weil [41], and an alternative proof is given in Schmidt [37].

Theorem 3.9 *Let χ be a character of order $d > 1$. Suppose that f is a polynomial with coefficients in F_q, and suppose that f is not a dth power—that is, $f(x)$ is not of the form $cg(x)^d$. If $f(x)$ has m distinct zeros in F_q, then*

$$\left| \sum_{x \in F} \chi(f(x)) \right| \le (m-1)q^{1/2}.$$

The trivial bound for this sum is q; Weil's result shows that there is so much cancellation taking place in the sum that we get a bound closer to the expected absolute value $q^{1/2}$ of a random walk in the plane with q steps of length 1. This result can be used to show that the Paley graph P_q has further properties that are close to those of the random graph (see [6, Chapter 13]).

Theorem 3.10 *Let U and W be disjoint sets of vertices of the Paley graph P_q, and let $v(U, W)$ be the number of vertices not in U or W and joined to each vertex of U and no vertex of W. Then*

$$|v(U, W) - 2^{-m}q| \le \tfrac{1}{2}[(m - 2 + 2^{1-m})q^{1/2} + m].$$

3.8 Quadratic residue tournaments

When $q = p^k$ with $p \equiv 3 \pmod 4$, the construction of the Paley graph P_q does not work. Since -1 is not a quadratic residue, $x - y$ is a quadratic residue if and only if $y - x$ is not. However, the construction can be modified to give a directed graph—in fact, a tournament. The *quadratic residue* (or *Paley*) *tournament* T_q has q vertices, identified with the elements of the field, with an arc from x to y precisely when $y - x$ is a quadratic residue (see Bollobás [6]); the quadratic residue tournament T_7 is shown in Fig. 3.3.

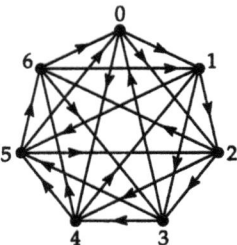

FIG. 3.3

An *automorphism* of a tournament is a permutation of the vertices that preserves the dominance relation on the vertices. The tournament T is called *vertex-homogeneous* when there is an automorphism T which sends u to v, for each pair of vertices u and v. Astie [2] proved the following result.

Theorem 3.11 *The number of vertex-homogeneous tournaments of prime order p is*

$$\frac{1}{p-1} \sum_{\substack{d \text{ odd} \\ d \mid (p-1)}} \phi(d) 2^{(p-1)/2d}.$$

A tournament T is *arc-homogeneous* if for each pair of arcs uv and xy, there is an automorphism of T taking u to x and v to y. With the trivial exception of the tournament of order 2, every arc-homogeneous tournament is also vertex-homogeneous. If T is a vertex-homogeneous tournament on n vertices, then n is odd. Further, for any vertex v, the subtournament induced on the neighbourhood of v is also vertex-homogeneous. It follows that $n \equiv 3 \pmod 4$. Goldberg [16] proved the following result, which was discovered independently by Berggren [3].

Theorem 3.12 *The automorphism group of the quadratic residue tournament T_q consists of all permutations π with*

$$\pi(x) = a^2 \alpha(x) + c,$$

where α is an automorphism of the field F_q, and a and c are elements of F_q with $a \neq 0$.

Fried [15] observed that for any arcs ij and hk in T_q, the permutation

$$\pi(x) = \frac{k-h}{j-i}x + h - \frac{k-h}{j-i}i$$

sends i to h and j to k. Further, it follows easily from the cyclic structure of F^* that $(k-h)/(j-i)$ is a quadratic residue, and that π is an automorphism of T. Thus every quadratic residue tournament is arc-homogeneous. Berggren [3] established that they are the only ones.

Theorem 3.13   *A tournament with n (≥ 3) vertices is arc-homogeneous if and only if it is a quadratic residue tournament.*

For further results on homogeneous tournaments, see Reid and Beineke [34].

3.9   Hadamard matrices and designs

Paley's original construction [30] was of tournaments rather than graphs. His interest arose through using them to construct Hadamard matrices. A square $n \times n$ matrix $\mathbf{H}$ is a *Hadamard matrix* of order n if each entry is 1 or -1, and $\mathbf{HH}^T = \mathbf{I}$. Let $\mathbf{A} = (a_{ij})$ be the adjacency matrix of the quadratic residue tournament T_q. Then

$$a_{ij} = \chi(i-j),$$

where χ is the quadratic residue character, and i and j are elements of F_q. Let $\mathbf{B} = \mathbf{A} - \mathbf{I}$, where $\mathbf{I}$ is the identity matrix of order q. A Hadamard matrix $\mathbf{H}$ of order $q+1$ can be formed by adding a border of 1s as the first row and column to the matrix $\mathbf{B}$. This gives an explicit construction of Hadamard matrices of order $q+1$, whenever q is a power of a prime $p \equiv 3 \pmod{4}$.

When $p \equiv 1 \pmod{4}$, the Paley graph P_q can be used to define a square $(q+1) \times (q+1)$ matrix that closely resembles a Hadamard matrix, in that it still satisfies $\mathbf{HH}^T = \mathbf{I}$ and all entries are 1 or -1, except that each entry on the main diagonal is 0. Starting with the adjacency matrix $\mathbf{A}$ of P_q, take $\mathbf{B} = 2\mathbf{A} + \mathbf{I} - \mathbf{J}$, where $\mathbf{J}$ is the $q \times q$ matrix with all entries 1. The $(q+1) \times (q+1)$ matrix $\mathbf{H}$ is formed by adding a border of 1s to the first row and column, except that the entry on the main diagonal is 0.

A (v, k, λ) *t-design* consists of a set V of v *varieties* and a collection of distinct k-element subsets, called *blocks*, such that any t-element subset of V occurs in exactly λ blocks (see, for example, Hughes and Piper [21]). A Hadamard matrix $\mathbf{H}$ is *normalized* if each entry in the first row and column is 1. Hadamard designs are obtained from normalized $n \times n$ Hadamard matrices by deleting the first row and column, and then replacing each -1 by 0. What is left is an $(n-1) \times (n-1)$ matrix that is the incidence matrix of an $(n-1, (n-2)/2, (n-4)/4)$ 2-design. Conversely, any Hadamard matrix of order $4(\lambda+1)$ gives a $(4\lambda+3, 2\lambda+1, \lambda)$ 2-design by reversing the procedure (see [21]). In particular, when the Hadamard matrix is constructed from a quadratic residue tournament in the manner described above, the design is called a *Paley design*.

Strongly regular graphs are closely related to designs. A k-regular graph G of order v is *strongly regular* with parameters (v, k, l, m) if there are integers l and m such that

if x and y are adjacent vertices, then there are exactly l vertices adjacent to both x and y;

if x and y are non-adjacent vertices, then there are exactly m vertices adjacent to both x and y.

For example, the Petersen graph is strongly regular with parameters $(10, 3, 0, 1)$, and when $q = p^k$ with $p \equiv 1 \pmod 4$, the Paley graph P_q is strongly regular with parameters $(q, (q-1)/2, (q-5)/4, (q-1)/4)$.

3.10 Ramanujan graphs

Telecommunications networks are frequently modelled using graph theory. Such networks have to transmit information quickly, so the question of how long it takes news (or gossip or a virus) to spread through the network is a central one. News spreads rapidly through the network if, for each subset X of the vertex set, there are many neighbours of the vertices of X that are not themselves in X. More formally, for a graph $G = (V, E)$ and a subset X of V, the *boundary* of X is the set

$$\partial X = \{v \in V - X : vx \in E \text{ for some } x \in X\}.$$

An (n, k, c)-*magnifier* is a graph G with n vertices and maximum degree k such that for each subset X of V with $|X| \leq n/2$, we have

$$|\partial X| \geq c|X|.$$

In the case of bipartite graphs, an (n, k, c)-*expander* is a bipartite graph on the sets I (inputs) and O (outputs), where $|I| = |O| = n$, the maximum degree is k, and for each subset X of I we have

$$|\partial X| \geq \{1 + c(1 - |X|/n)\}|X|.$$

If the bipartite graph G is an expander with $c \geq 0$, then G satisfies the conditions of Hall's Marriage Theorem.

The *magnification*, or *expansion*, *coefficient* of a graph is the maximum value of c for which the above inequalities are satisfied. The larger the value of this coefficient, the more quickly news spreads through the network. On the other hand, the maximum degree k gives some indication of the cost of the network. Probabilistic counting arguments show that almost all random graphs have large c and small k. The Ramanujan graphs are nearly optimal in terms of the parameters c and k; proving that they have the required properties depends on Ramanujan's conjectures on the Fourier coefficients of certain cusp forms (see Sarnak [36]). These conjectures follow from Deligne's proof of the Riemann–Weil Hypothesis [11].

Let $G_{n,k}$ be a k-regular graph of order n, and consider the eigenvalues of its adjacency matrix $\mathbf{A}$. The largest eigenvalue of $G_{n,k}$ is k; let μ be the absolute value of the next to largest (in absolute value) eigenvalue. An asymptotic lower bound for μ (see Lubotzky, Phillips and Sarnak [25]) states that

$$\liminf_{n \to \infty} \mu(G_{n,k}) \geq 2\sqrt{k-1}.$$

This lower bound gives rise to the following definition. A k-regular graph G is a *Ramanujan graph* if

$$\mu \leq 2\sqrt{k-1}.$$

These Ramanujan graphs are close to optimal in minimizing μ, and graphs with small μ have a large magnification coefficient. One example of a Ramanujan graph is the k-regular infinite tree. Finite examples were constructed by Lubotzky, Phillips and Sarnak [25] and Margulis [26]. Morgenstern [28] constructed new infinite families of $(q+1)$-regular graphs, for any prime power q.

Erdős [13] proved that there exist graphs with large girth and large chromatic number (see Chapter 13). His proof was probabilistic and does not give an explicit construction. Lovász [24] gave an explicit construction of such graphs using setsystems. Cook [9] showed that the graphs constructed by Lovász also have large connectivity. The Ramanujan graphs have large connectivity (see Bien [4]), while the constructions of Lubotzky, Phillips and Sarnak [25] and Margulis [26] give bipartite Ramanujan graphs with large girth. The following result is taken from Bien [4].

Theorem 3.14 *Let p and m be distinct odd primes such that p is not a square (modulo m). Then there exists a bipartite Ramanujan graph $X_{p,m}$ which is $(p+1)$-regular with order $n = (m^3 - m)/2$ and whose girth $g(X)$ satisfies*

$$\liminf_{m \to \infty} g(X_{p,m}) \geq \tfrac{4}{3} \log_p n.$$

This shows that the Ramanujan graphs have larger girth than any previously known family of graphs. Chung [8] established a bound for the diameter of k-regular graphs, using an estimate of Katz [22] for character sums.

Theorem 3.15 *The diameter $d(G)$ of a k-regular graph G of order n satisfies the inequality*

$$d(G) \leq \left\lceil \frac{\log(n-1)}{\log(k/\mu)} \right\rceil.$$

Since, for fixed k, μ is asymptotically minimal (as $n \to \infty$) for Ramanujan graphs, these graphs also minimize the diameter. Since the diameter of a graph corresponds to the worst-case transmission time in the corresponding telecommunications network, the Ramanujan graphs provide models of communications networks that are extremely good for both average and worst-case transmission times.

3.11 Negative Pell equations

Cremona and Odoni [10] considered the distribution of the values d for which the negative Pell equation
$$x^2 - dy^2 = -1$$
has a solution in integers (x, y). We call d *negative Pellian* when the equation has a solution in integers. Congruence considerations show that the equation is insoluble if $4|d$ or $p|d$ for some prime p with $p \equiv 3 \pmod{4}$. Furthermore, if d is negative Pellian, then the maximum square-free divisor d_0 of d is also negative Pellian. On the other hand, Rédei [32] gave a simple procedure to pass from the case of square-free d_0 to the general case. Let

$$D = \{d > 1 : d \text{ is square-free and has no prime factor } p \equiv 3 \,(\text{mod}\, 4)\},$$

and, for each positive integer n, let D_n be the subset of D consisting of those d with exactly n prime factors. Let p and q be primes in D_1; then p and q are not congruent to $3 \pmod{4}$. We say that p and q are *related*, denoted by pRq, if and only if $p \neq q$ and p^3 is not congruent to a square $(\text{mod } q^3)$. Note that R is a symmetric relation. Let

$$d = p_1 p_2 \ldots p_n \text{ be in } D_n, \text{ where } p_1 < \cdots < p_n.$$

We form a graph $H(d)$ with vertex set $\{1, 2, \ldots, n\}$, where the vertices i and j are adjacent if and only if $p_i R p_j$. A graph G with vertex set $\{1, 2, \ldots, n\}$ is *odd* when it has the following property: whenever $\{1, 2, \ldots, n\}$ is partitioned into the disjoint union $A \cup B$ of two non-empty sets A and B, then either there exists $a \in A$ joined (in G) to an odd number of vertices in B, or there exists $b \in B$ joined (in G) to an odd number of vertices in A. The graph on one vertex is taken to be odd. The class O_n of odd graphs on n vertices was enumerated, in a different context, by Thomason [39]. Cremona and Odoni [10] proved the following result.

Theorem 3.16 *If $d \in D_n$ and $H(d)$ is an odd graph, then d is negative Pellian.*

Using the Matrix-Tree Theorem (see Harary [17]) and the fact that a graph is odd if and only if it has an odd number of spanning trees, Cremona and Odoni showed that the number of odd graphs on n vertices is

$$2^{\binom{n}{2}} \prod_{2j < n+1} (1 - 2^{1-2j});$$

this result also appears, in a different version, in Thomason [39]. Using estimates on the distribution of primes in arithmetic progressions, they then showed that $H(d)$ is asymptotically uniformly distributed amongst the $2^{\binom{n}{2}}$ labelled graphs of order n, in the sense that for all fixed n and G,

$$\lim_{x\to\infty} \frac{|\{d \in D_n : d \leq x, H(d) = G\}|}{|\{d \in D_n : d \leq x\}|} = 2^{-\binom{n}{2}}.$$

From this estimate they obtained the following result.

Theorem 3.17 *For each positive integer n, let*

$$b_n = \prod_{j=1}^{\lfloor n/2 \rfloor} (1 - 2^{1-2j}).$$

Then, for each n,

$$\liminf_{x\to\infty} \frac{|\{d \in D_n : d \leq x \text{ and } d \text{ is negative Pellian}\}|}{|\{d \in D_n : d \leq x\}|} \geq b_n.$$

Clearly the sequence b_n decreases, and it converges to the limit

$$b = 0.4194\ldots.$$

Cremona and Odoni conjectured that

$$\lim_{x\to\infty} \frac{|\{d \in D : d \leq x \text{ and } d \text{ is negative Pellian}\}|}{|\{d \in D : d \leq x\}|} \geq b.$$

References

1. T. M. Apostol, *Introduction to Analytic Number Theory*, Springer-Verlag, 1976; *MR* **55**#7892.
2. A. Astie, Groupes d'automorphismes des tournois sommet-symmetriques d'ordre premier et dénombrement de ces tournois, *C. R. Acad. Sci. Paris (A/B)* **275** (1972), A167–A169; *MR* **46**#8888.
3. J. L. Berggren, An algebraic characterization of finite symmetric tournaments, *Bull. Austral. Math. Soc.* **6** (1972), 53–59; *MR* **45**#118.
4. F. Bien, Constructions of telephone networks by group representations, *Notices Amer. Math. Soc.* **36** (1989), 5–22; *MR* **90a**: 90052.
5. N. L. Biggs, E. K. Lloyd and R. J. Wilson, *Graph Theory 1736–1936*, paperback edn, Clarendon Press, 1986; *MR* **88e**: 01035.
6. B. Bollobás, *Random Graphs*, Academic Press, 1985; *MR* **87f**: 05152.
7. A. Cayley, On the mathematical theory of isomers, *Phil. Mag.* **47** (1874), 444–446 = *Math. Papers* **9** (1896), 202–204.
8. F. R. K. Chung, Diameters and eigenvalues, *J. Amer. Math. Soc.* **2** (1989), 187–196; *MR* **89k**: 05070.
9. R. J. Cook, On a construction of Lovász, *J. London Math. Soc. (2)* **12** (1975/6), 351–355; *MR* **53**#7838.

10. J. E. Cremona and R. W. K. Odoni, Some density results for negative Pell equations; an application of graph theory, *J. London Math. Soc. (2)* **39** (1989), 16–28; *MR* **90b**: 11019.
11. P. Deligne, La conjecture de Weil, I, *Publ. Math. I.H.E.S.* **43** (1974), 273–307; *MR* **49**#5013.
12. J. Edmonds and E. L. Johnson, Matching, Euler tours and the Chinese postman, *Math. Programming* **5** (1973), 88–124; *MR* **48**#168.
13. P. Erdős, Graph theory and probability, *Canad. J. Math.* **11** (1959), 34–38; *MR* **21**#876.
14. H. Fleischner, Eulerian graphs, *Selected Topics in Graph Theory 2* (ed. L. W. Beineke and R. J. Wilson), Academic Press, 1983, pp. 17–53; *MR* **86k**: 05079.
15. E. Fried, On homogeneous tournaments, *Combinatorial Theory and its Applications II* (ed. P. Erdős et al.), North-Holland, 1970, pp. 467–476; *MR* **45**#6688.
16. M. Goldberg, The group of the quadratic residue tournament, *Canad. Math. Bull.* **13** (1970), 51–54; *MR* **41**#6818.
17. F. Harary, *Graph Theory*, Addison-Wesley, 1969; *MR* **41**#1566.
18. F. Harary and E. M. Palmer, *Graphical Enumeration*, Academic Press, 1973; *MR* **50**#9682.
19. F. Harary, G. Prins and W. T. Tutte, The number of plane trees, *Indag. Math.* **26** (1964), 319–329; *MR* **29**#4049.
20. G. H. Hardy and E. M. Wright, *An Introduction to the Theory of Numbers*, 5th edn, Oxford University Press, 1978; *MR* **16**–673.
21. D. R. Hughes and F. C. Piper, *Design Theory*, Cambridge University Press, 1985; *MR* **87e**: 05022.
22. N. M. Katz, An estimate for character sums, *J. Amer. Math. Soc.* **2** (1989), 197–200; *MR* **90b**: 11081.
23. M.-K. Kwan (= Meigu Guan), Graphic programming using odd or even points, *Acta Math. Sinica* **10** (1960), 263–266 = *Chinese Mathematics* **1** (1962), 273–277; *MR* **28**#5828.
24. L. Lovász, On chromatic number of finite set-systems, *Acta Math. Acad. Sci. Hungar.* **19** (1968), 59–67; *MR* **36**#3673.
25. A. Lubotzky, R. S. Phillips and P. Sarnak, Ramanujan graphs, *Combinatorica* **8** (1988), 261–277; *MR* **89m**: 05099.
26. G. A. Margulis, Explicit group-theoretic constructions of combinatorial schemes and their applications in the construction of expanders and concentrators, *J. Problems Inform. Transmission* **24** (1988), 39–46; *MR* **89f**: 68054.
27. J. W. Moon, *Topics on Tournaments*, Holt, Rinehart and Winston, 1968; *MR* **41**#1574.
28. M. Morgenstern, Existence and explicit construction of $q+1$ regular Ramanujan graphs for every prime power q, *J. Combin. Theory (B)* **62** (1994), 44–62; *MR* **95h**: 05089.

29. R. Otter, The number of trees, *Ann. of Math. (2)* **49** (1948), 583–599; *MR* **10**–53c.
30. R. E. A. C. Paley, On orthogonal matrices, *J. Math. Phys.* **12** (1933), 311–320.
31. G. Pólya, Kombinatorische Anzahlbestimmungen für Gruppen, Graphen und chemische Verbindungen, *Acta Math.* **68** (1937), 145–254.
32. L. Rédei, Über die Pellsche Gleichung $t^2 - du^2 = -1$, *J. Reine Angew. Math.* **173** (1935), 193–211.
33. J. H. Redfield, The theory of group reduced distributions, *Amer. J. Math.* **49** (1927), 433–455.
34. K. B. Reid and L. W. Beineke, Tournaments, *Selected Topics in Graph Theory* (ed. L. W. Beineke and R. J. Wilson), Academic Press, 1978, pp. 169–204; *MR* **81e**: 05059.
35. G.-C. Rota, On the foundations of combinatorial theory. I. Theory of Möbius functions, *Z. Wahrscheinlichkeitstheorie* **2** (1964), 340–368; *MR* **30**#4688.
36. P. Sarnak, *Some Applications of Modular Forms*, Cambridge University Press, 1990; *MR* **92k**: 11045.
37. W. M. Schmidt, *Equations Over Finite Fields. An Elementary Approach*, Lecture Notes in Math. **536**, Springer-Verlag, 1976; *MR* **55**#2744.
38. M. R. Schroeder, *Number Theory in Science and Communication*, Springer-Verlag, 1986; *MR* **87d**: 11022.
39. A. G. Thomason, A graph property not satisfying a "zero-one law", *European J. Combin.* **9** (1988), 517–521; *MR* **90e**: 05051.
40. D. W. Walkup, The number of plane trees, *Mathematika* **19** (1972), 200–204; *MR* **51**#12587.
41. A. Weil, On some exponential sums, *Proc. Nat. Acad. Sci. U.S.A.* **34** (1948), 204–207; *MR* **10**–234.
42. H. S. Wilf, The Möbius function in combinatorial analysis and chromatic graph theory, *Proof Techniques in Graph Theory* (ed. F. Harary), Academic Press, 1969, pp. 179–188; *MR* **40**#7126.

4
Partial Orders

GRAHAM BRIGHTWELL

We explore several connections between graphs and partially ordered sets. We define various graphs and digraphs associated with a partially ordered set, and explore how the properties of the partial order are reflected in these graphs. We consider the incidence order of a graph, and prove Schnyder's Theorem, that a graph is planar if and only if its incidence order has dimension at most 3.

4.1 Introduction

It is possible, and only moderately perverse, to think of a graph as a special kind of partially ordered set. Then again, a partially ordered set can be regarded as nothing more than a special kind of directed graph. One should be wary of concluding that the concerns of one subject are similar to those of the other, or that results from one area are necessarily of use in the other, but it is not unreasonable to expect that there will be connections. This chapter is devoted to some of those links that have proved interesting.

We begin with some definitions and notation. Then we discuss Dilworth's Theorem and its relationship with Hall's Marriage Theorem. The next two sections are concerned with various graphs derived from partially ordered sets. Finally we study Schnyder's Theorem on the dimension of the incidence order of a graph, and related results.

4.2 Preliminaries

We begin with some definitions. A *partial order* $<$ on a set X is a transitive irreflexive relation on X. To say that $<$ is *transitive* means that if a, b, c are elements of X with $a < b$ and $b < c$, then also $a < c$. To say that $<$ is *irreflexive* means that we never have $a < a$. It follows from these axioms that we never simultaneously have $a < b$ and $b < a$. If $<$ is a partial order on X, we call the pair $(X, <)$ a *partially ordered set* (or *poset*). The set X is sometimes called the *ground-set* of the partial order.

Examples of partially ordered sets include the familiar number systems **N**, **Z**, **Q** and **R**, with the standard order. These are all *linear orders*: they have the additional property that if a and b are distinct elements of the ground-set, then either $a < b$ or $b < a$.

More interesting from the point of view of the order structure is the plane $\mathbf{R}^2$, with $(x, y) \leq (u, v)$ if both $x \leq u$ and $y \leq v$. Here, as usual, we write $a \leq b$ to mean that either $a < b$ or $a = b$. In this example there are pairs, such as $a = (0, 1)$ and $b = (1, 0)$, with neither $a \leq b$ nor $b \leq a$. The definition extends in the obvious way to give the *coordinate order* on $\mathbf{R}^k$, for any positive integer k.

In all our examples so far, the ground-set X is infinite. In this chapter, we shall be concerned rather with the case where X is finite.

If $\mathcal{A}$ is any collection of sets, we can put a natural order, called the *containment order*, on $\mathcal{A}$ by setting $A < B$ if $A \subset B$. This is a very general prescription: any partial order $<$ on a set X is isomorphic to the containment order obtained by identifying each element $a \in X$ with the set $S_a = \{x \in X : x \leq a\}$. Clearly, $S_a = S_b$ if and only if $a = b$, and $S_a \subset S_b$ if and only if $a < b$.

Viewed formally, a graph $G = (V, E)$ is just a collection of sets: each vertex in V can be thought of as a one-element set, and each edge as a two-element subset of V. The containment order on this collection $\mathcal{P}(G) = V \cup E$ of sets captures the incidence relation of the graph. Thus, in the containment order on $\mathcal{P}(G)$, $a < b$ if and only if a is a vertex and b is an edge incident with a. The set $\mathcal{P}(G)$ with its containment order is usually called the *incidence order* of G. An example is shown in Fig. 4.1. Obviously, G is determined by its incidence order: hence the comment that a graph is just a partially ordered set. It is now unnatural *not* to ask how the properties of the graph G are reflected in the properties of its incidence order.

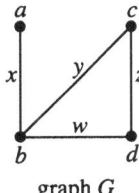
graph G

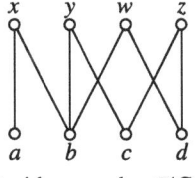
incidence order $\mathcal{P}(G)$

FIG. 4.1

We can also go the other way. Given a partially ordered set $P = (X, <)$, there are various graphs and digraphs naturally associated with P. Most obviously, we define the *directed comparability graph* $B(P)$ of P to be the digraph with vertex set X and an arc directed from a to b whenever $a < b$. There is obviously no great distinction between P and $B(P)$. Note that $B(P)$ is *acyclic*—that is, there are no directed cycles. The undirected graph obtained by ignoring the directions of the arcs of $B(P)$ is called the *comparability graph* $C(P)$; its complement is the *incomparability graph* $I(P)$.

When drawing a partially ordered set, it is often confusing and unattractive to include all the arcs of its directed comparability graph $B(P)$. Fortunately, most partial orders can be drawn unambiguously with many fewer edges. For instance, if we record that $a < b$ and $b < c$, then there is no need to add the information that $a < c$, as that follows from the axioms for partial orders. Following this idea, we define an ordered pair (a, b) of elements of X to be a *covering pair* of the partially ordered set $(X, <)$ if $a < b$ and there is no $c \in X$ with $a < c < b$. Thus, if we are given the set of covering pairs of a finite partially ordered set, we can recover the partial order. Alternatively, the set of covering pairs is the minimal relation on X whose *transitive closure* is $(X, <)$.

The *diagram* $D(P)$ of a partially ordered set $P = (X, <)$ is the digraph with vertex set X formed by drawing an arc from a to b whenever (a, b) is a covering pair of P. Note that $D(P)$ is a subdigraph of $B(P)$, and so the diagram is also an acyclic digraph. If X is finite, then the partially ordered set P is uniquely determined by its diagram. The undirected graph formed by ignoring the directions of arcs of $D(P)$ is called the *covering graph* $G(P)$.

When drawing a partially ordered set, it is conventional to draw just the covering edges (a, b), with an upward slope from a to b to indicate that $a < b$. Such a representation is called an *upward drawing*. Given an upward drawing of $P = (X, <)$, we see that $x < y$ if and only if there is an upward path in the drawing from x to y.

Figure 4.2 shows the directed comparability graph and the diagram of a simple partially ordered set P, and an upward drawing of P. The comparability graph and covering graph are obtained by ignoring the directions of the arcs of the directed comparability graph and the diagram, respectively.

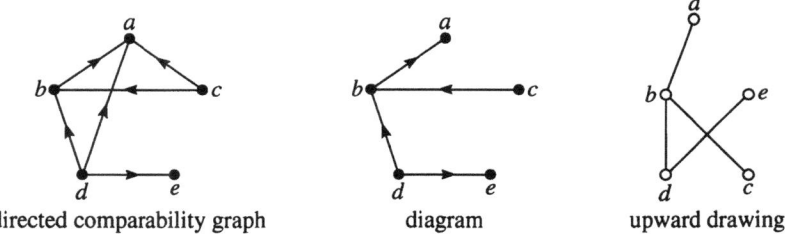

directed comparability graph   diagram   upward drawing

FIG. 4.2

Various questions now spring to mind. How are order-theoretic properties of P reflected in graph-theoretic properties of $C(P)$ or of $G(P)$? Which graphs are comparability graphs, or covering graphs, of some partially ordered set? It is not hard to produce pairs of non-isomorphic partially ordered sets with the same comparability graph, or with the same covering graph: must such pairs share similar properties? We begin to address these questions in the next section, but first we give a few more of the standard definitions connected with partial orders.

A *maximal element* in $P = (X, <)$ is an element a of X such that there is no $b \in X$ with $a < b$. Similarly, a *minimal element* is an element of X with no other elements below it.

A *chain* in a partially ordered set $P = (X, <)$ is a subset of X that is linearly ordered by $<$. Equivalently, a chain is a set of elements of X forming a clique in

$C(P)$. At the opposite extreme, an *antichain* in P is a subset of X in which no two elements are related by $<$. An antichain is thus an independent set in $C(P)$, or a clique in $I(P)$. The *height* of P is the size of a longest chain in P; this is just the clique number $\omega(C(P))$. The *width* of P is the size of a largest antichain; this is the independence number $\beta(C(P)) = \omega(I(P))$. Height and width are generally agreed to be the most fundamental parameters of partial orders, together with the number $|X|$ of elements of P.

Another parameter that has attracted a great deal of interest is the *dimension* $\dim(P)$ of a partially ordered set P; this is the minimum integer k such that P can be embedded in $\mathbf{R}^k$, with the coordinate order. This definition explains the name 'dimension', but in most circumstances an alternative definition turns out to be easier to work with. A *realizer* of a partially ordered set $P = (X, <)$ is a set S of linear orders on X whose intersection is P—that is, $x < y$ in P if and only if x is below y in each of the linear orders in S. The dimension of P is the minimum size of a realizer. It is easy to check that these two definitions are equivalent.

4.3 Dilworth's Theorem

We have already mentioned that the height and width of a partially ordered set $P = (X, <)$ are equal to the clique number and independence number of $C(P)$. A graph theorist might now ask whether there are results linking the height and width to the chromatic numbers of $C(P)$ and $I(P)$. Indeed there are, and these results are the most that one might hope to prove.

Recall that a graph G is *perfect* if for each induced subgraph H of G, the chromatic number $\chi(H)$ of H is equal to its clique number $\omega(H)$.

Theorem 4.1 *For any finite partially ordered set P, the graphs $C(P)$ and $I(P)$ are perfect.*

Let us see what Theorem 4.1 actually means. Note that every induced subgraph of $C(P)$ is itself the comparability graph of some partially ordered set, so to prove that $C(P)$ is perfect for every partially ordered set P, it is enough to show that $\omega(C(P)) = \chi(C(P))$ for every P. The same applies to $I(P)$, so Theorem 4.1 is equivalent to the following.

Theorem 4.2 *Let $P = (X, <)$ be a finite partially ordered set with height h and width w. Then*

(a) the ground-set X can be partitioned into h antichains;
(b) the ground-set X can be partitioned into w chains.

Either part of this result implies the simple but useful fact that, in the notation of the theorem, $|X| \leq hw$. As we shall now see, Theorem 4.2(a) is fairly simple. Theorem 4.2(b), known as *Dilworth's Theorem* [7], is a much more substantial result.

Proof of Theorem 4.2(a) Define the height $h(x)$ of an element x of X to be the size of a longest chain in P with top element x. Note that $1 \leq h(x) \leq h$, for all $x \in X$. For $j = 1, \ldots, h$, define $A_j = \{x \in X : h(x) = j\}$—for instance, A_1 is the set of minimal elements of P, and A_2 is the set of minimal elements of $P - A_1$. Clearly, the h sets A_j form a partition of X. To complete the proof, observe that each A_j is an antichain, since if $y > x$ and $h(x) = j$, then there is a chain of size $j + 1$ with top element y, and so $h(y) \geq j + 1$. □

This proves that $C(P)$ is perfect. One way to prove that $I(P)$ is perfect is to invoke the *Perfect Graph Theorem*—a famous result of Lovász [17, 18] stating that a graph is perfect if and only if its complement is. It is, however, substantially easier to prove Dilworth's Theorem (Theorem 4.2(b)) than it is to prove the Perfect Graph Theorem, so we give another proof.

Dilworth's Theorem is perhaps the most important result in the theory of partial orders. It belongs to the same family of minimax results as Hall's Marriage Theorem, the Max-Flow Min-Cut Theorem, Menger's Theorem and the König-Egerváry Theorem. All of these are 'if and only if' results in which one direction is trivial, whereas the other is far from obvious. Here, it is trivial that if there is a partition of P into w chains, then there is no antichain of size $w + 1$, but it is far from obvious that if there is no antichain of size $w + 1$, then there is a partition into w chains. Furthermore, the non-trivial parts of the results are all 'equivalent', in the sense that it seems to be easier to deduce them from each other than to prove them directly. Moreover, all involve some parameter that can be evaluated in polynomial time, but not by trivial means, and an algorithm for one of the problems can typically be converted to algorithms for others.

There are several short self-contained proofs of Dilworth's Theorem—see [2] or [32], for example. Here, we give a proof, due to Dantzig and Fulkerson [6], that depends on the 'defect version' of Hall's Marriage Theorem. Unlike the other proofs mentioned above, this proof forms the basis for a polynomial algorithm for finding the width of a partially ordered set. Let us state the version of Hall's Theorem that we shall use. Here, $\Gamma(U)$ is the set of neighbours of U.

Theorem 4.3 *Let G be a bipartite graph with vertex classes X and Y, with $|X| = |Y| = n$. Then there is a matching of size $n - k$ in G if and only if there is no set $U \subseteq X$ with $|\Gamma(U)| < |U| - k$.*

The case $k = 0$ is the usual form of Hall's Theorem, and it is easy to deduce Theorem 4.3 from this special case—see, for instance, [2]. As usual, the condition in Theorem 4.3 is trivially necessary, and the punch of the theorem is that it is sufficient. Incidentally, it is an easy, but pointless, exercise to derive Theorem 4.3 from the Perfect Graph Theorem.

Proof of Theorem 4.2(b) Given a partially ordered set $P = (X, <)$ with $|X| = n$ and width w, we form a bipartite graph $K(P)$ with vertex classes X and Y of size n, as follows. The class X is the ground-set of P, and Y is a copy of X, so for each $x \in X$ there is a corresponding vertex x' in Y. We put an edge of $K(P)$ between x and y' if and only if $x < y$ in P (see Fig. 4.3).

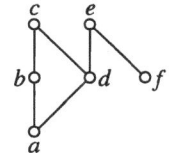 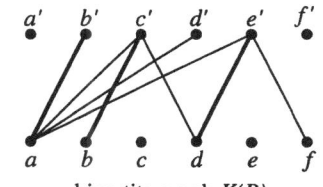

partially ordered set P   bipartite graph K(P)

FIG. 4.3

The key claim is that a matching M in $K(P)$ of size k corresponds to a partition of P into $n-k$ chains, with an edge of M between x and y' indicating that x and y are consecutive elements of the same chain. This is illustrated in Fig. 4.3, where the partially ordered set P has a decomposition into three chains, indicated by the vertical groups, and the associated matching in $K(P)$ consists of the three thick lines. To prove the claim, consider the graph obtained by taking the edges of M together with edges between each pair (x, x'). This graph has maximum degree 2, so each component is a path or a cycle. A cycle would correspond to a cycle in the directed comparability graph $B(P)$, so there are none. Each path component corresponds to a chain in P, and the number of path components is $n-k$.

The rest of the proof is automatic: the partition of P into fewest chains corresponds to the largest matching in $K(P)$, and an application of Hall's Theorem is almost bound to complete the proof.

Suppose then that there is no partition of P into w chains. This implies that there is no matching in $K(P)$ of size $n-w$. By Hall's Theorem, there is some set $U \subseteq X$ with $|\Gamma(U)| < |U| - w$. If x is an element of U that is not minimal in U, then $x' \in \Gamma(U)$. Thus $|\Gamma(U)| \geq |U| - |M(U)|$, where $M(U)$ is the set of minimal elements of U. We conclude that U has more than w minimal elements. But the set of minimal elements of U forms an antichain, contradicting our assumption that P has width w. This completes the proof. □

There are well-known polynomial-time algorithms for finding a maximum-size matching in a bipartite graph, and the above proof thus yields a fast algorithm to find the width of a partially ordered set.

We conclude this section with an easy argument reversing the above implication, enabling us to deduce Theorem 4.3 from Dilworth's Theorem. Given a bipartite graph G, with vertex classes X and Y each of size n, direct each edge from X to Y and consider this as the directed comparability graph $B(P)$ of a partially ordered set P. If there is a partition of P into $n+k$ chains, it must consist of $n-k$ 2-element chains and $2k$ singletons, and the 2-element chains constitute a matching of size $n-k$. If there is no such matching, then there is no such partition, and so, by Dilworth's Theorem, the partially ordered set P has an antichain A of size $n+k+1$. Let $U = X \cap A$, and $u = |U|$. Then $\Gamma(U) \cap A = \varnothing$, and so

$$|\Gamma(U)| \le n - |A \cap Y| = n - ((n+k+1) - u) = u - k - 1,$$

as required.

4.4 Comparability graphs

In the previous section, we saw that comparability graphs are rather special. One is led to ask which graphs are comparability graphs. A more modern approach might be to ask whether there is a polynomial algorithm to determine whether a graph is the comparability graph of some partially ordered set; providing such an algorithm might be regarded as tantamount to giving a characterization of comparability graphs, although not necessarily a very satisfying one. In this case there is no need to settle for half-measures, as we have the following pretty result of Ghouila-Houri [13] and Gilmore and Hoffman [14].

Theorem 4.4 *A finite graph G is a comparability graph if and only if, for each closed walk $x_1 x_2 \ldots x_{2k+1} x_1$ of odd length in G, there is an edge between x_j and x_{j+2} for some j, with the indices interpreted modulo $2k+1$.*

For instance, Fig. 4.4 shows a graph that is not a comparability graph. In order to check that the condition of Theorem 4.4 is not satisfied, it is necessary to consider the walk of length 9, with repeated vertices and edges, indicated in the figure.

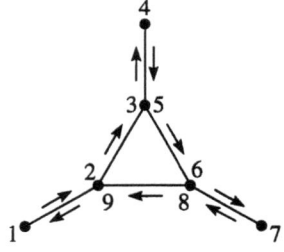

FIG. 4.4

It is relatively easy to see that the condition of Theorem 4.4 is necessary. Indeed, suppose that there is a closed walk $x_1 x_2 \ldots x_{2k+1} x_1$ of odd length in G such that no x_j is adjacent to x_{j+2}. Suppose that G can be oriented so that it is the directed comparability graph of some partially ordered set P. As the walk has odd length, any orientation gives the same direction to two consecutive edges $x_j x_{j+1}$ and $x_{j+1} x_{j+2}$ of the walk. Then x_j and x_{j+2} are comparable in P, but not adjacent in G, which is a contradiction.

To prove that the condition is sufficient requires rather more work, and we confine ourselves to outlining an approach. Given a graph G, we define the graph G' whose vertex set is the set of ordered pairs (x, y) with $xy \in E(G)$, with (x, y) adjacent to (y, z) if x is *not* adjacent to z in G—in particular, (x, y) is always adjacent to (y, x). Then Theorem 4.4 can be interpreted as saying that G is a comparability graph if and only if G' is bipartite. This is only the start of the proof, and there is a lot of work left to do: each bipartition of G' gives rise to an orientation of the edges of G in an obvious way, but this is *not* necessarily a diagram orientation. For a full proof, see Ghouila-Houri [13], Kelly [16] or Fishburn [9]. Note that it is a simple matter to construct G' and check whether it is bipartite, so the problem of testing whether a graph is a comparability graph can be solved in polynomial time; see Spinrad [29, 30] for more efficient algorithms for this problem.

Suppose that two partially ordered sets have the same comparability graph. How are they related? The next result, essentially due to Gallai [11], shows that the relationship is fairly close; this 1967 paper of Gallai is widely regarded as a classic in this area.

An *autonomous set* in a partially ordered set $P = (X, <)$ is a subset A of X such that, for each element $x \in X - A$, x is above all elements of A, or x is below all elements of A, or x is incomparable with all elements of A. Examples of autonomous sets are the entire ground-set X, and all singleton sets. If P has an autonomous set A that is not an antichain, then we can obtain another partial order on X from P by reversing the direction of all the comparabilities inside A: this is called *flipping* A. One can readily check that this does give another partially ordered set $\mathbf{F}_A P$, and that P and $\mathbf{F}_A P$ have the same comparability graph. As a special case, if $A = X$, then $\mathbf{F}_X P$ is the *dual* of P, the partially ordered set obtained from P by reversing all comparabilities. If P is a partially ordered set with comparability graph G, then any partially ordered set obtained from P by repeatedly flipping autonomous sets also has comparability graph G. Gallai's result states that there are no other partially ordered sets with comparability graph G.

Theorem 4.5 *If P and Q are two partially ordered sets with $C(P) = C(Q)$, then there is a sequence $A_1, \ldots, A_k$ such that each A_j is an autonomous set in the partially ordered set $\mathbf{F}_{A_{j-1}} \ldots \mathbf{F}_{A_1} P$, and $Q = \mathbf{F}_{A_k} \ldots \mathbf{F}_{A_1} P$.*

Gallai [11] used Theorem 4.5 to provide a 'forbidden suborder' characterization of comparability graphs. He gave a list of eight infinite families and eleven small exceptional graphs, such that a graph is a comparability graph if and only if it contains no graph from the list as an induced subgraph. This list can also be found in Trotter's book [32].

There are other, arguably more informative, ways of presenting results that are essentially equivalent to Theorem 4.5. However, this is the form that is needed if we wish to prove that some parameter f of partial orders is a 'comparability invariant'—that is, that $f(P) = f(Q)$ whenever $C(P) = C(Q)$. Indeed, to do this we need only check that $f(\mathbf{F}_A P) = f(P)$ whenever A is an autonomous set in P. In many cases—for example, if $f(P)$ is the dimension of P—it is straightforward to check this, even when the definition of f cannot obviously be couched so as to depend only on the comparability graph. Details of other comparability invariants can be found in Habib [15].

4.5 Covering graphs and diagrams

The results in Section 4.4 provide evidence that the comparability graph is an informative auxiliary object to a partial order. Unfortunately, the covering graph of a partially ordered set does not seem to capture essential features of the partial order in the same way; in particular, hardly any interesting parameters of partial orders are covering graph invariants. Thus the study of covering graphs, while not uninteresting for its own sake, may turn out to be something of a dead end. The situation for the diagram, the directed analogue of the covering graph, is more positive: for a start, a partially ordered set is determined by its diagram. Starting from a partially ordered set, it takes one step to reach the comparability graph or the diagram, but two to reach the covering graph, so perhaps one should not expect the covering graph to be so closely associated with the original partially ordered set.

We begin this section by considering the characterization problem: given a graph or digraph, when is it the covering graph or diagram of a partially ordered set? The situation for the covering graph is bad: the problem is NP-complete. This result is due to Nešetřil and Rödl, who in [21] presented a corrected version of an earlier paper [20]. A more elementary proof appears in [3]. The fact that a problem is NP-complete implies that unless P = NP, there is no algorithm to solve the problem in polynomial time. The P $\neq$ NP conjecture, widely believed to be true, would imply that testing whether a graph is a covering graph can be regarded in general as prohibitively time-consuming—for more on NP-completeness see, for instance, [12].

This does not rule out partial results about the family of covering graphs. For a start, it is easy to see that every bipartite graph is a covering graph, and that every covering graph is triangle-free. Two triangle-free graphs that are not covering graphs are shown in Fig. 4.5. One can check by case analysis that these are not covering graphs; a more reliable and informative approach was developed by Pretzel and Youngs [23].

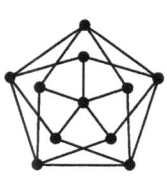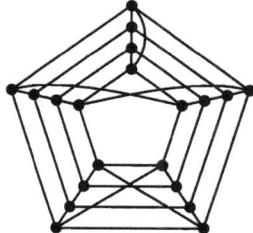

FIG. 4.5

There is an easy characterization of when a digraph is the diagram of some partially ordered set. Before giving this result, we introduce some notation. Given an undirected graph G, and an orientation O of G, we define G_O to be the directed graph obtained by directing each edge of G according to O. For a cycle $C = x_1 x_2 \ldots x_k x_1$ in G, equipped with a direction around C, an edge $x_j x_{j+1}$ is *positive* for an orientation O if it is directed from x_j to x_{j+1} by O, and *negative* otherwise. The *flow-difference* of O around C is the number of positive edges of C minus the number of negative edges. Thus a directed cycle in G_O is a cycle of length k and flow-difference k. A diagram cannot contain a directed cycle, and furthermore it cannot contain a cycle of length k and flow-difference $k-2$—that is, a cycle with just one negative edge—since the relation in the partial order indicated by the single negative edge can be deduced from the relations indicated by the positive edges. A cycle with just one negative edge is called a *bypass*. The following easy result is folklore.

Theorem 4.6 *A directed graph is the diagram of some partially ordered set if and only if it contains no directed cycle or bypass.*

Theorem 4.6 gives us another positive result about covering graphs.

Corollary 4.7 *Suppose that G has chromatic number k and no cycle of length k or less. Then G is a covering graph.*

Proof Take a colouring of G with k colour classes $C_1, \ldots, C_k$, and define an orientation of G by orienting each edge of G from the lower-numbered colour class to the higher. Each cycle contains at most $k-1$ consecutive positive edges in this orientation, so, since all cycles have at least $k+1$ edges, there must be at least two negative ones. So the orientation contains no directed cycles or bypasses and, by Theorem 4.6, is therefore a diagram. □

We deduce from Corollary 4.7 that a graph of chromatic number 3 is a covering graph if and only if it is triangle-free; this is an easy condition to check. However, the next case is already hard; as shown by Brightwell [3], it is NP-complete to decide whether a 4-chromatic graph is a diagram.

4.6 Schnyder's Theorem

This section is concerned with the incidence order $\mathcal{P}(G)$ of a graph G. We shall prove the following theorem of Schnyder [27].

Theorem 4.8 (Schnyder's Theorem) *A graph G is planar if and only if the dimension of $\mathcal{P}(G)$ is at most 3.*

This is a remarkable result in several ways. Planarity is primarily a geometric rather than a combinatorial concept, whereas partial order dimension, although motivated by geometric considerations, is usually thought of as an abstract parameter. Yet Theorem 4.8 links the two in as neat a fashion as could be wished for.

Compare Theorem 4.8 with Kuratowski's Theorem: *a graph is planar if and only if it contains no K_5 or $K_{3,3}$ minor*; I would suggest that Schnyder's Theorem is at least as elegant. On the algorithmic side, there are very fast (linear-time) algorithms to test whether a graph is planar, but these have no obvious connection with Theorem 4.8. It is NP-complete in general to decide whether a partially ordered set has dimension 3, but it is not known whether this remains true for partially ordered sets of height 2.

For convenience, we define the *dimension* $\dim(G)$ of a graph G to be the dimension $\dim(\mathcal{P}(G))$ of its incidence order.

We begin our proof of Theorem 4.8 with the relatively easy direction. This result, together with the proof given here, is due (in a different context) to Babai and Duffus [1].

Lemma 4.9 *If G is a graph with dimension at most 3, then G is planar.*

Proof Take an embedding p of $\mathcal{P}(G)$ in $\mathbf{R}^3$. Then each vertex v and edge e of G is associated with a point $p(v)$ or $p(e)$ of $\mathbf{R}^3$. Now consider the projection q of each point onto the plane A defined by $x + y + z = 1$, so that each vertex v and edge e is represented by a point $q(v)$ or $q(e)$ in A. For each edge $e = ab$, take straight lines from $q(a)$ to $q(e)$ and from $q(e)$ to $q(b)$. This gives an arc in A from $q(a)$ to $q(b)$, which we take as representing the edge e.

Suppose that e and f are two edges with no common end-points whose corresponding arcs cross: say, the line joining $q(e)$ to $q(a)$ crosses the line between $q(f)$ and $q(b)$ at a point $x \in A$, where a is an end-point of e and b is an end-point of f. Going back up to $\mathbf{R}^3$, we see that, without loss of generality, a point u on the line between $p(e)$ and $p(a)$ is directly over, or equal to, a point w on the line between $p(f)$ and $p(b)$. We deduce that, in the coordinate order on $\mathbf{R}^3$, $p(e) \geq u \geq w \geq p(b)$. Since we have a representation of $\mathcal{P}(G)$, we conclude that b is an end-point of e, which contradicts our assumptions.

Hence the only possible occasions when two arcs cross are those when the corresponding edges have a common end-point. Such crossings can be eliminated by interchanging the sections of arcs between the common end-point and the intersection, giving a plane drawing of the graph G. $\square$

In [27], Schnyder gave a slightly more complicated argument which yields a straight line drawing of G. Combined with the other half of Theorem 4.8, this gives a wonderfully indirect proof of the well-known fact, due to Wagner [33] and Fáry [8], that every planar graph has a straight line drawing.

To complete the proof of Theorem 4.8, we need to show that every planar graph G has dimension at most 3. To do this, we may (and shall) assume that G is a *maximal* planar graph, since adding edges can only increase the dimension. We also assume that G has at least three vertices, and that we are given a specific representation of G in the plane, in which each face is a triangle. Let v_1, v_2 and v_3 be the three vertices on the outside face, in clockwise order.

To prove that G has dimension at most 3, it is enough to find a triple of partial orders $<_1, <_2, <_3$ on $V(G)$ with the following property. We say that such a triple has *property Q* if, for each vertex v and each edge e not incident with v, there is an $i \in \{1, 2, 3\}$ such that v appears above both end-points of e in $<_i$. Indeed, given a triple with property Q, we may extend to linear orders on $V(G)$ arbitrarily, and then insert the edges of G into each linear order on $V(G)$

as low as possible—that is, just above the higher of its two end-points. Property Q ensures that, for each vertex v and each edge e not incident with v, v appears above e in one of the three resulting linear orders. It is then easy to check that these three linear orders form a realizer of $\mathcal{P}(G)$.

The construction of a triple of partial orders with property Q is based on a remarkable scheme called a *Schnyder labelling* of a triangulation. In such a scheme, a label from $\{1, 2, 3\}$ is assigned to each of the three angles of each internal triangle, satisfying the following conditions (see Fig. 4.6):

(a) each angle at vertex v_i is labelled i, for $i = 1, 2, 3$;
(b) around each internal triangle, the three labels used for its angles can be read clockwise as 1, 2, 3;
(c) around each internal vertex, the labels can be read clockwise as a non-empty string of 1s, followed by a non-empty string of 2s, and finally a non-empty string of 3s.

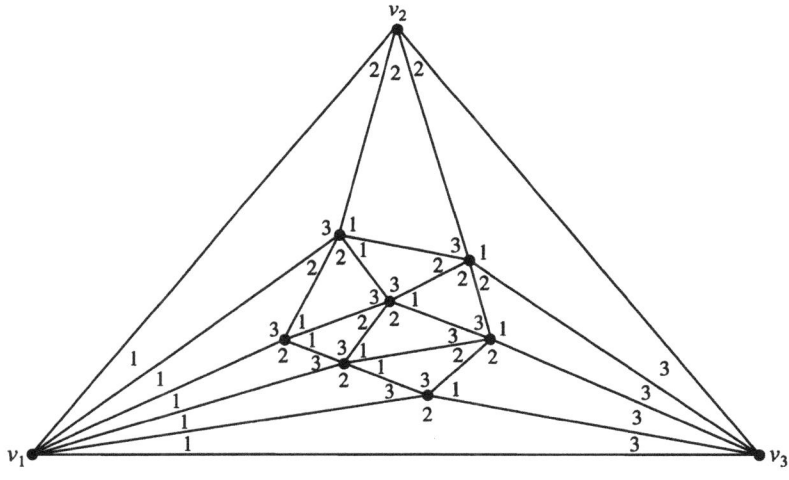

FIG. 4.6

Lemma 4.10 *Every plane triangulation admits a Schnyder labelling.*

Proof We work by induction on the number of vertices, the result being trivial for the triangulation consisting of one triangle.

Suppose that we have a triangulation G with $n \geq 4$ vertices, and that the result is true for all triangulations with fewer than n vertices. Let x be any neighbour of v_1 other than v_2 and v_3, such that v_1 and x have exactly two neighbours a and b in common. Deleting x from G leaves a face F whose vertices are exactly the neighbours of x in G. Three consecutive vertices around F are a, v_1 and b. We form a triangulation G' by sending an edge from v_1 across F to every other vertex on the face. By the induction hypothesis, G' admits a Schnyder labelling. We now recover a Schnyder labelling of G, as shown in Fig. 4.7. □

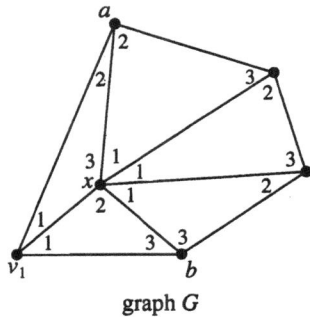 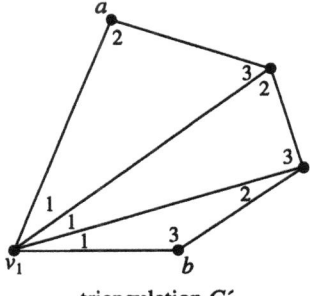

graph G triangulation G'

FIG. 4.7

We now explore the beautiful properties of Schnyder labellings, and the structures that can be derived from them. Our main aim is to show how a labelling can be used to give a 3-dimensional representation of $\mathcal{P}(G)$.

Suppose then that we have a Schnyder labelling of a plane triangulation G. Consider any internal edge e of G, and look at the four labels used for the angles adjoining it. A few moments' thought reveals that, by conditions (b) and (c), one of its ends (say, a) has both adjoining angles receiving the same label (say, i), while the other end features the other two labels. Consider the edge e as receiving the label i, and as being directed towards a. By condition (c), at each internal vertex x there is exactly one edge of each label directed away from x.

Thus we can follow the directed path labelled i from any vertex x; we call this the *i-path from* x. Note that as we follow the i-path, the $(i+1)$-path always heads off to the right, and the $(i+2)$-path heads off to the left; here, and in what follows, we interpret the labels modulo 3. We claim that i-paths all terminate at an external vertex, which must be v_i. The only other possibility is that there is a directed cycle with all edges labelled i.

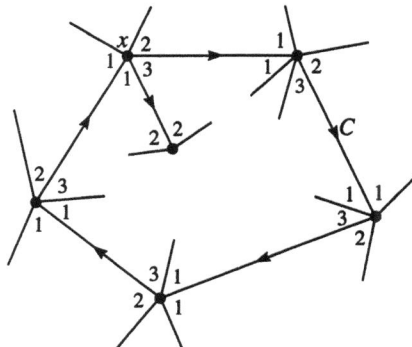

FIG. 4.8

Suppose that there is a cycle in which all edges have the same label, and let C be one enclosing the fewest faces. Without loss of generality, suppose that all the edges of C are labelled 1, and that the cycle runs clockwise, as in Fig. 4.8. Observe that from each vertex x of C, the edge labelled 2 heads off to the right, and so to the inside of C. Thus following the 2-path from any x on C must lead us to a cycle in the interior of C, which contradicts the minimality of C. We conclude that for $i = 1, 2, 3$, the edges labelled i form a directed tree with sink v_i.

One consequence of this is that the edges of a planar graph can be partitioned into three forests, and the tree consisting of edges labelled i can be augmented by any one or two of the three external edges. Using standard terminology, we say that every planar graph has *arboricity* at most 3; this result is originally due to Nash-Williams [19].

Now consider any internal vertex x of G, and its 1-path, 2-path and 3-path. These are three paths from x to the three external vertices, and it is easy to check that they do not cross. So the interior of the triangulation is split into three regions. For $i = 1, 2, 3$, let $S_i(x)$ be the closed region spanned by the $(i+1)$-path and the $(i+2)$-path from x and the external edge between v_{i+1} and v_{i+2}. Now define a partial order $<_i$ on $V(G)$ by $x <_i y$ if and only if $S_i(x) \subset S_i(y)$. Figure 4.9 shows the regions $S_i(x)$ and a vertex y with $S_2(y) \subset S_2(x)$, so that $y <_2 x$.

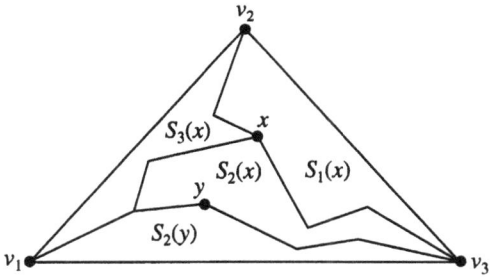

FIG. 4.9

Our aim is to prove that this triple of partial orders has property Q. Indeed, we shall show that if the edge e lies in the closed region $S_i(x)$, and if x is not an end-point of e, then x appears above both the end-points of e in $<_i$. Thus we must check that if the vertex y is in the region $S_i(x)$, then $S_i(y)$ is strictly contained in $S_i(x)$, and so $y <_i x$. Observe that the $(i+1)$-path from y cannot cross the $(i+1)$-path from x, since once the two paths meet, they coincide. Furthermore, the $(i+1)$-path from y cannot cross the $(i+2)$-path from x, since for each vertex z on the $(i+2)$-path from x, the edge labelled $i+1$ is directed into the region $S_i(x)$. Similarly, the $(i+2)$-path from y lies entirely in the region $S_i(x)$, and so $S_i(y)$ is contained in $S_i(x)$. Finally, we cannot have $S_i(x) = S_i(y)$, as this would entail a path between y and x consisting of edges labelled both $i+1$ and $i+2$. This completes the proof of Theorem 4.8. □

4.7 The incidence order

We conclude by considering two possible directions in which one might hope to extend Schnyder's Theorem.

If G is a planar 3-connected graph, then there is essentially only one way that it can be embedded in the plane, in the sense that the faces of G are determined. We can augment the incidence order $\mathcal{P}(G)$ by including the faces, putting a face above all the vertices and edges on its boundary: what can one say about the dimension of *this* partially ordered set? When G is a *maximal* planar graph, a minor adaptation of the proof of Theorem 4.8 shows that if the outside face is omitted, then the dimension of the resulting partially ordered set is still at most 3. However, it is not so simple to generalize to the case where G is not maximal planar.

More generally, let M be a plane map—that is, a set of points in the plane, with Jordan arcs drawn between various pairs of points and with no two arcs or points meeting except at common end-points. Multiple edges and loops are allowed. A *face* is then a component of the plane minus the points and arcs. We define a partially ordered set $\mathcal{Q}(M)$ on the set of points, arcs and faces by setting each edge above its one or two end-points, and each face above all the points and arcs in its boundary. Brightwell and Trotter [4, 5] proved the following results, extending Schnyder's Theorem.

Theorem 4.11 *For any plane map M, the dimension of $\mathcal{Q}(M)$ is at most 4.*

Theorem 4.12 *Suppose that M is the map of a 3-connected planar graph with no loops or multiple edges, and let F be the outside face. Then $\dim(\mathcal{Q}(M)) = 4$ and $\dim(\mathcal{Q}(M) - F) = 3$.*

What about dimensions other than 3? Which graphs have dimension 4, or 5, or 2? The last of these is easy: $\dim(G) \leq 2$ if and only if each component of G is a path. However, no characterization is known for any value of the dimension greater than 3.

It is tempting to think up attractive conjectures extending Schnyder's Theorem, but the next result scotches most of them. Note that if G is a subgraph of H, then $\mathcal{P}(G)$ is an induced suborder of $\mathcal{P}(H)$, and so $\dim(G) \leq \dim(H)$. In particular, if G has n vertices, then $\dim(G) \leq \dim(K_n)$. Our next result shows that $\dim(K_n)$ grows only very slowly with n.

Theorem 4.13 $\dim(K_n) = \log_2 \log_2 n + \left(\frac{1}{2} + o(1)\right) \log_2 \log_2 \log_2 n.$

A weaker version of Theorem 4.13, including the upper bound for $\dim(K_n)$, was proved by Spencer [28]. The full result appears in Trotter [32], based on work by Füredi, Hajnal, Rödl and Trotter [10] on the chromatic number of the *double shift graph*, and the dimension of *interval orders*. The connections between these subjects are explored fully in Trotter's book [32], and are well worth studying.

Our final results bound dim(G) in terms of the chromatic number of G. This supports the rather vague theory that the dimension of partially ordered sets is in some ways analogous to the chromatic number of graphs. The next theorem is again due to Schnyder [27].

Theorem 4.14 *If G is a graph with chromatic number at most 4, then dim(G) $\leq$ 4.*

A consequence of Theorem 4.8 is that there are graphs, such as $K_{3,3}$, with chromatic number 2, yet with dimension 4.

The smallest complete graph of dimension 5 is K_{13}. Subdividing each edge of K_{13} gives a bipartite graph which, by Theorem 4.14, has dimension at most (indeed, equal to) 4. However, this graph has a K_{13}-minor with dimension 5. So the property of having dimension at most r is not minor-closed for $r = 4$, and indeed not for any $r \geq 4$, although Theorem 4.8 implies that it is minor-closed for $r = 3$. Sadly, it seems that there is nothing as pretty as Theorem 4.8 to be said about the class of graphs of dimension r, for $r \geq 4$.

Finally, it is not hard to extend the above results to show that if G has chromatic number k, then dim(G) $\leq$ dim(K_k) + 3 = (1 + $o(1)$) $\log_2 \log_2 k$.

References

There is no one book that covers the topic of partially ordered sets thoroughly. Two books that I recommend are those by Fishburn [9], which concentrates on the special class of *interval orders*, and Trotter [32], which is mostly devoted to dimension theory. There are also many good survey articles on various aspects of the subject—for instance, Pretzel [22] on orientations, flow-differences and the like, and Kelly [16] on comparability graphs. Many surveys are collected in the conference volumes [24, 25, 26]. An earlier survey on connections between graphs and partially ordered sets is by Trotter [31]. Finally, the journal *Order* has attracted many of the important recent papers on partially ordered sets.

1. L. Babai and D. Duffus, Dimension and automorphism groups of lattices, *Algebra Universalis* **12** (1981), 279–289; *MR* **83a**: 06006.
2. B. Bollobás, *Graph Theory: An Introductory Course*, Springer-Verlag, 1979; *MR* **80j**: 05053.
3. G. Brightwell, On the complexity of diagram testing, *Order* **10** (1993), 297–303; *MR* **95d**: 05058.
4. G. Brightwell and W. T. Trotter, The order dimension of convex polytopes, *SIAM J. Discrete Math.* **6** (1993), 230–245; *MR* **94d**: 52013.
5. G. Brightwell and W. T. Trotter, The order dimension of planar maps, *SIAM J. Discrete Math.*, to appear.
6. G. Dantzig and D. Fulkerson, Minimizing the number of tankers to meet a fixed schedule, *Naval Res. Logist. Quart.* **1** (1954), 217–222.
7. R. P. Dilworth, A decomposition theorem for partially ordered sets, *Ann. of Math. (2)* **51** (1950), 161–166; *MR* **11**–309f.

8. I. Fáry, On straight line representation of planar graphs, *Acta Univ. Szeged Sect. Sci. Math.* **11** (1948), 229–233; *MR* **10**–136f.
9. P. C. Fishburn, *Interval Orders and Interval Graphs*, Wiley, 1985; *MR* **86m**: 06001.
10. Z. Füredi, P. Hajnal, V. Rödl and W. T. Trotter, Interval orders and shift graphs, *Sets, Graphs and Numbers* (ed. G. O. H. Katona), Colloq. Math. Soc. János Bolyai, 1992, pp. 297–313; *MR* **94e**: 06002.
11. T. Gallai, Transitiv orientierbare Graphen, *Acta Math. Acad. Sci. Hungar.* **18** (1967), 25–66; *MR* **36** #5026.
12. M. R. Garey and D. S. Johnson, *Computers and Intractability: A Guide to the Theory of NP-completeness*, Freeman, 1979; *MR* **80g**: 68056.
13. A. Ghouila-Houri, Caractérisation des graphes non orientés dont on peu orienter les arêtes de manière à obtenir le graphe d'une relation d'ordre, *C. R. Acad. Sci. Paris* **254** (1962), 1370–1371; *MR* **30**#2495.
14. P. C. Gilmore and A. J. Hoffman, A characterization of comparability graphs and of interval graphs, *Canad. J. Math.* **16** (1964), 539–548; *MR* **31**#87.
15. M. Habib, Comparability invariants, *Ann. of Discrete Math.* **23** (1984), 371–384; *MR* **86i**: 05069.
16. D. Kelly, Comparability graphs, *Graphs and Order* (ed. I. Rival), Reidel, 1985, pp. 3–40.
17. L. Lovász, Normal hypergraphs and the perfect graph conjecture, *Discrete Math.* **2** (1972), 253–267; *MR* **46**#1624.
18. L. Lovász, A characterization of perfect graphs, *J. Combin. Theory (B)* **13** (1972), 95–98; *MR* **46**#8885.
19. C. St. J. A. Nash-Williams, Decomposition of finite graphs into forests, *J. London Math. Soc.* **39** (1964), 12; *MR* **28**#4541.
20. J. Nešetřil and V. Rödl, Complexity of diagrams, *Order* **3** (1987), 321–330; *MR* **88h**: 05080.
21. J. Nešetřil and V. Rödl, More on complexity of diagrams, manuscript, 1993.
22. O. Pretzel, Orientations and edge functions on graphs, *Surveys in Combinatorics 1991* (ed. A. D. Keedwell), London Math. Soc. Lecture Notes **166**, Cambridge University Press, 1991, pp. 161–185; *MR* **93e**: 05039.
23. O. Pretzel and D. Youngs, Balanced graphs and noncovering graphs, *Discrete Math.* **88** (1991), 279–287; *MR* **92f**: 06006.
24. I. Rival (ed.), *Ordered Sets*, NATO ASI Series **83**, Reidel, 1982; *MR* **83e**: 06003.
25. I. Rival (ed.), *Graphs and Order*, NATO ASI Series **147**, Reidel, 1985; *MR* **86j**: 05006.
26. I. Rival (ed.), *Algorithms and Order*, NATO ASI Series **255**, Reidel, 1989; *MR* **90j**: 06002.
27. W. Schnyder, Planar graphs and poset dimension, *Order* **5** (1989), 323–343; *MR* **91b**: 06008.
28. J. Spencer, Minimal scrambling sets of simple orders, *Acta Math. Acad. Sci. Hungar.* **22** (1971/72), 349–353; *MR* **45**#1805.

29. J. P. Spinrad, On comparability and permutation graphs, *SIAM J. Comput.* **14** (1985), 658–670.
30. J. Spinrad, Dimension and algorithms, *Orders, Algorithms and Applications* (ed. V. Bouchitté and M. Morvan), Lecture Notes in Computer Science **831**, Springer-Verlag, 1994, pp. 33–52; *MR* **95m**: 06012.
31. W. T. Trotter, Graphs and partially ordered sets, *Selected Topics in Graph Theory 2* (ed. L. W. Beineke and R. J. Wilson), Academic Press, 1983, pp. 237–268.
32. W. T. Trotter, *Combinatorics and Partially Ordered Sets—Dimension Theory*, Johns Hopkins University Press, 1992; *MR* **94a**: 06001.
33. K. Wagner, Bemerkungen zum Vierfarbenproblem, *Jahresber. Deutsch. Math.-Verein.* **46** (1936), 26–32.

5
First-order Logic

PETER J. CAMERON

This chapter concerns some connections between graph theory and first-order logic. After an introduction to first-order logic and a discussion of which graph properties are first-order, we consider logical concepts such as $\aleph_0$-categoricity and homogeneity for graphs, and present some theorems about finite graphs requiring logical techniques in their proofs. The chapter concludes with a discussion of a recent construction method of Hrushovski related to sparse random graphs.

5.1 Introduction

Model theory concerns itself with the relationship between mathematical structures and the logical formulas they satisfy. Structures are normally required to have a 'finitary' description in terms of relations, functions, etc.; graphs form a particularly natural class of examples. There is some choice about the kind of formulas we allow. The most commonly used is first-order logic, where formulas are finite and quantification is allowed only over the elements of the structure.

In this chapter, we survey first-order properties of graphs. Many of the results are folklore, and should not be regarded as new, even if no references are given. At first glance, model theory does not appear very useful for the graph theorist: any property of finite graphs is the restriction of a first-order property; and on the other hand, many simple and natural properties of arbitrary graphs (such as connectedness) are not first-order. However, there is some confusion among graph theorists about what is a first-order property, which this chapter may help to dispel.

We do more than just list first-order (and non-first-order) properties. One important theme of logic is the relationship between finite and infinite structures. There are several recent results about finite graphs whose proofs depend on the techniques of model theory and substantial results about infinite graphs. Theorems 5.10 and 5.15 below are examples of this.

5.2 First-order logic

First-order logic describes mathematical structures which consist of sets carrying various functions and relations, and having various distinguished elements. We use a language appropriate to the kind of structure to be considered. Accordingly, the syntax is specified by a set of *relation symbols*, a set of *function symbols*, and a set of *constant symbols*. Each relation or function symbol is equipped with its *arity*, the number of arguments it takes. There is also a countable set of *variables*, together with the standard relation =, the connectives ∨, ∧, ¬, and the quantifiers ∀, ∃. Expressions in the language are built up recursively:

a *term* is either a constant symbol or a variable, or a function symbol with the correct number of terms as arguments;

an *atomic formula* is a relation symbol (possibly =) with the correct number of terms as arguments;

a *formula* is either an atomic formula, a combination of formulas using the connectives in the usual way, or a quantified formula $(\forall x)\phi$ or $(\exists x)\phi$, where x is a variable and ϕ is a formula.

For example, graphs can be described in a language with one binary relation symbol for adjacency in addition to equality. As usual, we write this relation symbol as ∼. Thus the following are formulas of first-order graph theory:

(a) $(\exists x_1)(\exists x_2)(\exists x_3)((x_1 \sim x_2) \wedge (x_1 \sim x_3) \wedge (x_2 \sim x_3))$;

(b) $(\forall x_1)(\forall x_2)((x_1 = x_2) \vee (x_1 \sim x_2) \vee ((\exists x_3)((x_3 \sim x_1) \wedge (x_3 \sim x_2))))$.

The word *first-order* refers to the facts that

quantification can be done only over variables (which will stand for elements of a structure), and not over subsets, functions, natural numbers, formulas, etc.;

conjunction and disjunction of infinite sets of subformulas are not permitted.

Higher-order logics relax these conditions, but usually fail to have the important properties of first-order logic, notably Theorems 5.1 and 5.2 below.

In a subformula $(\forall x)\phi$ or $(\exists x)\phi$ of a formula, the occurrence of x in the quantifier and all occurrences of x in ϕ are *bound*. Any unbound occurrence of a variable is *free*. A *sentence* is a formula with no free occurrences of variables. The formulas (a) and (b) above are both sentences.

A *structure* $\mathcal{M}$ over a first-order language consists of a non-empty set M supporting constants, relations and functions (of the appropriate arity) corresponding to the symbols in the language, where the relation symbol = corresponds to the relation of equality on M. Given any assignment of values in M to the variables, the inductive rules allow us to associate an element of M with each term, and a truth value with each formula, in the expected manner: the connectives ∨ ('or'), ∧ ('and') and ¬ ('not') have their familiar truth tables; $(\forall x)\phi$ is **true** if and only if ϕ is **true** for all values of the variables which coincide with the given one at each variable except x; $(\exists x)\phi$ is interpreted similarly.

The truth value of a formula depends only on the values assigned to the variables with free occurrences in the formula. In particular, a sentence σ has a truth value in $\mathcal{M}$ independent of the values of the variables. We write $\mathcal{M} \models \sigma$ (read '$\mathcal{M}$ models σ') if σ has the value **true**. If Σ is a set of sentences, we say that $\mathcal{M} \models \Sigma$ if $\mathcal{M} \models \sigma$ for all $\sigma \in \Sigma$. The *first-order theory* of $\mathcal{M}$, written $\text{Th}(\mathcal{M})$, is the set of all sentences σ such that $\mathcal{M} \models \sigma$. It is a *complete* theory—that is, for any sentence σ, it contains either σ or $\neg\sigma$. It follows that if $\mathcal{N}$ is a model of $\mathrm{Th}(\mathcal{M})$, then $\mathrm{Th}(\mathcal{M}) = \mathrm{Th}(\mathcal{N})$.

For example, a graph G models the sentence (a) above if and only if it contains a triangle; and G models the sentence (b) if and only if it has diameter at most 2.

The next two results can be regarded as the portals of model theory. Both follow from the proof of Gödel's Theorem, asserting that there is a finite deduction system that is complete for first-order logic—that is, any logically valid formula can be proved.

Theorem 5.1 (Compactness Theorem) *A set Σ of sentences has a model if and only if each finite subset of Σ has a model.*

Theorem 5.2 (Downward Löwenheim–Skolem Theorem) *If a set of sentences has an infinite model, then it has one whose cardinality is no greater than that of the language.*

A consequence of the Compactness Theorem is the following.

Theorem 5.3 (Upward Löwenheim–Skolem Theorem) *If a set of sentences has an infinite model, then it has arbitrarily large infinite models.*

We consider structures over a fixed first-order language. A property $\mathcal{P}$ of structures is *first-order* if there is a set Σ of first-order sentences such that a structure $\mathcal{M}$ has the property $\mathcal{P}$ if and only if $\mathcal{M} \models \Sigma$. A first-order property $\mathcal{P}$ is *finitely axiomatizable* if Σ can be taken to be finite. In this case, we can take Σ to contain just one sentence, the conjunction of the finitely many given sentences. So if $\mathcal{P}$ is finitely axiomatizable (by a sentence σ), then not-$\mathcal{P}$ is also finitely axiomatizable (by the sentence $\neg\sigma$).

First-order properties behave like closed sets in a topological space.

Theorem 5.4

(a) Any conjunction of first-order properties is first-order.

(b) A disjunction of finitely many first-order properties is first-order.

Proof (a) If Σ_i is a set of sentences axiomatizing $\mathcal{P}_i$ for each $i \in I$, then the conjunction of the properties $\mathcal{P}_i$ is defined by $\bigcup_{i \in I} \Sigma_i$.

(b) It suffices to consider two properties $\mathcal{P}_1$ and $\mathcal{P}_2$, defined by sets Σ_1 and Σ_2 of sentences. Let $\Sigma = \{\sigma_1 \vee \sigma_2 : \sigma_1 \in \Sigma_1,\ \sigma_2 \in \Sigma_2\}$. Clearly, any structure having either of the properties $\mathcal{P}_1$ and $\mathcal{P}_2$ models Σ. Conversely, suppose that $\mathcal{M} \models \Sigma$ and $\mathcal{M}$ does not satisfy $\mathcal{P}_1$. Then $\mathcal{M} \models \neg\sigma_1$ for some $\sigma_1 \in \Sigma_1$. Since $\mathcal{M} \models \sigma_1 \vee \sigma_2$, we conclude that $\mathcal{M} \models \sigma_2$ for all $\sigma_2 \in \Sigma_2$; thus $\mathcal{M}$ satisfies $\mathcal{P}_2$. So $\mathcal{P}_1 \vee \mathcal{P}_2$ is axiomatized by Σ. □

The Compactness Theorem is analogous to topological compactness. The role of open-and-closed sets is played by the finitely axiomatizable properties.

Theorem 5.5 *The properties $\mathcal{P}$ and not-$\mathcal{P}$ are both first-order if and only if they are both finitely axiomatizable.*

Proof Suppose that Σ and T axiomatize $\mathcal{P}$ and not-$\mathcal{P}$, respectively. Then $\Sigma \cup T$ is unsatisfiable. By the Compactness Theorem, some finite subset $\Sigma_0 \cup T_0$ is unsatisfiable. Then Σ_0 axiomatizes $\mathcal{P}$. For, clearly, a structure with property $\mathcal{P}$ satisfies Σ_0. Also, if Σ_0 is satisfied, then some sentence in T_0 must fail, so that not-$\mathcal{P}$ does not hold, and so $\mathcal{P}$ does. The reverse implication is trivial. $\square$

Infiniteness is a first-order property of sets: for each n, let σ_n be a sentence saying 'there are at least n elements'; the set of all these sentences axiomatizes infiniteness. Also, finiteness is not finitely axiomatizable: for, if it were axiomatized by ϕ, then $\{\phi, \sigma_1, \sigma_2, \ldots\}$ would be unsatisfiable, whereas each of its finite subsets is satisfiable, contradicting compactness. Hence, by Theorem 5.5, finiteness cannot be first-order.

The following consequence of compactness is very useful.

Theorem 5.6 *If a set of first-order sentences has arbitrarily large finite models, then it has an infinite model.*

This follows by applying the Compactness Theorem to the union of the given set and the set of sentences σ_n defined above.

5.3 First-order properties of graphs

For any finite graph G, we can write a sentence ψ_G in the language of graph theory, any model of which is isomorphic to G. For example, if G is the complete graph K_2, then ψ_G could be taken to be

$$(\exists x_1)(\exists x_2)(\neg(x_1 = x_2) \wedge (x_1 \sim x_2) \wedge (\forall x_3)((x_3 = x_1) \vee (x_3 = x_2))).$$

Hence, for any finite set of finite graphs, there is a sentence, the disjunction of the corresponding sentences ψ_G, having just these graphs and their isomorphs as models. However, it follows from Theorem 5.6 that if Σ is any set of sentences having infinitely many non-isomorphic finite graphs as models, then Σ has an infinite model.

For any set $\mathcal{C}$ of isomorphism classes of finite graphs, there is a set of sentences whose *finite* models are precisely the graphs in $\mathcal{C}$. For example, this holds for the set of sentences $\neg \psi_G$ for all graphs $G \notin \mathcal{C}$. This particular set admits all infinite graphs as models! Thus first-order logic is useful for studying properties of not only *finite* graphs: infinite graphs play an essential role.

One type of graph property is necessarily first-order.

Theorem 5.7 *Let $\mathcal{P}$ be a property of graphs which is closed under taking induced subgraphs, and such that a graph has the property $\mathcal{P}$ if and only if all its finite subgraphs do. Then $\mathcal{P}$ is first-order. If there are infinitely many minimal excluded subgraphs for $\mathcal{P}$, then $\mathcal{P}$ is not finitely axiomatizable.*

Proof Such a property $\mathcal{P}$ is characterized by its minimal excluded subgraphs, each of which is finite. If $H_1, H_2, \ldots$ are the minimal excluded subgraphs for $\mathcal{P}$, and ϕ_n is the sentence saying that H_n is not embeddable in a graph G, then the set $\{\phi_1, \phi_2, \ldots\}$ axiomatizes $\mathcal{P}$. Suppose that $\mathcal{P}$ is finitely axiomatizable, and let σ be an axiom for it. Then the set $\{\neg\sigma, \phi_1, \phi_2, \ldots\}$ is not satisfiable; so some finite subset is not satisfiable, by compactness. This subset must include every ϕ_n, since if it does not include a particular ϕ_n, then the corresponding H_n is a model for the set of formulas. So there are only finitely many H_n. □

In particular, the property of bipartiteness is first-order, but not finitely axiomatizable, and so its negation is not first-order. A similar remark applies to perfectness, where an infinite graph is said to be *perfect* if each *finite* induced subgraph has equal clique number and chromatic number.

Connectedness is not a first-order property. For, suppose that Σ is a set of axioms for connectedness. Extend the language with two constant symbols c_1 and c_2; then Σ axiomatizes connected graphs with two distinguished vertices. Let ϕ_n be the sentence saying that no path of length n connects c_1 and c_2. Then the set $\Sigma \cup \{\phi_0, \phi_1, \phi_2, \ldots\}$ is finitely satisfiable: take c_1 and c_2 as the end-vertices of a path longer than the index of any ϕ_n in the finite set. But this set is clearly not satisfiable, contradicting compactness.

Another proof of this fact uses Theorem 5.3. Suppose that connectedness is a first-order property. Then, for fixed $k > 1$, the property of being connected and k-regular is first-order. There are infinite graphs with this property, and hence there are such graphs of arbitrarily large cardinality, which is impossible.

Disconnectedness is not a first-order property, but this fact is more subtle, and the proof is postponed until Section 5.7. It does satisfy the weaker condition of being *pseudo-elementary*—that is, of being equivalent to a first-order property in a larger language. To see this, we add to the language a unary relation symbol R. A graph supports a relation satisfying the sentence

$$(\forall x)(\forall y)(R(x) \wedge (x \sim y) \to R(y)) \wedge (\exists x)R(x) \wedge (\exists x)\neg R(x)$$

if and only if it is disconnected. This fact indicates that no simple compactness argument can show that it is not first-order, since such an argument would presumably work in the expanded language.

The Löwenheim–Skolem theorems imply that the properties 'Hamiltonian' and 'non-Hamiltonian' are not first-order properties: a Hamiltonian graph cannot be larger than countable, but any countable complete graph is Hamiltonian. For a similar reason, planarity, which might seem to be a first-order property since it is defined by excluding topological K_5 or $K_{3,3}$ subgraphs, is actually not first-order, since a planar graph has bounded cardinality.

We conclude this section with a catalogue classifying some graph properties. For k-colourable graphs, see Section 5.4. In many cases we omit the proof, since it is similar to one that is given.

Finitely axiomatizable properties include being complete, null, complete bipartite, a disjoint union of paths and cycles, a line graph or a series-parallel graph, or having clique number k (or at most k, or at least k) or bounded order, degree or diameter.

First-order but not finitely axiomatizable properties include being acyclic (a forest), a union of paths, bipartite, perfect or a comparability graph, or having infinite order, degree, diameter or clique number, or having chromatic number k (or at most k).

Non-first-order properties include being connected, disconnected, planar, non-planar, Hamiltonian, non-Hamiltonian or non-bipartite, or having finite order, degree, diameter or clique number, or having trivial (or non-trivial) automorphism group, or infinite (or finite) chromatic number, or chromatic number at least k.

Note that a few of these non-first-order properties (such as being disconnected, or having non-trivial automorphism group) are pseudo-elementary.

5.4 Applications of compactness

We have already seen some typical applications of compactness, in showing that certain properties are not first-order properties. Other applications involve transferring properties from the finite to the infinite. One of the best-known applications shows that the chromatic number of a graph is finitely determined.

Theorem 5.8 *If every finite induced subgraph of a graph G is k-colourable, then G is k-colourable.*

This theorem is usually proved as an application of the Compactness Theorem for propositional (Boolean) logic (see Barwise [1], for example). The following is a 'first-order' proof.

Proof Let $G = (V, E)$ be a graph, every finite subgraph of which is k-colourable. Let $L_{G,k}$ be the language of graph theory, augmented with unary relation symbols $C_1, \ldots, C_k$ and constant symbols c_v for $v \in V$. Let Σ be the following set of sentences:

a sentence asserting that for any x, exactly one of $C_1(x), \ldots, C_k(x)$ is true;

for $1 \leq i \leq k$, the sentence $(\forall x)(\forall y)(x \sim y \to \neg(C_i(x) \wedge C_i(y)))$;

for each pair v, w of distinct vertices, the sentence $c_v \sim c_w$ if $vw \in E$, or $\neg(c_v = c_w) \wedge \neg(c_v \sim c_w)$ if $vw \notin E$.

Now any finite subset Σ_0 of Σ is satisfiable, as we can see by choosing a k-colouring of the subgraph on all vertices v for which c_v is mentioned in Σ_0, setting $C_i(c_v)$ true if v has colour i and false otherwise, and setting the remaining relations arbitrarily. Hence Σ is satisfiable. Then the relations C_i on the vertices c_v define a k-colouring of G. □

Corollary 5.9 *For any k, the property of k-colourability is first-order.*

Proof k-colourability is defined by excluding all finite minimal non-k-colourable graphs. □

A similar proof technique applies to properties such as being a line graph or a comparability graph of a partial order.

A different application of compactness is a result of Cameron [4] about finite distance-transitive graphs; see Chapter 6 or 9 for further discussion of these graphs. A connected graph G is *distance-transitive* if whenever x, y, u and v are vertices with $d(x,y) = d(u,v)$, there is an automorphism of G carrying x to u and y to v. Note that a distance-transitive graph is regular.

Theorem 5.10 *For any given $k > 2$, there are only finitely many finite k-regular distance-transitive graphs.*

A permutation group is *primitive* if it leaves invariant no equivalence relation other than equality and the 'universal' relation, and a graph is *primitive* if its automorphism group is. Now, by a reduction due to Smith [23], it suffices to prove Theorem 5.10 for primitive graphs. The argument uses two subsidiary results. The first is a conjecture of Sims proved by Cameron, Praeger, Saxl and Seitz [7], and the second is a theorem of Macpherson [18].

Theorem 5.11 *There is a function f such that $|H| \leq f(k)$ for any finite primitive permutation group G with vertex stabilizer H having an orbit of size $k > 1$.*

Theorem 5.12 *If G is an infinite distance-transitive graph of finite degree, then, for some integers s and t, each block is the complete graph K_s and each vertex lies in t blocks.*

Figure 5.1 shows part of Macpherson's graph with $s = 4$, $t = 2$.

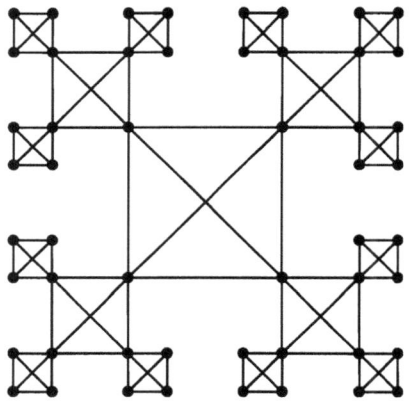

FIG. 5.1

Now distance-transitivity is not a first-order property of graphs, but we can describe in first-order language a structure consisting of a graph and a group acting on it, and write sentences asserting that the action is distance-transitive. Suppose that there are infinitely many primitive finite distance-transitive graphs of degree k. By Theorem 5.11, the stabilizer of a vertex has order at most $f(k)$. For any d, the vertices lying at distance d from v form an orbit of the stabilizer of v, so there are at most $f(k)$ such vertices. All of these numerical conditions are first-order, so, by Theorem 5.6, there is an infinite k-regular distance-transitive graph with at most $f(k)$ vertices at any given distance d from a given vertex.

But this contradicts Theorem 5.12 if $k > 2$, since in one of Macpherson's graphs, $k = t(s-1)$, and the number of vertices at distance d from v is $t(t-1)^{d-1}(s-1)^d$, which grows exponentially with d.

The proof of Theorem 5.11 uses the classification of finite simple groups. Subsequently, Weiss [27] found a proof of Theorem 5.10 avoiding the classification. It uses the Compactness Theorem and Macpherson's Theorem more carefully, to find structural information about large distance-transitive graphs of degree k, and then group-theoretic analysis to show that such graphs are impossible.

5.5 The random graph

There is a remarkable countable graph R which 'controls' the first-order theory of finite graphs. It is called *Rado's graph*, since Rado [20] gave the first explicit construction of it, or the *random graph*, since Erdős and Rényi [10] showed that a random countable graph chosen by selecting edges independently with probability $\frac{1}{2}$ is almost surely isomorphic to R. The remarkable properties of R are surveyed in Cameron [5], and we concentrate here on just one feature. In the course of the argument of Erdős and Rényi, the following result is shown.

Theorem 5.13 *Let m and n be non-negative integers, and let $\phi_{m,n}$ be the sentence in the language of graph theory asserting that for any distinct vertices $u_1, \ldots, u_m, v_1, \ldots, v_n$, there is a vertex z joined to all of $u_1, \ldots, u_m$ and to none of $v_1, \ldots, v_n$. Then a countable graph is isomorphic to R if and only if it satisfies all the sentences $\phi_{m,n}$.*

The sense in which the first-order theory of finite random graphs is determined by R is shown by the following result (see Glebskii *et al.* [14], Fagin [11] and Blass and Harary [2]). Recall that a property is satisfied in almost all finite graphs if the probability that it holds in an n-vertex random graph tends to 1 as $n \to \infty$. As before, edges are chosen independently with probability $\frac{1}{2}$, so that all n-vertex labelled graphs are equally likely.

Theorem 5.14 *Let σ be a sentence in the language of graph theory. Then σ holds in almost all finite graphs if and only if it holds in R. In particular, either σ holds in almost all finite graphs, or $\neg \sigma$ does.*

The proof is simple enough to be outlined. Suppose that $R \models \sigma$. By Theorem 5.13, σ is a logical consequence of the sentences $\phi_{m,n}$. By compactness, it is a logical consequence of finitely many of these sentences. A simple calculation shows that each $\phi_{m,n}$ holds in almost all finite graphs; the same is true for any finite set of these sentences, and hence for the sentence σ.

The 'zero-one law' for finitely axiomatizable properties expressed by the third sentence of Theorem 5.14 is a result about finite graphs whose proof involves a detour via the infinite graph R, although it is possible to rephrase the argument with no reference to R. Another result about finite graphs using R in its proof, where, however, no finite proof is known, is due to Cameron and Martins [6]. The statement requires some notation.

Let $\mathcal{F}$ be a finite set of finite graphs. For any graph G, let $\mathcal{F}(G)$ denote the hypergraph whose vertex set is that of G, and whose hyperedges are the subsets for which the induced subgraph is isomorphic to a member of $\mathcal{F}$. The general question is: *to what extent does $\mathcal{F}(G)$ determine G?*

Switching a graph G with respect to a set Y of vertices is defined as the operation of interchanging adjacency and non-adjacency between each pair of vertices, just one of which lies in Y (see Chapter 6 or 9). We now define five equivalence relations on the set of finite graphs: $\equiv_1$, equality; $\equiv_2$, equality or complementation; $\equiv_3$, switching; $\equiv_4$, switching, or complementation and switching; and $\equiv_5$, having the same vertex set.

Theorem 5.15 *Let $\mathcal{F}$ be a finite set of finite graphs. Then there is a number $i \in \{1,2,3,4,5\}$ such that for almost all finite graphs G, we have $\mathcal{F}(G) = \mathcal{F}(G')$ if and only if $G \equiv_i G'$.*

The proof uses Theorem 5.14, a theorem of Engeler, Ryll-Nardzewski and Svenonius (see Section 5.7) and a theorem of Thomas [25] on reducts of R.

5.6 Homogeneous graphs

Homogeneity is the highest degree of symmetry that a graph can possess. The graph G is *homogeneous* if each isomorphism between finite induced subgraphs of G can be extended to an automorphism of G. Thus a homogeneous graph is vertex-transitive, edge-transitive, non-edge-transitive, and so on.

It is not obvious that homogeneity of a countable graph can be expressed as a first-order property. We say that G has the *extension property* if, whenever A and B are finite graphs embeddable in G, with $A \subseteq B$ and $|B| = |A| + 1$, then any embedding of A into G can be extended to an embedding of B into G.

Theorem 5.16 *A finite or countable graph G is homogeneous if and only if it has the extension property.*

If the extension property holds, then any isomorphism between finite subgraphs of G can be extended so that its domain or codomain contains any prescribed additional vertex; then by the 'back-and-forth' method, it can be extended to an automorphism. Conversely, if G is homogeneous, then all embeddings of A in G are equivalent under the automorphism group of G; so if some embedding of A extends to B—that is, if B is embeddable—then all embeddings extend.

The extension property is a first-order property. (For each pair A, B, let $\theta_{A,B}$ be the sentence asserting the property for that pair: this sentence says that for all $|A|$-tuples of vertices inducing the subgraph A, there exists a vertex z extending the given A to a copy of B.) Now the axioms $\phi_{m,n}$ of the last section assert that the random graph R has the extension property, and we conclude that R is homogeneous.

Using Theorem 5.16, Fraïssé [12] gave a necessary and sufficient condition for a class $\mathcal{C}$ of finite graphs to consist of all finite graphs embeddable in a given countable homogeneous graph. The class $\mathcal{C}$ of finite graphs has the *amalgamation property* if, whenever $A \in \mathcal{C}$ and embeddings of A in two members B_1 and B_2 of $\mathcal{C}$ are given, then there is a graph $C \in \mathcal{C}$ with embeddings of B_1 and B_2 in C extending the given embeddings. (For logicians, we point out that A is permitted to be empty.) The *age* of a graph G is the class of all finite graphs that are embeddable in G.

Theorem 5.17 *A class $\mathcal{C}$ of finite graphs is the age of some countable homogeneous graph if and only if it is closed under isomorphism, closed under taking induced substructures, and has the amalgamation property. Moreover, any two countable homogeneous graphs with the same age are isomorphic.*

The unique countable homogeneous graph whose age is the class $\mathcal{C}$, if it exists, is called the *Fraïssé limit* of $\mathcal{C}$.

Determination of the countable homogeneous graphs is thus reduced to the determination of all classes of finite graphs satisfying Fraïssé's hypotheses. This was achieved by Lachlan and Woodrow [17], who proved the following theorem.

Theorem 5.18 *A countable homogeneous graph is one of the following:*

(a) *the disjoint union of m complete graphs of order n, where at least one of m and n is infinite;*

(b) *the complement of a graph in (a);*

(c) *the Fraïssé limit of the class of finite graphs containing no complete subgraph of fixed order $n \geq 3$;*

(d) *the complement of a graph in (c);*

(e) *the random graph R.*

Gardiner [13] had earlier determined the finite homogeneous graphs. As well as those of type (a) and (b) in the theorem (with m and n finite), there are only the 5-cycle and the line graph of $K_{3,3}$.

There is also a connection between homogeneity and the logical notion of *elimination of quantifiers*. Let G be a homogeneous graph and let $\phi(x_1, \ldots, x_n)$ be an n-variable formula. If $\phi(a_1, \ldots, a_n)$ holds in G, then $\phi(a'_1, \ldots, a'_n)$ also holds if $(a'_1, \ldots, a'_n)$ is any n-tuple isomorphic to $(a_1, \ldots, a_n)$. This means that $a'_i = a'_j$ if and only if $a_i = a_j$, and $a'_i \sim a'_j$ if and only if $a_i \sim a_j$. Thus the truth of ϕ depends only on the equality and adjacency relations among its arguments. Hence there is a quantifier-free formula $\psi(x_1, \ldots, x_n)$ (a Boolean combination of atomic formulas $x_i = x_j$ or $x_i \sim x_j$) that is equivalent to ϕ modulo the theory of G. This relation between homogeneity and quantifier elimination actually holds much more generally; see Kaye and Macpherson [16].

Fraïssé's Theorem is also much more general than described above. In particular, it holds for arbitrary relational structures (that is, structures over a language with no function or constant symbols). It is necessary to add the condition that there are at most countably many non-isomorphic finite structures in the class $\mathcal{C}$ to ensure that there is a countable homogeneous structure with age $\mathcal{C}$: this is automatically satisfied if there are only a finite number of relation symbols, but not in general. A further extension is described in Section 5.8.

5.7 $\aleph_0$-categorical graphs

According to Theorem 5.3, first-order sentences cannot specify an infinite structure completely, since they can put no restriction on its cardinality. The most they can do is to specify the structure when its cardinality is given. Accordingly, a set Σ of sentences is called λ-*categorical*, where λ is an infinite cardinal, if it has a unique model of cardinality λ (up to isomorphism), and a structure of cardinality λ is λ-*categorical* if its theory is. In fact, there are only two different instances of this concept, in view of a theorem of Morley [19].

Theorem 5.19 *Let λ and μ be uncountable cardinals. Then a theory in a countable language is λ-categorical if and only if it is μ-categorical.*

Thus a given theory may be $\aleph_0$-categorical (or *countably categorical*), or λ-categorical for all uncountable λ (*uncountably categorical*), or both, or neither.

It turns out that the two types of categoricity have profound implications for a structure. Uncountable categoricity implies a kind of 'structure theory', typified by vector spaces over countable fields. (A single cardinal number, the dimension, completely determines such a vector space up to isomorphism; any space of finite or countable dimension is countable, but an uncountable space has dimension equal to its cardinality.) An example of an uncountably categorical graph property is that of being 2-regular without cycles. A graph with this property is determined by the number of components it has, so there are a countable infinity of countable graphs, but only one of each uncountable cardinality. We do not discuss uncountable categoricity further.

On the other hand, countable categoricity implies a high degree of symmetry, as is shown by the following theorem, due independently to Engeler [9], Ryll-Nardzewski [21] and Svenonius [24]. A permutation group on an infinite set X is *oligomorphic* if it has only finitely many orbits on X^n, for each positive integer n.

Theorem 5.20 *A countable structure $\mathcal{M}$ is $\aleph_0$-categorical if and only if its automorphism group is oligomorphic.*

For example, let G be a countable homogeneous graph. Then two n-tuples lie in the same orbit of the automorphism group if and only if they induce isomorphic labelled subgraphs—this is just a re-statement of the definition of homogeneity—and so Aut(G) is oligomorphic. By Theorem 5.20, there is a set of first-order sentences whose only countable model is G. In this case, we can give such a set explicitly. Let ϕ_A express the fact that the finite graph A occurs as an induced subgraph. Let Σ consist of

(a) all ϕ_A for which A is in the age of G;

(b) all $\neg \phi_A$ for which A is not in the age of G;

(c) the sentences $\theta_{A,B}$ expressing the extension property of G.

Then any countable model of Σ has the same age as G, by (a) and (b), and is homogeneous, by (c). By the uniqueness statement in Fraïssé's Theorem, it is isomorphic to G.

Although there are only countably many countable homogeneous graphs, Droste and Macpherson [8] showed that there are $2^{\aleph_0}$ countable $\aleph_0$-categorical graphs. Note that a connected $\aleph_0$-categorical graph has finite diameter, and this does not exceed the number of orbits of its automorphism group on pairs of vertices, since pairs of vertices at different distances lie in different orbits.

Theorem 5.20 allows us to show that disconnectedness is not a first-order property. Suppose that G is a connected graph that is not $\aleph_0$-categorical but is the unique *connected* model of its first-order theory. Then there must be a disconnected graph H which models the theory of G.

The two-way infinite path G satisfies these requirements. It is connected and has infinite diameter, so by the above remark it is not $\aleph_0$-categorical. But G is the unique connected 2-regular graph without cycles, and all these properties except connectedness are first-order.

Further analysis shows that the models of the theory of G are just the disjoint unions of two-way infinite paths.

5.8 Sparse graphs

In the theory of sparse random graphs, we take a function p with $0 < p(n) < 1$ and (usually) $p(n) \to 0$ as $n \to \infty$, and choose a graph on n vertices by selecting edges independently with probability $p(n)$. Let $\mathcal{P}$ be a graph property closed under taking spanning subgraphs. We say that the function f is a *threshold* for $\mathcal{P}$ if the probability that the random graph has property $\mathcal{P}$ tends to 1 or 0 according as $p(n)/f(n) \to 0$ or $p(n)/f(n) \to \infty$. It is well known that many properties, both first-order (being a forest, 3-colourability, ...) and not (disconnectedness, being non-Hamiltonian, ...), have thresholds.

In two recent lines of development, due to Shelah and Spencer [22] and Hrushovski [15], this theory is related to first-order logic and infinite graphs. The results are technical, and the subject is developing rapidly, so only a brief sketch is given here. The articles of Winkler [28] and Wagner [26] shed further light, and Bollobás [3] is a good general reference for finite random graphs. This section is based on Wagner's account.

If the edge-probability tends to 0 as $n \to \infty$, the naive approach to constructing an infinite random graph fails: the limiting edge-probability is 0, and the graph is almost surely null. What we must do instead is find a suitable property holding with high probability in finite graphs, and then show that there is a particularly nice countable graph with this property, playing the role of Rado's graph R for constant edge-probability. This countable graph is likely to resemble R in having a lot of symmetry and being interesting to model theorists.

One such property that has been particularly useful is the following. Let ϕ be an integer function. The graph G is ϕ-*bounded* if every finite induced subgraph H of G satisfies $|E(H)| \leq \phi(|V(H)|)$; note that ϕ-boundedness is first-order. It can be shown that random finite graphs with small edge-probability $p(n)$ are almost surely ϕ-bounded, for some suitable function ϕ. We shall be concerned mainly with the second step: what does the 'nice' countable universal ϕ-bounded graph look like, if there is one?

For an illustrative example, let $\phi(n) = n - 1$ for $n \geq 1$, and $\phi(0) = 0$. Then a graph is ϕ-bounded if and only if it is a forest. The obvious candidate for the countable universal graph is the forest M, the disjoint union of countably many trees each of countable degree. Certainly, every finite forest can be embedded in M. However, the embeddings of a given finite forest G are not all equivalent. The best-behaved embeddings are those in which different components of G go into different components of M; these are all equivalent under automorphisms of M. But there is another problem: if G is contained in another finite forest H, the embedding of G into M may not lift to an embedding of H; indeed, it will do so only if different components of G are not contained in the same component of H. This can be guaranteed by requiring that the number of components of any G' with $G \subseteq G' \subseteq H$ is not smaller than that of G. Noting that the number of components of G' is one more than the excess of $\phi(|V(G')|)$ over $|E(G')|$, we come to the definition of a *self-sufficient* embedding.

Let α be a positive real number. We define the *predimension* $\delta(G)$ of a graph G by $\delta(G) = |V(G)| - \alpha|E(G)|$. For any non-negative function f, let $\mathcal{C}_f$ be the class of graphs G with $\delta(G') \geq f(|V(G')|)$ for all induced subgraphs G' of G. This class is clearly closed under isomorphism and under induced subgraphs. Our 'sparse' graphs are those in $\mathcal{C}_0$, and we consider $\mathcal{C}_f$ and various related subclasses $\mathcal{C}$. In our earlier terminology, the graphs in $\mathcal{C}_f$ are ϕ-bounded, where $\phi(x) = (x - f(x))/\alpha$. We require $f(x) \leq x$ for all x, to ensure that $\mathcal{C}_f$ contains arbitrarily large graphs. We have

$$\delta(G \cup H) + \delta(G \cap H) \leq \delta(G) + \delta(H),$$

with equality if and only if there are no edges between $G \setminus H$ and $H \setminus G$; if equality holds, we say that $G \cup H$ is the *free amalgamation* of G and H over $G \cap H$.

Suppose that $G, H \in \mathcal{C}_0$. We write $G \subseteq H$ to mean that G is an induced subgraph of H. We say that G is *self-sufficient* in H (written $G \leq H$) if $\delta(G) \leq \delta(G')$ for all G' with $G \subseteq G' \subseteq H$. Now we have:

$\emptyset \leq H$, for any H;

if $G \subseteq H' \subseteq H$ and $G \leq H$, then $G \leq H'$;

if $G_1, G_2 \subseteq H$ and $G_2 \leq H$, then $G_1 \cap G_2 \leq G_1$.

A class $\mathcal{C}$ is *good* if there is no infinite chain $G_0 \subseteq G_1 \subseteq G_2 \subseteq \ldots$ of graphs in $\mathcal{C}$ such that $\delta(G_i)$ is strictly decreasing. Let M be an infinite graph whose age is contained in $\mathcal{C}$. For finite $G \subseteq M$, we write $G \leq M$ if $G \leq H$ for all finite H with $G \subseteq H \subseteq M$. If $\mathcal{C}$ is good, then, for any $G \subseteq M$, there is a unique smallest $G' \leq M$ containing G; we call G' the *closure* of G in M, written $G' = \mathrm{cl}_M(G)$. Hrushovski [15] proved the following analogue of Fraïssé's Theorem (Theorem 5.17).

Theorem 5.21 *Suppose that $\mathcal{C}$ is a good subclass of $\mathcal{C}_0$ that is closed under free amalgamation of two graphs over a self-sufficient subgraph. Then there is a countable graph M with age $\mathcal{C}$, having the property that if $G \leq H$ are finite* *graphs, then any self-sufficient embedding of G in M extends to a self-sufficient embedding of H. Moreover, M is unique up to isomorphism, and any isomorphism between finite closed subgraphs of M extends to an automorphism of M.*

In practice, there are two situations in which $\mathcal{C}$ is good. If α is rational, then the values of $\delta(G)$ form a discrete set, and $\mathcal{C}$ is automatically good. For example, suppose that $\alpha = 1$ and $\mathcal{C} = \mathcal{C}_1 \cup \{\varnothing\}$. This is our earlier example: $\mathcal{C}$ is the class of finite forests. In this case, $\delta(G)$ is the number of connected components of G; we have $G \le H$ if and only if distinct components of G are contained in distinct components of H; and $\mathrm{cl}_M(G)$ is the union of the paths in M joining vertices of G. The graph M given by Theorem 5.21 is the disjoint union of countably many trees of countably infinite degree.

A more subtle situation occurs when α is an irrational number. In this case, the class $\mathcal{C}_f$ is good if and only if f is unbounded. However, the requirement that $\mathcal{C}_f$ is closed under free amalgamation puts a constraint on f. This is most easily seen geometrically. We represent a graph G by the point $P(G) = (|V(G)|, \delta(G))$ in the plane. Each graph in $\mathcal{C}_f$ is represented by a point between the curve $y = f(x)$ and the line $y = x$. The point representing the free amalgam of G_1 and G_2 is the fourth point of the parallelogram through $P(G_1 \cap G_2)$, $P(G_1)$ and $P(G_2)$. Now closure under free amalgamation requires that f is convex, and that if $G \le H$, then $f'(|H|)$ does not exceed the slope of the segment from $P(G)$ to $P(H)$.

Hrushovski showed, using partial fractions and the Baire Category Theorem, that for continuum many irrational numbers α, there is a function $f = f_\alpha$ for which f is unbounded and $\mathcal{C}_f$ is closed under free amalgamation.

Note that if these conditions hold, then the graph M given by Theorem 5.21 is $\aleph_0$-categorical. For, given $G \subseteq M$, there is a bound for $|V(\mathrm{cl}_M(G))|$ in terms of $|V(G)|$, since

$$f(|V(\mathrm{cl}_M(G))|) \le \delta(\mathrm{cl}_M(G)) \le \delta(G) \le |V(G)|,$$

and f is unbounded. Now $\mathrm{Aut}(M)$ has only finitely many orbits on self-sufficient subgraphs of bounded size, and any such subgraph contains only finitely many subgraphs of order $|V(G)|$; so $\mathrm{Aut}(M)$ is oligomorphic. This gives another proof of the existence of $2^{\aleph_0}$ countable $\aleph_0$-categorical graphs.

We conclude with a model-theoretic application of this construction. If $\frac{3}{4} < \alpha < 1$, then the class $\mathcal{C}_f$ has girth greater than 4. Hence if we take the open neighbourhoods of vertices of M as 'lines', we obtain an $\aleph_0$-categorical partial linear space with infinite line size. Thus Hrushovski refuted a conjecture of Lachlan asserting, in particular, that no such geometry could exist.

The construction is much more flexible than indicated; it can be extended by using other classes $\mathcal{C}$, by using relational structures other than graphs, or by further intricacies in the construction itself; see Wagner [26] for an account of these.

References

1. J. Barwise, An introduction to first-order logic, *Handbook of Mathematical Logic* (ed. J. Barwise), North-Holland, 1977, pp. 5–46.
2. A. Blass and F. Harary, Properties of almost all graphs and complexes, *J. Graph Theory* **3** (1979), 225–240; *MR* **81c**: 05081.
3. B. Bollobás, *Random Graphs*, Academic Press, 1985; *MR* **87f**: 05012.
4. P. J. Cameron, There are only finitely many finite distance-transitive graphs of given valency greater than two, *Combinatorica* **2** (1982), 9–13; *MR* **83k**: 05050.
5. P. J. Cameron, The random graph, *The Mathematics of Paul Erdős* (ed. R. L. Graham and J. Nešetřil), Springer-Verlag, to appear.
6. P. J. Cameron and C. Martins, A theorem on reconstruction of random graphs, *Combinatorics Probab. Comput.* **2** (1993), 1–9; *MR* **94f**: 05126.
7. P. J. Cameron, C. E. Praeger, J. Saxl and G. M. Seitz, On the Sims conjecture and distance transitive graphs, *Bull. London Math. Soc.* **15** (1983), 499–506; *MR* **85g**: 20006.
8. M. Droste and H. D. Macpherson, On k-homogeneous posets and graphs, *J. Combin. Theory (A)* **56** (1991), 1–15; *MR* **92e**: 03049.
9. E. Engeler, Äquivalenzklassen von n-Tupeln, *Z. Math. Logik Grundl. Math.* **5** (1959), 340–345; *MR* **25**#4996.
10. P. Erdős and A. Rényi, Asymmetric graphs, *Acta Math. Acad. Sci. Hungar.* **14** (1963), 295–315; *MR* **27**#6258.
11. R. Fagin, Probabilities on finite models, *J. Symbolic Logic* **41** (1976), 50–58; *MR* **57**#16042.
12. R. Fraïssé, Sur certains relations qui généralisent l'ordre des nombres rationnels, *C. R. Acad. Sci. Paris* **237** (1953), 540–542; *MR* **15**-192.
13. A. Gardiner, Homogeneous graphs, *J. Combin. Theory (B)* **20** (1976), 94–102; *MR* **54**#7316.
14. Y. V. Glebskii, D. I. Kogan, M. I. Liogon'kii and V. A. Talanov, Range and degree of realizability of formulas in the restricted predicate calculus, *Kybernetika* **2** (1969), 17–28.
15. E. Hrushovski, A new strongly minimal set, *Ann. Pure Appl. Logic* **62** (1993), 147–166; *MR* **94d**: 03064.
16. R. Kaye and H. D. Macpherson, Models and groups, *Automorphisms of First-Order Structures* (ed. R. Kaye and H. D. Macpherson), Oxford University Press, 1994, pp. 3–31; *MR* **96d**: 03048.
17. A. H. Lachlan and R. E. Woodrow, Countable ultrahomogeneous undirected graphs, *Trans. Amer. Math. Soc.* **262** (1980), 51–94; *MR* **82c**: 05083.
18. H. D. Macpherson, Infinite distance transitive graphs of finite valency, *Combinatorica* **2** (1982), 63–69; *MR* **84g**: 05075.
19. M. Morley, Categoricity in power, *Trans. Amer. Math. Soc.* **114** (1965), 514–538; *MR* **31**#58.
20. R. Rado, Universal graphs and universal functions, *Acta Arith.* **9** (1964), 331–340; *MR* **30**#2488.

21. C. Ryll-Nardzewski, On categoricity in power $\leq \aleph_0$, *Bull. Acad. Polon. Sci.* **7** (1959), 545–548; *MR* **22**#2543.
22. S. Shelah and J. Spencer, Zero-one laws for sparse random graphs, *J. Amer. Math. Soc.* **1** (1988), 97–115; *MR* **89i**: 05249.
23. D. H. Smith, Primitive and imprimitive graphs, *Quart. J. Math. Oxford (2)* **22** (1971), 551–557; *MR* **48**#5926.
24. L. Svenonius, $\aleph_0$-categoricity in first-order predicate calculus, *Theoria* **25** (1959), 82–94; *MR* **25**#1986a.
25. S. Thomas, Reducts of the random graph, *J. Symbolic Logic* **56** (1991), 176–181; *MR* **92m**: 05092.
26. F. O. Wagner, Relational structures and dimensions, *Automorphisms of First-Order Structures* (ed. R. Kaye and H. D. Macpherson), Oxford University Press, 1994, pp. 153–180; *MR* **96d**: 03048.
27. R. M. Weiss, On distance-transitive graphs, *Bull. London Math. Soc.* **17** (1985), 253–256; *MR* **87a**: 05082.
28. P. Winkler, Random structures and zero-one laws, *Finite and Infinite Combinatorics in Sets and Logic* (ed. N. W. Sauer, R. E. Woodrow and B. Sands), NATO Advanced Science Institutes Series, Kluwer, 1993, pp. 399–420; *MR* **95d**: 03051.

6
Linear Algebra

PETER ROWLINSON

We discuss the relationship between the structure of a graph and the spectrum of its adjacency matrix, and we describe a classic application of linear algebra to distance-regular graphs. The notions of graph angles and star partitions enable us to explore the influence on graph structure of an individual eigenvalue.

6.1 Introduction

The representation of a finite graph by a matrix provides an immediate link between linear algebra and graph theory. If the vertices of the graph G are labelled $1, 2, \ldots, n$, then the corresponding *adjacency matrix* $\mathbf{A}$ is the $n \times n$ matrix whose ij-entry is 1 if the vertices i and j are adjacent, and 0 if they are non-adjacent. Thus $\mathbf{A}$ is a symmetric matrix with zero diagonal; see Fig. 6.1 for a simple example.

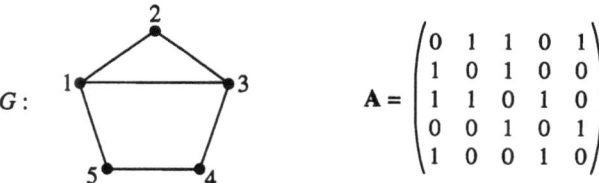

FIG. 6.1

Also commonly used to represent G are the *Seidel matrix* $\mathbf{S} = \mathbf{J} - 2\mathbf{A} - \mathbf{I}$, where $\mathbf{J}$ is the all-1 matrix, and the *Laplacian* (or *admittance*) *matrix* $\mathbf{L} = \mathbf{D} - \mathbf{A}$, where $\mathbf{D}$ is the diagonal matrix whose ith diagonal entry is the degree d_i of the vertex i; see [19] for a survey of Laplacian matrices and [24] for an application of the Seidel matrix.

We shall be concerned almost entirely with $\mathbf{A}$, regarded as a matrix with real entries. Then $\mathbf{A}$ is orthogonally diagonalizable and its eigenvalues are real. These eigenvalues are invariants of G because different vertex orderings yield similar adjacency matrices, and accordingly we speak of the *eigenvalues of the graph G*. In the next section we see what information about G can (and cannot) be extracted from the spectrum of its eigenvalues.

In Section 6.3 the multiplicities of eigenvalues feature in a classic application of linear algebra to distance-regular graphs. Such graphs illustrate further the limits of the information provided by the spectrum of a graph. These limitations motivate the study of additional algebraic invariants in Section 6.4. These in turn lead to the recent notion of a *star partition*, which is used in Section 6.5 to investigate the role of a single eigenvalue in the structure of a graph.

6.2 Graph spectra

Let G be a graph with adjacency matrix $\mathbf{A}$ and eigenvalues $\lambda_1, \lambda_2, \ldots, \lambda_n$. We begin our discussion of the relationship between structure and spectrum with the following result, which is easily proved by induction on k.

Theorem 6.1 *For each non-negative integer k, the ij-entry $a_{ij}^{(k)}$ of $\mathbf{A}^k$ is the number of ij-walks of length k in G.*

In particular, $a_{ii}^{(2)} = d_i$, and so the number of edges in G is equal to $\frac{1}{2}\mathrm{tr}(\mathbf{A}^2)$, or $\frac{1}{2}(\lambda_1^2 + \lambda_2^2 + \cdots + \lambda_n^2)$. Again, $a_{ii}^{(3)}$ is the number of 3-cycles $ijki$. Each triangle in G determines six such 3-cycles, because there are three choices of initial vertex and two possible orientations. It follows that the number of triangles in G is $\frac{1}{6}\mathrm{tr}(\mathbf{A}^3)$, or $\frac{1}{6}(\lambda_1^3 + \lambda_2^3 + \cdots + \lambda_n^3)$.

Although we can find the number of triangles in G, we cannot in general find the number of k-gons for $k > 3$, essentially because a closed walk of length $k > 3$ is not necessarily a k-cycle. For example, $C_4 \cup K_1$ and $K_{1,4}$ have the same spectrum (namely, $2, 0, 0, 0, -2$) but different numbers of quadrangles. The same pair of graphs shows that in general we can neither determine whether a graph is connected nor find its degree sequence. We can, however, find the length of a shortest cycle of odd length, as it is the least odd positive integer k such that $\lambda_1^k + \lambda_2^k + \cdots + \lambda_n^k$ is non-zero. This is because the subgraph induced by a closed walk of odd length necessarily contains a cycle of odd length.

Note that the connectedness of G is equivalent to the irreducibility of $\mathbf{A}$, for by definition the (symmetric) matrix $\mathbf{A}$ is reducible if and only if there exists a permutation matrix $\mathbf{P}$ such that $\mathbf{P}^{-1}\mathbf{A}\mathbf{P}$ has block-diagonal form $\mathrm{diag}(\mathbf{A}_1, \mathbf{A}_2)$. It follows from the Perron–Frobenius theory of matrices (see [16, Chapter 14]) that if G is connected, then its largest eigenvalue (or *index*) is a simple eigenvalue. The converse is false, as shown by the graph $C_4 \cup K_1$.

We can, however, find necessary and sufficient conditions in terms of the spectrum for G to be bipartite, regular, or strongly regular. Moreover, if G is regular then we can determine whether it is connected, and if G is strongly regular then we can determine its parameters. We discuss these three types of graph in turn. Additional results on graph spectra may be found in [12] and [11].

Bipartite graphs A graph G is bipartite if and only if its spectrum is symmetric about 0. For, if the spectrum has this property, then $\mathrm{tr}(\mathbf{A}^k) = 0$ for each odd positive integer k; thus G has no closed walks of odd length and is therefore bipartite. Conversely, if G is bipartite then, with an appropriate labelling of the vertices, $\mathbf{A}$ has the form

$$\begin{pmatrix} \mathbf{0} & \mathbf{B}^T \\ \mathbf{B} & \mathbf{0} \end{pmatrix},$$

and we have
$$\begin{pmatrix} 0 & B^T \\ B & 0 \end{pmatrix} \begin{pmatrix} x \\ y \end{pmatrix} = \lambda \begin{pmatrix} x \\ y \end{pmatrix}$$
if and only if
$$\begin{pmatrix} 0 & B^T \\ B & 0 \end{pmatrix} \begin{pmatrix} x \\ -y \end{pmatrix} = -\lambda \begin{pmatrix} x \\ -y \end{pmatrix}.$$

This shows that if λ is an eigenvalue, then $-\lambda$ is also an eigenvalue and has the same multiplicity; for, if the vectors
$$\begin{pmatrix} x_1 \\ y_1 \end{pmatrix}, \ldots, \begin{pmatrix} x_r \\ y_r \end{pmatrix}$$
are linearly independent, then so are the vectors
$$\begin{pmatrix} x_1 \\ -y_1 \end{pmatrix}, \ldots, \begin{pmatrix} x_r \\ -y_r \end{pmatrix},$$
since the matrices
$$\begin{pmatrix} x_1 & \cdots & x_r \\ y_1 & \cdots & y_r \end{pmatrix} \text{ and } \begin{pmatrix} x_1 & \cdots & x_r \\ -y_1 & \cdots & -y_r \end{pmatrix}$$
have the same row rank.

Regular graphs A necessary and sufficient condition for regularity is
$$\max\{\lambda_1, \ldots, \lambda_n\} = \frac{1}{n} \sum_{i=1}^{n} \lambda_i^2;$$

equivalently, the index μ_1 is equal to the mean degree $\bar{d}$. Indeed, by considering Rayleigh quotients, one can prove that for any graph G, $\mu_1 \geq \bar{d}$, with equality if and only if G is regular (see [12, Theorem 3.8]). This result was first proved by Collatz and Sinogowitz [6], and led them to suggest $\mu_1 - \bar{d}$ as a measure of irregularity. For a discussion of irregularity in this context, see [2].

The above remark on irreducible matrices shows that if G is regular of degree r, then the multiplicity of r as an eigenvalue of G is equal to the number of components of G. In particular, we can determine from the spectrum of a regular graph G whether G is connected. We note in passing that for a connected regular graph of degree r, the spectra of the matrices $\mathbf{A}$, $\mathbf{L}$ and $\mathbf{S}$ are determined by the spectrum of just one of them since, in each case, all but one of the n eigenvalues correspond to eigenvectors orthogonal to $(1, 1, \ldots, 1)^T$.

Strongly regular graphs A connected non-complete graph G is *strongly regular* with parameters r, s, t ($r > 0$, $s \geq 0$, $t > 0$) if G is regular of degree r, any two adjacent vertices have s common neighbours, and any two non-adjacent vertices have t common neighbours; equivalently, the adjacency matrix $\mathbf{A}$ of G satisfies the equation

$$\mathbf{A}^2 + (t-s)\mathbf{A} + (t-r)\mathbf{I} = t\mathbf{J}.$$

It follows that the minimal polynomial of $\mathbf{A}$ is

$$(x-r)\{x^2 + (t-s)x + (t-r)\}.$$

Thus a necessary and sufficient condition for a connected regular graph to be strongly regular is that it have exactly three distinct eigenvalues (see [12, Theorem 3.32]). Further, if the distinct eigenvalues are μ_1, μ_2, μ_3 in decreasing order, then

$$r = \mu_1, \quad s = \mu_1 + \mu_2 + \mu_3 + \mu_2\mu_3, \quad t = \mu_1 + \mu_2\mu_3.$$

The Petersen graph is an example with $r = 3$, $s = 0$, $t = 1$ and $\mu_1 = 3$, $\mu_2 = 1$, $\mu_3 = -2$.

6.3 Distance-regular graphs

A graph is *distance-regular* if, for each non-negative integer k, the number $a_{ij}^{(k)}$ of ij-walks of length k depends only on the distance $d(i,j)$ between the vertices i and j. The distance-regular graphs of diameter 2 are precisely the strongly regular graphs; in particular, for a strongly regular graph with parameters r, s, t, we have $a_{ij}^{(2)} = r$, s or t, according as $d(i,j) = 0$, 1 or 2. From Section 6.2 we know that the property of being a distance-regular graph of diameter 2 is identifiable from its spectrum. We shall see in this section that we cannot make an analogous statement for distance-regular graphs of diameter greater than 2.

For any graph G and any non-negative integer h, we define the *distance matrix* $\mathbf{A}_h$ as follows: the ij-entry of $\mathbf{A}_h$ is 1 if $d(i,j) = h$, and 0 otherwise; thus $\mathbf{A}_0 = \mathbf{I}$ and $\mathbf{A}_1 = \mathbf{A}$. In addition, we define the *adjacency algebra* $\mathcal{A}$ to be the algebra generated by all powers of $\mathbf{A}$. In this section, G denotes a distance-regular graph of diameter d, so that each power of $\mathbf{A}$ is a linear combination of the $d+1$ linearly independent matrices $\mathbf{A}_0, \mathbf{A}_1, \ldots, \mathbf{A}_d$. Since $\mathbf{I}, \mathbf{A}, \mathbf{A}^2, \ldots, \mathbf{A}^d$ are also linearly independent, both $\{\mathbf{A}_0, \mathbf{A}_1, \ldots, \mathbf{A}_d\}$ and $\{\mathbf{I}, \mathbf{A}, \ldots, \mathbf{A}^d\}$ are bases for $\mathcal{A}$. It follows that the minimal polynomial $m_\mathbf{A}$ of $\mathbf{A}$ has degree $d+1$, and so G has precisely $d+1$ distinct eigenvalues—say, $\mu_1, \mu_2, \ldots, \mu_{d+1}$ in decreasing order. For example, the skeleton of a cube (the cube graph Q_3) is a distance-regular graph of diameter 3, with eigenvalues $3, 1, -1, -3$ of multiplicities $1, 3, 3, 1$, respectively.

Now consider the linear transformation τ of $\mathcal{A}$ mapping $\mathbf{X}$ to $\mathbf{XA}$, for each $\mathbf{X} \in \mathcal{A}$. The matrix of τ with respect to $\{\mathbf{I}, \mathbf{A}, \ldots, \mathbf{A}^d\}$ is just the companion matrix of $m_\mathbf{A}$, but the matrix $\mathbf{B}$ of τ with respect to $\{\mathbf{A}_0, \mathbf{A}_1, \ldots, \mathbf{A}_d\}$ has an interesting tri-diagonal form, as we shall see. If $\mathbf{B} = (\beta_{ij})$, then

$$\mathbf{A}_i \mathbf{A} = \sum_{j=0}^{d} \beta_{ij} \mathbf{A}_j;$$

on equating uv-entries, where $d(u,v) = j$, we see that β_{ij} is the number of neighbours of v at distance i from u. It is clear that if $\beta_{ij} \neq 0$, then $|i - j| \leq 1$, and so **B** is a tri-diagonal matrix. In standard notation,

$$\mathbf{A}_i\mathbf{A} = b_{i-1}\mathbf{A}_{i-1} + a_i\mathbf{A}_i + c_{i+1}\mathbf{A}_{i+1} \quad (0 < i < d),$$

$$\mathbf{A}_0\mathbf{A} = 0\mathbf{A}_0 + 1\mathbf{A}_1, \quad \mathbf{A}_d\mathbf{A} = b_{d-1}\mathbf{A}_{d-1} + a_d\mathbf{A}_d. \tag{6.1}$$

The parameters a_j, b_j and c_j have the following interpretation, where u and v are any two vertices at distance j and $\Gamma_i(u)$ denotes the set of vertices at distance i from u:

$$a_j = |\Gamma_j(u) \cap \Gamma_1(v)|, \quad b_j = |\Gamma_{j+1}(u) \cap \Gamma_1(v)|, \quad c_j = |\Gamma_{j-1}(u) \cap \Gamma_1(v)|. \tag{6.2}$$

Thus a_j, b_j and c_j are the numbers of neighbours of v at distances j, $j+1$ and $j-1$ from u, respectively. We write $r = b_0$, so that G is regular of degree r, and

$$a_j + b_j + c_j = r \ (0 < j < d), \quad a_0 = 0, \quad c_1 = 1, \quad a_d + c_d = r.$$

For the skeleton of a cube, we have $a_1 = 0, b_1 = 2$; $a_2 = 0, b_2 = 1, c_2 = 2$; $a_3 = 0, c_3 = 3$. In the general case, we have

$$\mathbf{B} = \begin{pmatrix} 0 & 1 & & & & & \\ r & a_1 & c_2 & & & & \\ & b_1 & a_2 & \cdot & & & \\ & & b_2 & \cdot & \cdot & & \\ & & & \cdot & \cdot & \cdot & \\ & & & \cdot & \cdot & c_{d-1} & \\ & & & & \cdot & a_{d-1} & c_d \\ & & & & & b_{d-1} & a_d \end{pmatrix},$$

and **B** is determined by the *intersection array* $\{r, b_1, b_2, \ldots, b_{d-1}; 1, c_2, \ldots, c_d\}$.

A distance-regular graph is usually defined as one with such an intersection array, where the parameters are given by equations (6.2). The study of such graphs originated in a paper of Biggs [3], and a good introduction to the topic may be found in his monograph [4].

A general problem is to determine which arrays are the intersection arrays of distance-regular graphs, and a number of necessary conditions are known. For example, from simple combinatorial considerations (see [4, Proposition 20.4] and [26]), we have:

$1 \leq c_2 \leq c_3 \leq \cdots \leq c_d$,
$r \geq b_1 \geq b_2 \geq \cdots \geq b_{d-1}$,
$c_i \leq b_j$ whenever $i + j \leq d$,
for each $j \in \{2, \ldots, d\}$, $rb_1 \ldots b_{j-1}/c_2 c_3 \ldots c_j = |\Gamma_j(v)|$, which is an integer.

But linear algebra can provide additional constraints, and the description that follows is based on [4, Chapter 21].

First, note that the matrix $\mathbf{B}$ determines the distinct eigenvalues $\mu_1, \ldots, \mu_{d+1}$ of G because $m_\mathbf{B} = m_\tau = m_\mathbf{A}$. Secondly, if the polynomials $v_0(x), \ldots, v_d(x)$ are defined recursively by

$$v_0(x) = 1, \ v_1(x) = x, \ c_{i+1}v_{i+1}(x) + (a_i - x)v_i(x) + b_{i-1}v_{i-1}(x) = 0 \ (0 < i < d),$$

then it follows by induction from the identities (6.1) that $\mathbf{A}_i = v_i(\mathbf{A})$, for $i = 0, 1, \ldots, d$. In other words, $\mathbf{B}$ determines the transition matrix from $\{\mathbf{A}_0, \mathbf{A}_1, \ldots, \mathbf{A}_d\}$ to $\{\mathbf{I}, \mathbf{A}, \mathbf{A}^2, \ldots, \mathbf{A}^d\}$, and hence the inverse transition matrix (w_{hk}), where

$$\mathbf{A}^k = \sum_{h=0}^{d} w_{hk} \mathbf{A}_h \quad (k = 0, 1, \ldots, d).$$

Since $\text{tr}(\mathbf{A}_0) = n$, while $\text{tr}(\mathbf{A}_h) = 0$ for $h \in \{1, \ldots, d\}$, we have

$$\sum_{i=1}^{d+1} m(\mu_i) \mu_i^k = n w_{0k} \quad (k = 0, 1, \ldots, d),$$

where $m(\mu_i)$ denotes the multiplicity of μ_i as an eigenvalue of $\mathbf{A}$.

It is clear from these $d+1$ equations that the multiplicities $m(\mu_i)$, and hence the spectrum of G, are determined by the entries of $\mathbf{B}$. In fact, there is a neat way to obtain $m(\mu_i)$ in terms of the left and right eigenvectors of $\mathbf{B}$ (see [4, Proposition 21.4]), and then $m(\mu_i)$ is determined as a rational function of μ_i. Since $m(\mu_i)$ is a positive integer, this imposes a powerful restriction on the parameters of an intersection array. For example (see [4, p. 168]), there is no distance-regular graph with intersection array $\{3, 2, 1; 1, 1, 3\}$, an array that survives the above combinatorial feasibility conditions. On the other hand, there are four non-isomorphic distance-regular graphs with intersection array $\{7, 6, 4; 1, 3, 7\}$; they have spectrum $\{7^1, 2^{14}, -2^{14}, -7^1\}$, where the superscripts denote multiplicities, and they are the only graphs with this spectrum. Haemers and Spence [18] showed, in contrast, that while there is a unique distance-regular graph with intersection array $\{13, 6, 1; 1, 6, 13\}$, it is one of no fewer than 515 graphs of diameter 3 that share the spectrum $\{13^1, \sqrt{13}^7, -1^{13}, -\sqrt{13}^7\}$.

Further necessary conditions on the parameters of an intersection array arise from the fact that $\mathcal{A}$ is closed under elementwise matrix multiplication; for these and other constraints, see the excellent monograph by Brouwer, Cohen and Neumaier [5], which provides 800 references related to distance-regular graphs. For graphs in the following categories, all the arrays that pass all known feasibility tests for distance-regularity are listed:

graphs with diameter ≥ 5 and at most 4096 vertices,

non-bipartite graphs with diameter 4 and at most 4096 vertices,

primitive graphs with diameter 3 and at most 1024 vertices.

(A graph with diameter d is *primitive* if each of $\mathbf{A}_1, \mathbf{A}_2, \ldots, \mathbf{A}_d$ is the adjacency matrix of a connected graph.)

6.4 Other algebraic invariants

We have seen that the structural information provided by the spectrum of a graph is severely limited, and so it is natural to seek additional algebraic invariants. Such invariants should be computable in polynomial time—that is, in a number of steps bounded above by some polynomial function of the number n of vertices of the graph. The reason is that we would like a complete set of invariants that enable us to determine in polynomial time whether two graphs are isomorphic. It remains to be seen whether this goal can be achieved, but at the least any new invariants provide additional sufficient conditions for non-isomorphism of graphs.

A graph G is reconstructible from its spectrum and a basis $\{x_1, \ldots, x_n\}$ consisting of eigenvectors of the adjacency matrix $\mathbf{A}$. For, if $x_1, \ldots, x_n$ are taken as the columns of the matrix $\mathbf{P}$, then $\mathbf{A} = \mathbf{PDP}^{-1}$, where $\mathbf{D} = \mathrm{diag}(\lambda_1, \ldots, \lambda_n)$ and $\mathbf{A}x_i = \lambda_i x_i$ $(i = 1, \ldots, n)$. Moreover, such a basis can be found in polynomial time, but we are still faced with the problem of determining whether different adjacency matrices $\mathbf{A}, \mathbf{A}'$ with the same spectrum represent isomorphic graphs. At the very worst, for each permutation π of $\{1, \ldots, n\}$, we have to check whether $\mathbf{M}(\pi)^T \mathbf{A} \mathbf{M}(\pi) = \mathbf{A}'$, where $\mathbf{M}(\pi)$ is the permutation matrix determined by π. If, there is an absolute upper bound on the dimensions of eigenspaces (in particular, if there are n distinct eigenvalues), then graph isomorphism can be tested in polynomial time (see [1, 9]). In general, eigenspaces play a crucial role, yet eigenvectors themselves are not algebraic invariants because their coordinates are permuted when the vertices of G are renumbered. Accordingly, we look to geometric attributes of eigenspaces with algebraic interpretations that are essentially coordinate-free.

One example consists of the angles between eigenspaces and the all-1 vector $\mathbf{j}$. For further refinement we consider the angles between eigenspaces and the vectors $\mathbf{e}_1, \ldots, \mathbf{e}_n$, where $\mathbf{e}_1 = (1, 0, 0, \ldots, 0)^T$, etc. In order to investigate the implications for graph structure, we invoke the existence of an orthogonal matrix $\mathbf{U}$ such that $\mathbf{U}^T \mathbf{A} \mathbf{U} = \mathbf{D}$, where $\mathbf{D} = \mathrm{diag}(\lambda_1, \ldots, \lambda_n)$. The columns of $\mathbf{U}$ form an orthonormal basis for $\mathbf{R}^n$, and we order them so that repeated eigenvalues occur consecutively on the diagonal of $\mathbf{D}$. Thus if $k_i = \dim \mathcal{E}(\mu_i)$ $(i = 1, 2, \ldots, m)$, then

$$\mathbf{U}^T \mathbf{A} \mathbf{U} = \mu_1 \mathbf{E}_1 + \cdots + \mu_m \mathbf{E}_m,$$

where $\mu_1, \ldots, \mu_m$ are the distinct eigenvalues and

$$\mathbf{E}_1 = \begin{pmatrix} \mathbf{I}_{k_1} & 0 \\ 0 & 0 \end{pmatrix}, \ldots, \mathbf{E}_m = \begin{pmatrix} 0 & 0 \\ 0 & \mathbf{I}_{k_m} \end{pmatrix}.$$

On conjugating by $\mathbf{U}$, we obtain the *spectral decomposition*

$$\mathbf{A} = \mu_1 \mathbf{P}_1 + \cdots + \mu_m \mathbf{P}_m,$$

where $\mathbf{P}_i = \mathbf{U} \mathbf{E}_i \mathbf{U}^T$ $(i = 1, \ldots, m)$. Note that $\mathbf{P}_i \mathbf{P}_j = \mathbf{0}$ when $i \neq j$, and that $\mathbf{P}_i^2 = \mathbf{P}_i = \mathbf{P}_i^T$ and $\mathbf{P}_i \mathbf{A} = \mathbf{A} \mathbf{P}_i$ for each i.

A *eutactic star* in the subspace $\mathcal{U}$ of an inner product space $\mathcal{V}$ is, to within a scalar multiple, the orthogonal projection onto $\mathcal{U}$ of an orthonormal set of vectors in $\mathcal{V}$ (see [25]). Since $\mathbf{P}_i$ represents the orthogonal projection of $\mathbf{R}^n$ onto the eigenspace $\mathcal{E}(\mu_i)$, the vectors $\mathbf{P}_i\mathbf{e}_1, \mathbf{P}_i\mathbf{e}_2, \ldots, \mathbf{P}_i\mathbf{e}_n$ are the arms of a eutactic star $\mathcal{S}_i$ in $\mathcal{E}(\mu_i)$. Figure 6.2 illustrates the construction of $\mathcal{S}_i$ when $n = 3$ and $k_i = 2$. Generally speaking, geometric properties of the stars $\mathcal{S}_1, \mathcal{S}_2, \ldots, \mathcal{S}_m$ can be related to spectral properties of G and its induced subgraphs. Certain aspects of this relationship are explored here and in the next section.

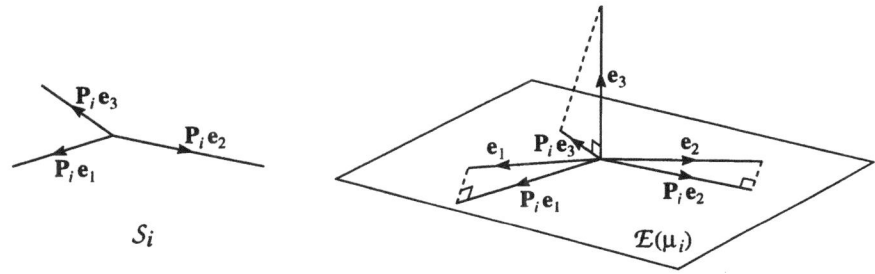

FIG. 6.2

If $\alpha_{ij} = \|\mathbf{P}_i\mathbf{e}_j\|$, then $\cos^{-1}(\alpha_{ij})$ is the angle between $\mathbf{e}_j$ and $\mathcal{E}(\mu_i)$, but it is the α_{ij} themselves that are customarily referred to as the *angles* of the graph G. The $m \times n$ matrix (α_{ij}) is an algebraic invariant, computable in polynomial time, when its rows are ordered so that $\mu_1 > \mu_2 > \ldots > \mu_m$ and its columns are ordered lexicographically; we call (α_{ij}) the *angle matrix* of G. It is not difficult to show (see [11]) that

$$\sum_{i=1}^{m} \alpha_{ij}^2 = 1 \quad \text{and} \quad \sum_{j=1}^{n} \alpha_{ij}^2 = k_i.$$

From the spectrum and angle matrix of G, we can find the degree sequence. This is the case $k = 2$ of the relation

$$a_{jj}^{(k)} = \sum_{i=1}^{m} \alpha_{ij}^2 \mu_i^k, \tag{6.3}$$

obtained by equating diagonal entries in the matrix equation

$$\mathbf{A}^k = \mu_1^k \mathbf{P}_1 + \mu_2^k \mathbf{P}_2 + \cdots + \mu_m^k \mathbf{P}_m.$$

We can also use equation (6.3) to express the number of quadrangles in G and the number of pentagons in G in terms of eigenvalues and angles; explicit formulas are given in [13, Theorems 3 and 4]. Moreover, as noted in [10], we can find from the spectrum and the angle matrix the spectrum of each vertex-deleted subgraph $G-j$. The result here is best expressed in terms of characteristic polynomials:

$$\phi_{G-j}(x) = \phi_G(x) \sum_{i=1}^{m} \frac{\alpha_{ij}^2}{x - \mu_i}, \tag{6.4}$$

where $\phi_G(x) = \det(xI - A)$. To see this, note that $\phi_{G-j}(x)$ is the jj-entry of the adjoint of $xI - A$; this adjoint is expressible as $(xI - A)^{-1}\det(xI - A)$, and hence as

$$\phi_G(x) \sum_{i=1}^{m} (x - \mu_i)^{-1} P_i.$$

From equations (6.3) and (6.4), we can see that, given the spectrum of G, the following are equivalent:

knowledge of the angles of G;

knowledge of $a_{jj}^{(k)}$, for $j = 1, \ldots, n$ and $k = 0, \ldots, m-1$;

knowledge of the polynomials $\phi_{G-j}(x)$, for $j = 1, \ldots, n$.

However, the angles and eigenvalues are far from sufficient to characterize a graph, for Cvetković [8] has shown how to construct an infinite family of pairs of non-isomorphic trees, each pair having the same spectrum and the same angle matrix.

Earlier, we mentioned the angles between the all-1 vector j and the eigenspaces $\mathcal{E}(\mu_i)$. Their cosines, called the *main angles* β_i of G, are given by $\beta_i = \|P_i j\|/\sqrt{n}$, for $i = 1, \ldots, m$, where (as before) we assume that $\mu_1 > \mu_2 > \ldots > \mu_m$. Note that a connected graph is regular if and only if its main angles are $1, 0, 0, \ldots, 0$. This is because G is a regular graph if and only if $j \in \mathcal{E}(\mu_1)$, while each $\mathcal{E}(\mu_i)$ ($i > 1$) is orthogonal to $\mathcal{E}(\mu_1)$. It is shown in [20] that, given the spectrum of G, any of the following can be obtained from any other:

the main angles of G;

the Seidel spectrum of G;

the spectrum of the complement $\overline{G}$ of G;

the spectrum of the graph obtained from G by adding one new vertex adjacent to each vertex of G.

Non-isomorphic graphs can have the same spectrum, angles and main angles, and examples may be found in the family of *walk-regular* graphs. Such a graph is one in which, for each positive integer k, $a_{jj}^{(k)}$ is independent of j. Thus walk-regular graphs are regular and include the distance-regular graphs. Figure 6.3 shows a walk-regular graph which is not distance-regular.

It is shown in [17] that a graph is walk-regular if and only if all the columns of its angle matrix are the same, a result which follows from the identities (6.1). It follows from this and earlier remarks that cospectral walk-regular graphs have the same angles and main angles. In particular, the four non-isomorphic distance-regular graphs with intersection array $\{7, 6, 4; 1, 3, 7\}$ have the same spectrum, angles and main angles.

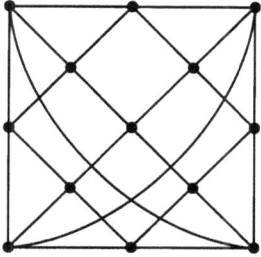

FIG. 6.3

6.5 Eigenvalues and star partitions

In the preceding sections, we have mainly considered properties of graphs related to the eigenvalues collectively. We now investigate the role of an individual eigenvalue. First note that, for each $i \in \{1,\ldots,m\}$, $\mathcal{E}(\mu_i)$ is spanned by the vectors $\mathbf{P}_i\mathbf{e}_1,\ldots,\mathbf{P}_i\mathbf{e}_n$, and so there exists a k_i-element subset X_i of $\{1,\ldots,n\}$ such that the vectors $\mathbf{P}_i\mathbf{e}_j$ ($j \in X_i$) form a basis of $\mathcal{E}(\mu_i)$. It turns out that the sets $X_1,\ldots,X_m$ can always be chosen to be pairwise disjoint (see [14] or [21]), and then the partition $V(G) = X_1 \dot\cup \ldots \dot\cup X_m$ is called a *star partition* of G, with *star cells* $X_1,\ldots,X_m$. In this situation, $\bigcup_{i=1}^{m}\{\mathbf{P}_i\mathbf{e}_j : j \in X_i\}$ is a basis for $\mathbf{R}^n$, called the corresponding *star basis*. Such a basis can always be found in polynomial time (see [14]).

As explained in [14, Section 5] and [9], it is possible to define recursively a *canonical star basis* corresponding to G—that is, a star basis with the property that two cospectral graphs are isomorphic if and only if they determine the same canonical star basis. It remains to be seen whether these canonical bases can be obtained in polynomial time.

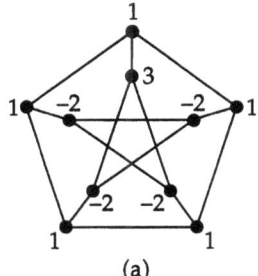
(a)

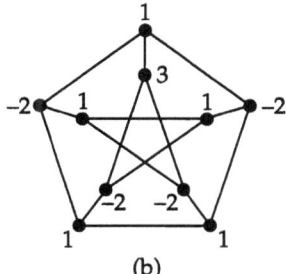
(b)

FIG. 6.4

Figure 6.4 shows two different star partitions of the Petersen graph, for which $m = 3$, $\mu_1 = 3$, $\mu_2 = 2$, $\mu_3 = -2$ and $k_1 = 1$, $k_2 = 5$, $k_3 = 4$. In each case, the vertices in X_i are labelled by μ_i ($i = 1, 2, 3$). One way to verify that these are star partitions is to use part (c) of the following result from [14].

Theorem 6.2 *The following statements are equivalent for each i:*
(a) $\{\mathbf{P}_i \mathbf{e}_j : j \in X_i\}$ *is a basis for* $\mathcal{E}(\mu_i)$;
(b) $\mathbf{R}^n = \mathcal{E}(\mu_i) \oplus \langle \mathbf{e}_j : j \notin X_i \rangle$;
(c) $|X_i| = k_i$, *and* μ_i *is not an eigenvalue of* $G - X_i$.

The next result, sometimes known as the *Reconstruction Theorem*, establishes a relation between an individual eigenvalue of G and the structure of G. We give the proof from [13] to demonstrate a simple application of linear algebra. In what follows, X_i denotes a set of vertices satisfying the conditions given in Theorem 6.2. We write $E(X_i)$ for the set of edges between vertices of X_i, and $E(X_i, \overline{X}_i)$ for the set of edges between X_i and its complement $\overline{X}_i$ in $V(G)$.

Theorem 6.3 (Reconstruction Theorem) *If* $|X_i| = k_i$ *and* μ_i *is not an eigenvalue of* $G - X_i$, *then the graph* G *is reconstructible from the graph* $G - X_i$, *the edge set* $E(X_i, \overline{X}_i)$ *and the eigenvalue* μ_i.

Proof With appropriate labelling of vertices, the adjacency matrix $\mathbf{A}$ of G has the form
$$\begin{pmatrix} \mathbf{A}_i & \mathbf{B}_i^T \\ \mathbf{B}_i & \mathbf{C}_i \end{pmatrix},$$
where $\mathbf{A}_i$ is the adjacency matrix of $G - \overline{X}_i$ and $\mathbf{C}_i$ is the adjacency matrix of $G - X_i$. We have to show that $\mathbf{A}_i$ is determined by $\mathbf{C}_i$, $\mathbf{B}_i$ and μ_i. Now
$$\mu_i \mathbf{I} - \mathbf{A} = \begin{pmatrix} \mu_i \mathbf{I} - \mathbf{A}_i & -\mathbf{B}_i^T \\ -\mathbf{B}_i & \mu_i \mathbf{I} - \mathbf{C}_i \end{pmatrix},$$
and the rank of this matrix is $n - k_i$. But $\mu_i \mathbf{I} - \mathbf{C}_i$ is invertible since μ_i is not an eigenvalue of $G - X_i$, and so the rows of $(-\mathbf{B}_i \mid \mu_i \mathbf{I} - \mathbf{C}_i)$ form a basis for the row-space of $\mu_i \mathbf{I} - \mathbf{A}$. Hence there exists a matrix $\mathbf{L}$ such that
$$(\mu_i \mathbf{I} - \mathbf{A}_i \longrightarrow -\mathbf{B}_i^T) = \mathbf{L}(-\mathbf{B}_i \mid \mu_i \mathbf{I} - \mathbf{C}_i).$$
Thus
$$\mu_i \mathbf{I} - \mathbf{A}_i = -\mathbf{L}\mathbf{B}_i \quad \text{and} \quad -\mathbf{B}_i^T = \mathbf{L}(\mu_i \mathbf{I} - \mathbf{C}_i).$$
It follows that
$$\mathbf{L} = -\mathbf{B}_i^T (\mu_i \mathbf{I} - \mathbf{C}_i)^{-1}$$
and
$$\mathbf{A}_i = \mu_i \mathbf{I} - \mathbf{B}_i^T (\mu_i \mathbf{I} - \mathbf{C}_i)^{-1} \mathbf{B}_i. \tag{6.5}$$
This completes the proof. □

Theorem 6.3 shows that G is reconstructible from μ_i and the two graphs $G - X_i$ and $G - E(X_i)$. If $E(X_i) \neq \varnothing$, then $G - E(X_i)$ has fewer edges than G, while $G - X_i$ always has fewer vertices. We remark in passing that the graphs in which $E(X_i) = \varnothing$ ($i = 1, 2, \ldots, m$) for all star partitions $X_1 \dot{\cup} X_2 \dot{\cup} \ldots \dot{\cup} X_m$ probably merit further investigation. The reason is that they obstruct a recursive ordering of graphs in which graphs are ordered first by the number of vertices, secondly by the number of edges, and thirdly lexicographically by spectrum. The graphs in question clearly include those in which every eigenvalue is simple, but as we have seen in Section 6.4, such graphs do not pose a problem in relation to the complexity of testing for graph isomorphism. In general, we need a better understanding of the relation between star cells and the structure of G, and we now show how linear algebra can give us further insights into this relationship.

Crucial to all subsequent results is the following basic relation between arms of the eutactic star $\mathcal{S}_i$:

$$\mu_i \mathbf{P}_i \mathbf{e}_j = \sum_{k \in \Gamma(j)} \mathbf{P}_i \mathbf{e}_k \quad (j \in V(G)), \tag{6.6}$$

where we write $\Gamma(j)$ for $\Gamma_1(j)$. Equation (6.6) is verified as follows:

$$\mu_i \mathbf{P}_i \mathbf{e}_j = \mathbf{A} \mathbf{P}_i \mathbf{e}_j = \mathbf{P}_i \mathbf{A} \mathbf{e}_j = \mathbf{P}_i \sum_{k \in \Gamma(j)} \mathbf{e}_k = \sum_{k \in \Gamma(j)} \mathbf{P}_i \mathbf{e}_k.$$

When $\Gamma(j) \neq \varnothing$, equation (6.6) says that the vertices $\mathbf{P}_i \mathbf{e}_j$ and $\mathbf{P}_i \mathbf{e}_k$ ($k \in \Gamma(j)$) are linearly dependent. On the other hand, the vectors $\mathbf{P}_i \mathbf{e}_j$ ($j \in X_i$) are linearly independent, and this enables us to prove the next two results. The first says that $\overline{X}_i$ is always a dominating set when G has no isolated vertices.

Theorem 6.4 *Let v be a vertex in X_i. If v is not isolated in G, or if $\mu_i \neq 0$, then v is adjacent to some vertex of $\overline{X}_i$.*

The second result says that when μ_i is neither -1 nor 0 (these are essential exceptions), then $\overline{X}_i$ is a 'location-dominating set'; this means that distinct vertices in X_i have distinct neighbourhoods in $\overline{X}_i$.

Theorem 6.5 *If u and v are distinct vertices in X_i, and $\mu_i \notin \{-1, 0\}$, then $\Gamma(u) \cap \overline{X}_i \neq \Gamma(v) \cap \overline{X}_i$.*

As noted in [23], it follows from Theorems 6.4 and 6.5 that if $\mu_i \notin \{-1, 0\}$, then there are only finitely many graphs G in which $\overline{X}_i$ has prescribed cardinality; for, if $|\overline{X}_i| = t$, then $|X_i| < 2^t$ and $|V(G)| < t + 2^t$. In certain circumstances, we can characterize a graph G by properties of $\overline{X}_i$ or by the graph $G - X_i$ induced by $\overline{X}_i$. The Reconstruction Theorem suggests that the properties likely to be useful in this context include those that tell us something about $E(X_i, \overline{X}_i)$. (There is an analogy here with the characterization of finite simple groups by the centralizer of an involution: there are only finitely many simple groups in which such a centralizer has prescribed order, and if the embedding of the centralizer is known, then the group can usually be determined uniquely; see [7, Chapter 2].)

We illustrate the idea in the case of a regular graph G, where we impose the condition that $G - X_i$ is itself regular. Then each vertex in $\overline{X}_i$ is adjacent to the same number of vertices in X_i, and in this situation we can extend our previous arguments to show that each vertex in X_i is adjacent to the same number of vertices in $\overline{X}_i$. Here we make use of the fact that if G is regular of degree r and

$\mu_i \neq r$, then the eigenspaces $\mathcal{E}(\mu_i)$, and $\mathcal{E}(r)$ are orthogonal. Since $\mathbf{j} \in \mathcal{E}(r)$, we have $\mathbf{P}_i \mathbf{j} = \mathbf{0}$; that is, $\mathbf{P}_i \mathbf{e}_1 + \mathbf{P}_i \mathbf{e}_2 + \cdots + \mathbf{P}_i \mathbf{e}_n = \mathbf{0}$, and we can exploit this relation in conjunction with the basic relation (6.6) on the arms of S_i (see [23]).

Theorem 6.6 *If G is regular of degree r, if $\mu_i \neq r$, and if $G - X_i$ is regular of degree s, then $G - \overline{X}_i$ is regular of degree $\mu_i + r - s$.*

In the situation of Theorem 6.6, for given r there are only finitely many possibilities for s and for the integer eigenvalue μ_i. The first non-trivial case to consider is $r = 3$, and when μ_i is not one of the exceptional values -1 and 0, we can use equation (6.5) to identify G (see [23]).

Theorem 6.7 *Let G be a connected cubic graph in which $\overline{X}_i$ induces a regular subgraph. If $\mu_i \neq \{-1, 0\}$, then one of the following holds:*

(a) $\mu_i = 3$, and $G = K_4$;

(b) $\mu_i = 1$, and G is a Petersen graph with X_i as in Fig. 6.4(a);

(c) $\mu_i = -2$, and G is a Petersen graph with X_i as in Fig. 6.4(b).

With more work, we can characterize the Petersen graph among all connected regular graphs (not just cubic graphs) as follows. Here, the property required of $\overline{X}_i$ is that it is a minimal dominating set, and this ensures (independently of the regularity of G) that $E(X_i, \overline{X}_i)$ is a perfect matching for G (see [22]).

Theorem 6.8 *Let G be a connected regular graph for which the dominating set $\overline{X}_i$ is minimal, with $\mu_i \notin \{-1, 0\}$. If $G - X_i$ has no isolated vertices, then $\mu_i = 1$ and G is a Petersen graph with X_i as in Fig. 6.4(a).*

In conclusion, we remark that Theorems 6.7 and 6.8 show that in certain circumstances it is possible to construct a graph directly from a single eigenspace $\mathcal{E}(\mu_i)$, given graph-theoretic properties of the set $\overline{X}_i$ associated with the eigenvalue μ_i. Taken together, the recent results presented in this section suggest that linear algebra will play a significant role in further investigation of the relationship between graph structure and star partitions. Some exploratory results in this direction may be found in [21].

References

1. L. Babai, D. Yu. Grigoryev and D. M. Mowat, Isomorphism of graphs with bounded eigenvalue multiplicity, *Proc. 14th ACM Symp. Theory of Computing*, ACM Press, 1982, pp. 310–324.
2. F. K. Bell, A note on the irregularity of graphs, *Linear Algebra Appl.* **161** (1992), 45–54; *MR* **92m**: 05096.
3. N. L. Biggs, Intersection matrices for linear graphs, *Combinatorial Mathematics and its Applications* (ed. D. J. A. Welsh), Academic Press, 1971, pp. 15–23; *MR* 44#2639.
4. N. L. Biggs, *Algebraic Graph Theory*, 2nd edn, Cambridge University Press, 1993; *MR* **95h**: 05015.
5. A. E. Brouwer, A. M. Cohen and A. Neumaier, *Distance-regular Graphs*, Springer-Verlag, 1989; *MR* **90e**: 05001.

6. L. Collatz and U. Sinogowitz, Spektren endlicher Grafen, *Abh. Math. Sem. Univ. Hamburg* **21** (1957), 63–77; *MR* **19**–443b.
7. M. J. Collins (ed.), *Finite Simple Groups II*, Academic Press, 1980; *MR* **82h**: 20021.
8. D. Cvetković, Constructing trees with given eigenvalues and angles, *Linear Algebra Appl.* **105** (1988), 1–8; *MR* **89i**: 05195.
9. D. Cvetković, Star partitions and the graph isomorphism problem, *Linear and Multilinear Algebra* **39** (1995), 109–132.
10. D. Cvetković and M. Doob, Developments in the theory of graph spectra, *Linear and Multilinear Algebra* **18** (1985), 153–181; *MR* **87f**: 05111.
11. D. Cvetković, M. Doob, I. Gutman and A. Torgašev, *Recent Results in the Theory of Graph Spectra*, North-Holland, 1988; *MR* **89d**: 05130.
12. D. Cvetković, M. Doob and H. Sachs, *Spectra of Graphs*, Academic Press, 1980; *MR* **81i**: 05054.
13. D. Cvetković and P. Rowlinson, Further properties of graph angles, *Scientia A* **1** (1988), 41–51.
14. D. Cvetković, P. Rowlinson and S. K. Simić, A study of eigenspaces of graphs, *Linear Algebra Appl.* **182** (1993), 45–66; *MR* **83d**: 05068.
15. D. Cvetković, P. Rowlinson and S. K. Simić, Some algorithmic investigations of star partitions, *Discrete Applied Math.* **62** (1995), 119–130.
16. F. R. Gantmacher, *The Theory of Matrices*, Chelsea, 1971.
17. C. D. Godsil and B. D. McKay, Feasibility conditions for the existence of walk-regular graphs, *Linear Algebra Appl.* **30** (1980), 51–61; *MR* **83d**: 05068.
18. W. Haemers and E. Spence, Graphs cospectral with distance-regular graphs, *Linear and Multilinear Algebra* **39** (1995), 9–107.
19. R. Merris, A survey of graph Laplacians, *Linear and Multilinear Algebra* **39** (1995), 19–31.
20. P. Rowlinson, Characteristic polynomials of modified graphs, *Discrete Appl. Math.*, to appear.
21. P. Rowlinson, Eutactic stars and graph spectra, *Combinatorial and Graph-Theoretical Problems in Linear Algebra* (ed. R. A. Brualdi, S. Friedland and V. Klee), Springer, 1993, pp. 153–164; *MR* **94i**: 05061.
22. P. Rowlinson, Dominating sets and eigenvalues of graphs, *Bull. London Math. Soc.* **26** (1994), 248–254; *MR* **95f**: 05081.
23. P. Rowlinson, Star partitions and regularity in graphs, *Linear Algebra Appl.* **226–228** (1995), 247–265.
24. J. J. Seidel, Strongly regular graphs with $(-1, 1, 0)$-adjacency matrix having eigenvalue 3, *Linear Algebra Appl.* **1** (1968), 281–298; *MR* **38**#3175.
25. J. J. Seidel, Eutactic stars, *Combinatorics* (ed. A. Hajnal and V. Sós), North-Holland, 1974, pp. 983–999.
26. D. E. Taylor and R. Levingston, Distance-regular graphs, *Combinatorial Mathematics* (ed. D. A. Holton and J. Seberry), Lecture Notes in Math. **686**, Springer-Verlag, 1978, pp. 257–274; *MR* **80i**: 05077.

7
Matroids

JAMES OXLEY

A matroid consists of a collection of subsets of a finite set which, loosely speaking, behave like the edge sets of cycles in a graph. Matroids also arise naturally from matrices and projective geometries. This chapter provides some examples, mainly in the context of connectivity, of how the interplay between graphs and matroids has led to new results for both matroids and graphs.

7.1 Introduction

Matroids were introduced by Hassler Whitney [22] in 1935, with the aim of capturing the fundamental properties of dependence that are common to graphs and matrices. Consider the graph G in Fig. 7.1, where we have labelled the edges $1, 2, \ldots, 8$. Now list the collection $\mathcal{C}$ of edge sets of cycles of G. Evidently,

$$\mathcal{C} = \{\{1\}, \{2,3\}, \{2,4,5\}, \{3,4,5\}, \{5,6,7\}, \{2,4,6,7\}, \{3,4,6,7\}\}.$$

The pair $(E(G), \mathcal{C})$ is an example of a matroid. We call this matroid the *cycle matroid* of the graph G.

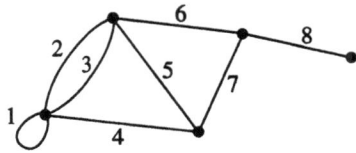

FIG. 7.1

In general, a matroid M consists of a finite set E and a collection $\mathcal{C}$ of subsets of E, called the *circuits* of M, such that the members of $\mathcal{C}$ behave like the edge sets of cycles in a graph. Obviously, this is a rather vague description of a matroid; a formal definition will be given in the next section. For the moment, we are content to note that, from every graph, we can obtain a matroid in the manner described. However, as we shall see, there are also matroids that do not arise in this way.

The purpose of this chapter is to discuss some of the links that exist between graphs and matroids. Certain graph-theoretic results have generalizations or analogues for matroids, while others do not; we shall give examples of both

Matroids

types of results. Our main goal is to describe some instances where the interaction between graphs and matroids has been exploited to give new results for both matroids and graphs. In many places in this chapter, the details of an argument have been merely sketched, or even omitted completely; a complete treatment of such arguments appears in [12]. Throughout this chapter, graphs are allowed to have loops and multiple edges.

7.2 Definitions and examples

In this section, we formalize the loose description of a matroid given in the last section, and we introduce some basic examples of matroids.

A *matroid* M is a pair $(E, \mathcal{C})$ consisting of a finite set E, called the *ground set*, and a collection $\mathcal{C}$ of subsets of E, called *circuits*, such that the following conditions hold:

(C1) $\varnothing \notin \mathcal{C}$;

(C2) if C_1 and C_2 are members of $\mathcal{C}$, and if $C_1 \subseteq C_2$, then $C_1 = C_2$;

(C3) if C_1 and C_2 are distinct members of $\mathcal{C}$, and if $e \in C_1 \cap C_2$, then there is a member C_3 of $\mathcal{C}$ such that $C_3 \subseteq (C_1 \cup C_2) - \{e\}$.

The third of these conditions, the most powerful of the three, is called the *circuit elimination axiom*.

It is easy to verify that every graph yields a matroid in the manner described in the Introduction. Formally, we have the following result.

Theorem 7.1 *Let G be a graph with edge set $E(G)$, and let $\mathcal{C}$ be the collection of edge sets of cycles of G. Then $(E(G), \mathcal{C})$ is a matroid $M(G)$.*

It is natural to ask whether the cycle matroid $M(G)$ of G uniquely determines G, and it is not difficult to answer this question in the negative. The graphs in Figs 7.1 and 7.2 have identical cycle matroids: they have the same ground sets and the same collections of edge sets of cycles. It follows that we cannot determine from $M(G)$ whether G has any isolated vertices; even if G has no isolated vertices, we may not be able to determine whether G is connected; and, even if it is, we still may not be able to determine G from $M(G)$.

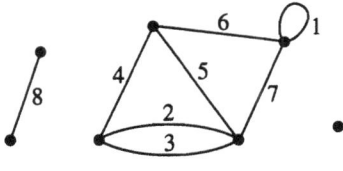

FIG. 7.2

The relationship between the graphs in Figs 7.1 and 7.2 hints at the essence of Whitney's 2-isomorphism theorem (see [21]), which characterizes precisely when two graphs have isomorphic cycle matroids. Two matroids M_1 and M_2 are *isomorphic*, written $M_1 \cong M_2$, if there is a bijection $\phi : E(M_1) \to E(M_2)$ such that $C \in \mathcal{C}(M_1)$ if and only if $\phi(C) \in \mathcal{C}(M_2)$. We call a matroid M *graphic* if M is isomorphic to the cycle matroid of some graph.

Is every matroid graphic? To answer this question, consider the following example.

Example 7.1 Let $E = \{1, 2, 3, 4\}$, and let $\mathcal{C}$ be the collection of 3-element subsets of E. It is routine to check that $(E, \mathcal{C})$ is a matroid; we denote it by $U_{2,4}$. It is not graphic, because there is no graph with four edges every three of which form a cycle. In fact, $U_{2,4}$ is the unique smallest non-graphic matroid, but showing this requires quite a bit more effort. We can represent $U_{2,4}$ geometrically as in Fig. 7.3, with the elements of the matroid corresponding to the four labelled points, and the circuits of the matroid being all sets of three collinear points. Note that this diagram is not a graph.

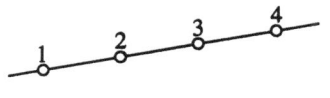

FIG. 7.3

Next, we extend this method of representing a matroid to the plane.

Example 7.2 Figure 7.4(a) is a geometric representation of the non-Fano matroid F_7^-. The elements of this matroid are $1, 2, \ldots, 7$, and the circuits are all sets of three collinear points, together with all sets of four points no three of which are collinear. Figure 7.4(b) is a geometric representation of the Fano matroid F_7. The circle through the elements 4, 5 and 6 indicates a collinearity. Hence $\{4, 5, 6\}$ is a circuit of F_7, although it is clearly not a circuit of F_7^-. The verification that each of F_7^- and F_7 is actually a matroid is left to the reader. In fact, one can obtain a matroid from any finite set of points in the plane by distinguishing certain sets of at least three points as lines, provided that any two distinct such lines share at most one common point. Neither F_7^- nor F_7 is graphic, but we delay the proof of this until Section 7.4.

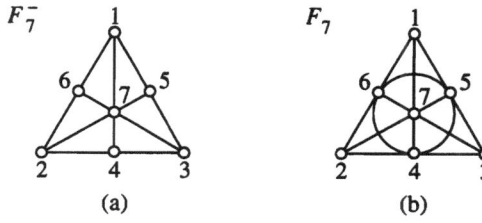

FIG. 7.4

It is clear that a matroid M on a fixed set E is uniquely determined by its collection $\mathcal{C}$ of circuits. It is not difficult to see that M is also uniquely determined by the collection $\mathcal{I}$ of subsets I of E such that I contains no member of $\mathcal{C}$. The members of $\mathcal{I}$ are called the *independent sets* of M. For the matroid $U_{2,4}$ in Example 7.1, the independent sets are all the subsets of $\{1, 2, 3, 4\}$ with at most two elements; $U_{2,4}$ is called the rank-2 *uniform matroid* on $\{1, 2, 3, 4\}$. For the cycle matroid of a graph G, the independent sets of $M(G)$ are all the subsets X of $E(G)$ for which the subgraph $G[X]$ induced by X has no cycles. In any matroid, a set that is not independent is called *dependent*, and it is straightforward to show that the circuits are precisely the minimal dependent sets.

In the Introduction, we noted that the two fundamental classes of matroids considered in Whitney's founding paper on the subject arose from graphs and from matrices. Let us now consider how to derive a matroid from any matrix $\mathbf{A}$ over a field F. For instance, take $\mathbf{A}$ to be the matrix

$$\begin{array}{cccc} 1 & 2 & 3 & 4 \end{array}$$
$$\mathbf{A} = \begin{pmatrix} 1 & 0 & 1 & 1 \\ 0 & 1 & 1 & -1 \end{pmatrix}$$

over the field $\mathbf{R}$. Let E be the set of column labels of $\mathbf{A}$; in our example, $E = \{1, 2, 3, 4\}$. Now let $\mathcal{I}$ be the set of subsets X of E for which the multiset of columns labelled by X is linearly independent. In our example, $\mathcal{I} = \{X : X \subseteq \{1, 2, 3, 4\}, |X| \leq 2\}$, and so $\mathcal{I}$ is the collection of independent sets of the matroid $U_{2,4}$ from Example 7.1. More generally, for all choices of the matrix $\mathbf{A}$, the set $\mathcal{I}$ is the collection of independent sets of a matroid on E. We call this matroid the *vector matroid* $M[\mathbf{A}]$ of $\mathbf{A}$. For the particular matrix $\mathbf{A}$ above, $M[\mathbf{A}]$ is isomorphic to the matroid $U_{2,4}$, and we say that $U_{2,4}$ is *represented over* $\mathbf{R}$ by $\mathbf{A}$.

Suppose now that we view the above matrix $\mathbf{A}$ over the field $GF(2)$—that is, we view its entries as residues modulo 2. Then $-1 = 1$, and $\mathbf{A}$ no longer represents $U_{2,4}$. Instead, $\mathbf{A}$ now represents the cycle matroid of the graph obtained from K_3 by adding an extra edge between two of its vertices. In fact, as can be easily checked, $U_{2,4}$ is not *binary*—that is, there is no matrix $\mathbf{A}_2$ over $GF(2)$ for which $M[\mathbf{A}_2] \cong U_{2,4}$.

7.3 Basic matroid operations

In this section, we introduce the three most basic matroid operations. We show that each is an extension of a natural graph operation, and that there is an attractive and fundamental link between the three.

Two graphs that are related to the graph G in Fig. 7.1 are shown in Fig. 7.5. The first of these, denoted by $G \backslash 5$, is obtained from G by *deleting* the edge 5; the second, denoted by $G/5$, is obtained by *contracting* 5—that is, by shrinking the edge 5 to a single vertex, or (equivalently) by deleting 5 and then identifying its end-vertices. In particular, contracting a loop is the same as deleting it; for example, $G/1 = G \backslash 1$.

FIG. 7.5

To extend these definitions to matroids, we consider how the cycles of $G \backslash e$ and G/e are related to those of G. In general, if e is an element of a matroid M, then $M \backslash e$, the *deletion* of e from M, is the matroid with ground set $E - \{e\}$ whose set of circuits is $\{C \in \mathcal{C}(M) : e \notin C\}$; the *contraction* of e from M, denoted by M/e, also has $E - \{e\}$ as its ground set, but its circuits are the minimal non-empty members of $\{C - \{e\} : C \in \mathcal{C}(M)\}$. It is straightforward to check that both $M \backslash e$ and M/e are actually matroids, and that if e is an edge of a graph G, then $M(G) \backslash e = M(G \backslash e)$ and $M(G)/e = M(G/e)$.

For the above matrix $\mathbf{A}$, $M[\mathbf{A}] \backslash 1$ is represented over $\mathbf{R}$ by the matrix obtained from $\mathbf{A}$ by deleting the first column. On the other hand, $M[\mathbf{A}]/1$ is represented over $\mathbf{R}$ by the matrix $(1\ 1\ -1)$ obtained by deleting both the first row and the first column from $\mathbf{A}$. This last operation is not as trivial as it may appear. In particular, to obtain the representation for $M[\mathbf{A}]/1$, it is important that the element 1 corresponds to a unit vector. As an exercise, the reader may wish to construct a matrix representing $M[\mathbf{A}]/3$ over $\mathbf{R}$. A discussion of how to find such a representation appears in Section 7.7.

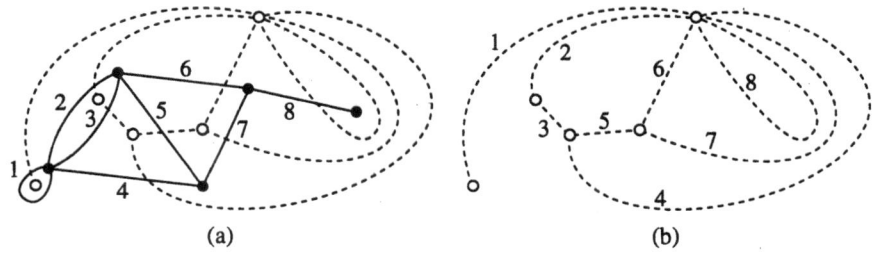

FIG. 7.6

For both graphs and matroids, the operations of deletion and contraction can be linked through a third operation, that of taking duals. The graph G in Fig. 7.1 is a plane graph—that is, it is drawn in the plane so that edges do not cross. In Fig. 7.6(a) we have indicated the construction for the dual graph G^* of G, and in Fig. 7.6(b) G^* has been drawn in isolation. In Fig. 7.7, we have constructed $(G/5)^*$. We observe that $(G/5)^* = G^* \backslash 5$. This illustrates the general principle that, for plane graphs, deletion and contraction are dual operations.

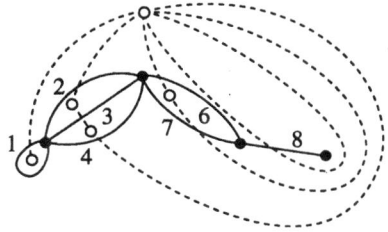
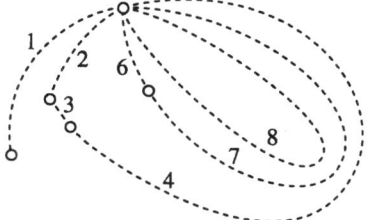

FIG. 7.7

To see how this observation can be extended to matroids, we first need to define matroid duality. We noted earlier that a matroid with a fixed ground set is uniquely determined by its collection of independent sets. Clearly, every independent set of a matroid is contained in some maximal independent set or *basis*, and every subset of a basis is independent. Thus a matroid M on a fixed set is uniquely determined by its collection of bases. If $M = M(G)$ and G is connected, then the bases of M are precisely the edge sets of the spanning trees of G. Moreover, just as all spanning trees of a graph have the same number of edges, all bases of an arbitrary matroid M have the same number of elements. This number $r(M)$ is the *rank* of M.

For the graphs G and G^* in Fig. 7.6, it is not difficult to check that the edge sets of spanning trees of G^* coincide precisely with the complements of edge sets of spanning trees of G. In fact, this relationship holds in general for every connected plane graph G and its dual G^*. For a matroid M with ground set E and set of bases $\mathcal{B}$, define $\mathcal{B}^* = \{E - B : B \in \mathcal{B}\}$. Then it can be shown that $\mathcal{B}^*$ is the set of bases of a matroid M^* on E. We call M^* the *dual* of M; evidently $(M^*)^* = M$. Moreover, for all e in E, we have the following basic link between deletion, contraction and duality:

$$M^* \backslash e = (M/e)^*, \quad \text{or (equivalently)} \quad M \backslash e = (M^*/e)^*.$$

For a graph G, the dual of the cycle matroid $M(G)$ is written as $M^*(G)$. How can we distinguish the circuits of $M^*(G)$ within the graph G? We noted that when G is a plane graph with dual G^*, the matroid $M^*(G)$ is the cycle matroid of G^*. Hence the circuits of $M^*(G)$ are the cycles of G^*. In the example in Fig. 7.6, these cycles are $\{8\}$, $\{6,7\}$, $\{2,3,4\}$, $\{4,5,6\}$, $\{4,5,7\}$, $\{2,3,5,6\}$ and $\{2,3,5,7\}$. The feature that characterizes these sets is that they are precisely the minimal edge-cuts of G—that is, those minimal sets of edges of G whose deletion increases the number of connected components of the graph. Since, in our example, G is connected, these minimal edge-cuts are the minimal sets of edges whose removal disconnects G.

For an arbitrary graph G, whether plane or not, $M^*(G)$ is the matroid on the edge set of G whose circuits are the minimal edge-cuts of G. For an arbitrary matroid M, the circuits of M^* are called the *cocircuits* of M. Thus the cocircuits of $M(G)$ are the minimal edge-cuts of G, and, in an arbitrary graph, we may expect some similarities in the behaviour of cycles and minimal edge-cuts.

7.4 Connectivity

One area in which the ties between graphs and matroids have been very successfully exploited is the study of connectivity. In this section, we show how the concepts of 2-connectedness and 3-connectedness for graphs have been extended to matroids.

The examples in Figs 7.1 and 7.2 show that the notion of connectedness for graphs does not carry over to matroids. Indeed, we have the following result.

Theorem 7.2 *Every graphic matroid is isomorphic to the cycle matroid of a connected graph.*

Proof Let M be a graphic matroid. Then M is isomorphic to the cycle matroid of some graph G. Suppose that G has components $G_1, G_2, \ldots, G_n$. Then it is straightforward to check that $M(G)$ equals $M(G')$, where G' is obtained by choosing one vertex in each G_i and identifying all the chosen vertices. □

We are now able to prove a result delayed from Section 7.2.

Theorem 7.3 *Neither the Fano nor the non-Fano matroid is graphic.*

Proof Assume that M is graphic, where M is F_7 or F_7^-. Then $M \cong M(G)$, for some connected graph G with seven edges. The bases in M have three elements, so the spanning trees in G have three edges. Thus G has four vertices. But M has no circuits with fewer than three elements. Hence G is simple, and so the number of edges of G cannot exceed 6, the number of edges of K_4. This contradiction completes the proof. □

Higher connectivity for graphs is usually defined in terms of the number of vertices that must be deleted from the graph in order to disconnect it. In trying to extend this concept to matroids, we first observe that when we derive a matroid from a graph, it is the edges of the graph that play the crucial role, with the vertices being largely ignored. This suggests that as a first step towards finding this extension, we should seek a re-statement of the graph definition that does not mention vertices. It is well known that if G is a loopless graph of order at least 3 without isolated vertices, then G is 2-connected if and only if every two edges of G lie on a cycle.

Generalizing this, we define a matroid M to be *2-connected* if every two distinct elements of M are in a common circuit. Somewhat confusingly from this standpoint, a 2-connected matroid is frequently referred to simply as a *connected matroid*.

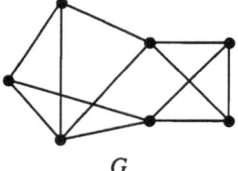
G

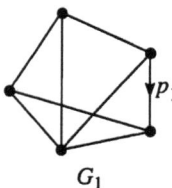
G_1

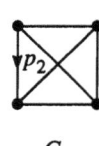
G_2

FIG. 7.8

Next we consider 3-connectedness. The graph G in Fig. 7.8 is 2-connected, but not 3-connected. It can be formed from the graphs G_1 and G_2 by identifying the directed edges p_1 and p_2 and then deleting the resulting edge. We say that G is the *2-sum* of G_1 and G_2. From a matroid standpoint, the appeal of this operation is that we can specify all of the cycles of G in terms of those of G_1 and G_2. The matroid definition of 2-sum, which we give next, embodies this specification.

Let M_1 and M_2 be 2-connected matroids on disjoint sets, each with at least three elements. Let p_1 and p_2 be elements of M_1 and M_2, respectively. The *2-sum* of M_1 and M_2 with respect to p_1 and p_2 is the matroid $M_1 \oplus_2 M_2$ with ground set $(E(M_1) - \{p_1\}) \cup (E(M_2) - \{p_2\})$ for which the circuits are the following:

all circuits of M_1 avoiding p_1;

all circuits of M_2 avoiding p_2;

all sets of the form $(C_1 - \{p_1\}) \cup (C_2 - \{p_2\})$, where C_i is a circuit of M_i containing p_i.

A matroid is *3-connected* if it is 2-connected and cannot be written as a 2-sum.

This matroid concept of 3-connectedness was introduced by Tutte [18] in 1966, and its link with the operation of 2-sum was noted by several authors (see, for example, [15]). Tutte sought not only to generalize the graph concept of 3-connectedness, but also to incorporate invariance under duality. He achieved the latter aim: *a matroid is 3-connected if and only if its dual is.* It is well known, and is easily checked, that a 3-connected graph has no edge-cuts with fewer than three edges. Thus, for a graph G, we expect that if $M(G)$ is 3-connected, then G is simple. In fact, it is not difficult to prove the following results.

Theorem 7.4 *Let G be a graph of order n without isolated vertices. Then*

(a) for $n \geq 3$, $M(G)$ is 2-connected if and only if G is loopless and 2-connected;

(b) for $n \geq 4$, $M(G)$ is 3-connected if and only if G is simple and 3-connected.

Tutte also defined k-connectedness for matroids, for all $k \geq 4$. Under this definition, if G is a graph of order at least $k + 1$ without isolated vertices, then $M(G)$ is k-connected if and only if G is k-connected and has no cycles with fewer than k edges. Clearly, as k increases, this matroid concept of k-connectedness and its graphical namesake diverge. For this reason, we focus on the cases when k is 2 or 3.

There are other notions of k-connectedness in the literature, including one that is a truer generalization of the graph notion for all values of k (see [3] or [12, Section 8.2]). For the matroid theorist, the disadvantage of this alternative concept, called *vertical* (or *Whitney*) *k-connectedness*, is that it is not invariant under duality. With both vertical k-connectedness and (Tutte) k-connectedness, most of the substantial matroid results that have been proved restrict to the cases when k is 2 or 3. For these cases, the two versions of k-connectedness are identical for *simple matroids*, those matroids with no circuits of size less than three.

We noted in Section 7.2 that an arbitrary graph is not determined by its cycle matroid. One attractive feature of 3-connected graphs without loops is that such graphs *are* determined by their cycle matroids (see [21]).

7.5 Wheels and whirls

Numerous authors have studied *minimally k-connected graphs*, the k-connected graphs for which no single-edge deletion yields a k-connected graph. This work has been surveyed by Mader [8, 9]. A number of these graph results have attractive extensions to matroids, although almost all such results have been proved only when $k \leq 3$, whereas many of the graph results hold without this restriction. From now on, we focus attention on the case $k = 3$, remarking that there is a correspondingly well-developed body of matroid results when $k = 2$ (see, for example, [12, Section 10.2]).

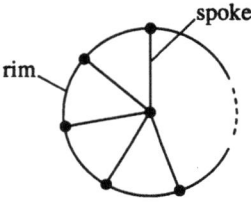

FIG. 7.9

One important example of a minimally 3-connected graph is the *wheel* W_t formed from a t-cycle C_t ($t \geq 3$) by adding a new vertex and joining this by a single edge (a *spoke*) to every vertex of (the *rim*) C_t (see Fig. 7.9). The next result, *Tutte's Wheels Theorem* (see [17]), can be used to give a recursive construction for all simple 3-connected graphs.

Theorem 7.5 (Tutte's Wheels Theorem) *A minimally 3-connected graph G has no edge e for which G/e is both simple and 3-connected if and only if G is a wheel.*

The main result of the paper [18] that introduced higher connectivity for matroids is a generalization of the last theorem. This matroid result, *Tutte's Wheels and Whirls Theorem*, is of crucial importance in the study of 3-connected matroids. A *minimally 3-connected matroid* is a 3-connected matroid with the property that none of its single-element deletions is 3-connected. Theorem 7.5 translates directly to matroids as follows: *a minimally 3-connected graphic matroid M has no element e such that M/e is 3-connected if and only if M is isomorphic to the cycle matroid of a wheel.*

To state the generalization of this result to all matroids, we need another definition. For all $r \geq 3$, the *rank-r whirl* W^r is the matroid whose ground set is the edge set of the wheel W_r, and whose set of circuits consists of all cycles of W_r except the rim, together with all sets of edges consisting of the rim plus a single spoke. None of the whirls is a graphic matroid. Figure 7.10 compares geometric representations for the rank-3 whirl and the cycle matroid of the 3-spoked wheel, the spokes of the latter being 1, 2 and 3.

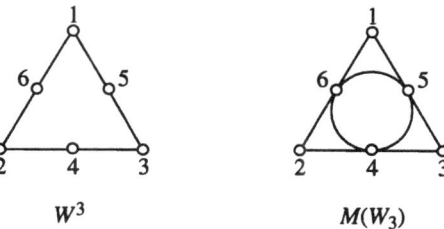

FIG. 7.10

We now state Tutte's Theorem. Its applications include providing a recursive construction for all 3-connected matroids.

Theorem 7.6 (Tutte's Wheels and Whirls Theorem) *A minimally 3-connected matroid M has no element e for which M/e is 3-connected if and only if M is isomorphic to a whirl or the cycle matroid of a wheel.*

7.6 Minimally 3-connected graphs and matroids

For all $t \geq 3$, the complete bipartite graph $K_{3,t}$, like the wheel W_t, is simple and 3-connected with each edge meeting a vertex of degree 3. It is easy to see that all simple 3-connected graphs with this property are minimally 3-connected. While it is not true in general that all edges of a minimally 3-connected graph meet some vertex of degree 3, it is true that minimally 3-connected graphs have an abundance of such vertices. In this section, we make this assertion more precise and discuss matroid analogues of it.

Halin [6] has established that, by two separate criteria, every minimally 3-connected graph has many vertices of degree 3.

Theorem 7.7 *Let G be a minimally 3-connected graph of order n. Then*

(a) each cycle of G meets at least two vertices of degree 3;

(b) G has at least $(2n+6)/5$ vertices of degree 3.

If we are seeking an analogue of this result for minimally 3-connected matroids, we first need to decide what corresponds to a vertex of degree 3. In a loopless 2-connected graph G, the set of edges incident with a vertex of G forms a minimal edge-cut, a cocircuit of $M(G)$. Of course, $M(G)$ may have cocircuits that do not arise in this way, but one matroid analogue of a vertex of degree 3 is a *triad*—that is, a 3-element cocircuit. The following is a matroid analogue of the first part of Halin's Theorem (see [10]).

Theorem 7.8 *In a minimally 3-connected matroid with at least four elements, each circuit meets at least two triads.*

This theorem was extended by Lemos [7] as follows.

Theorem 7.9 *Let M be a 3-connected matroid with at least four elements, and let C be a circuit of M. If, for each e in C, the matroid $M\backslash e$ is not 3-connected, then C meets at least two triads of M.*

The last result was motivated by a result of Halin for graphs. If we apply it directly to graphs, we deduce that, if C is a cycle of a simple 3-connected graph G and, for each edge e of C, the graph $G\backslash e$ is not 3-connected, then C meets at least two minimal edge-cuts of G of size 3. The natural question now is whether this conclusion can be strengthened to assert that C meets at least two degree-3 vertices of G. Lemos [7] noted that it can indeed, and Mader [9] gave a short proof of this fact.

Theorem 7.10 *Let G be a simple 3-connected graph and let C be a cycle of G. If, for each edge e of C, the graph $G\backslash e$ is not 3-connected, then C meets at least two degree-3 vertices of G.*

The graph technique used to prove Theorem 7.7(b) does not extend readily to matroids. However, by using a matroid argument based on Theorem 7.9, one can prove the following result (see [11]). Recall that the rank $r(M)$ is the number of elements in a basis of M.

Theorem 7.11 *A minimally 3-connected matroid M with at least four elements has at least $(|E(M)| - r(M) + 2)/2$ triads.*

Although this result is not an exact analogue of Halin's result, it does maintain the spirit of that result by showing that a minimally 3-connected matroid has many triads.

As before, if we apply the last theorem to graphs, we obtain a bound on the number of minimal edge-cuts of size 3. We would prefer a result about vertices of degree 3 and, by an argument based on Theorem 7.7(a), such a result can indeed be proved (see [11]). Note that for a connected graph G of order n, the rank of $M(G)$ is $n - 1$, the number of edges in a spanning tree of G.

Theorem 7.12 *In a minimally 3-connected graph G with n vertices and m edges, there are at least $(m - n + 3)/2$ vertices of degree 3.*

Theorems 7.7 and 7.12 give lower bounds for the number of vertices of degree 3 in a minimally 3-connected graph with n vertices. Before comparing these bounds, we note another result of Halin [5] that has an attractive extension to matroids.

Theorem 7.13 *Let G be a minimally 3-connected graph with n vertices and m edges. Then*
$$m \leq 2n - 2, \quad \text{if } n \leq 6,$$
$$m \leq 3n - 9, \quad \text{if } n \geq 7.$$
Moreover, the only graphs attaining equality in these bounds are W_t for $3 \leq t \leq 6$ and $K_{3,t}$ for $t \geq 4$.

The next result (see [10]) shows that precisely the same bounds hold for arbitrary minimally 3-connected matroids.

Theorem 7.14 *Let M be a minimally 3-connected matroid with at least four elements. Then*
$$|E(M)| \leq 2r(M), \quad \text{if } r(M) \leq 5,$$
$$|E(M)| \leq 3r(M) - 6, \quad \text{if } r(M) \geq 6.$$

From Theorem 7.13, we know that the bounds in the preceding result are best possible. The matroids attaining equality have been characterized (see [10]). Although this characterization is somewhat cumbersome in general, the only *binary* matroids attaining equality are the cycle matroids of the graphs that attain equality in the bounds of Theorem 7.13.

The next result (see [11]) is obtained by comparing the bounds in Theorems 7.7 and 7.12 and, for each graph, choosing the better bound. The relative sharpness of these bounds depends on the density of the graph. A small amount of additional argument enables Halin's bound to be improved slightly in the given range. We observe that for a minimally 3-connected graph G with n vertices and m edges, one can easily check that $m \geq 3n/2$.

Theorem 7.15 *The number ν_3 of vertices of degree 3 in a minimally 3-connected graph G with n vertices and m edges satisfies*
$$\nu_3 \geq (2n+7)/5, \quad \text{if } 3n/2 \leq m < (9n-3)/5,$$
$$\nu_3 \geq (m-n+3)/2, \quad \text{if } (9n-3)/5 \leq m \leq 3n - 9.$$

Since Theorem 7.11 applies to all minimally 3-connected matroids, we may apply it to $M^*(G)$ when G is simple and 3-connected, to obtain the following result.

Corollary 7.16 *Let G be a simple 3-connected graph of order n, for which no single-edge contraction is both simple and 3-connected. Then the number of 3-cycles in G is at least $(n+1)/2$.*

To conclude this section, we broadly summarize its contents. We began with several graph results, and these motivated a search for matroid analogues. In turn, these matroid results suggested new graph results which, when proved, sometimes sharpened the initiating graph results. Evidently, exploiting the link between graphs and matroids has been very fruitful in this area.

7.7 Excluded minors

In this section, we characterize graphic matroids. It was noted in Section 7.2 that, for a graphic matroid M, all single-element deletions and all single-element contractions are also graphic. It follows that every minor of M is graphic, where a *minor* of a matroid N is any matroid that can be obtained from N by a sequence of single-element deletions and single-element contractions. We say that the class of graphic matroids is *minor-closed*.

In Example 7.1, we observed that the matroid $U_{2,4}$ is not graphic. It is not difficult to check that all of the single-element deletions of $U_{2,4}$ are isomorphic to $M(K_3)$, while all of its single-element contractions are isomorphic to $M(G_3)$, where G_3 is the graph consisting of two vertices joined by three edges. We conclude that $U_{2,4}$ is an *excluded minor* for the class of graphic matroids because it is not in the class, yet all of its proper minors are in the class.

Can we specify all of the excluded minors for the class of graphic matroids? Theorem 7.3 shows that neither F_7^- nor F_7 is graphic. The first of these has a $U_{2,4}$-minor, but the second is another excluded minor. A third excluded minor is F_7^*, the dual of the second. Tutte [16] showed that there are just two other excluded minors for the class of graphic matroids.

Theorem 7.17 *A matroid is graphic if and only if it has no minor isomorphic to any of the matroids $U_{2,4}$, F_7, F_7^*, $M^*(K_5)$ or $M^*(K_{3,3})$.*

The appearance of both of the graphs K_5 and $K_{3,3}$ in this result is reminiscent of Kuratowski's famous characterization of planar graphs. Indeed, Kuratowski's Theorem can be deduced from this theorem.

On combining Theorem 7.15 with its dual and using Kuratowski's Theorem, we obtain the following result.

Theorem 7.18 *A matroid is isomorphic to the cycle matroid of a planar graph if and only if it has no minor isomorphic to any of the matroids $U_{2,4}$, F_7, F_7^*, $M(K_5)$, $M^*(K_5)$, $M(K_{3,3})$ or $M^*(K_{3,3})$.*

Recall from Section 7.2 that a matroid is binary if it is isomorphic to the vector matroid of some matrix over $GF(2)$. We noted earlier that $U_{2,4}$ is not binary. In contrast, one can check that F_7 is binary, being represented over $GF(2)$ by the 3×7 matrix whose columns are all of the non-zero ordered vectors of zeros and ones.

We also noted in Section 7.2 that every deletion of a binary matroid $M[\mathbf{A}]$ is binary. To see that every contraction is also binary, we argue as follows.

If the column labelled by e is zero, then $M[\mathbf{A}]/e = M[\mathbf{A}]\backslash e$, and so the contraction of e is certainly binary. If the column labelled by e is non-zero, then we choose a non-zero entry of this column—say, the entry in row i. We now add suitable multiples of row i to the other rows of $\mathbf{A}$ to produce a matrix $\mathbf{A}'$ in which the column labelled by e is the ith unit vector. It is easily checked that $M[\mathbf{A}'] = M[\mathbf{A}]$. Moreover, $M[\mathbf{A}']/e$ is represented by the matrix obtained from $\mathbf{A}'$ by deleting row i and column e. We conclude that the class of binary matroids is minor-closed. The excluded minor characterization of this class was also proved by Tutte [16].

Theorem 7.19 *A matroid is binary if and only if it has no minor isomorphic to $U_{2,4}$.*

The last result has several interesting consequences. First, since $U_{2,4}$ is isomorphic to its dual, it follows that a matroid is binary if and only if its dual is binary. Indeed, if M is represented over $GF(2)$ by the $r \times n$ matrix $[\mathbf{I}_r \,|\, \mathbf{D}]$, then M^* is represented by the matrix $[\mathbf{D}^T \,|\, \mathbf{I}_{n-r}]$.

A second consequence of Theorem 7.19 is that, since the set of excluded minors for the class of graphic matroids contains the set of excluded minors for the class of binary matroids, every graphic matroid is binary. One can see this directly by checking that the $n \times \binom{n}{2}$ matrix over $GF(2)$ that consists of all columns with exactly two ones represents $M(K_n)$.

7.8 Infinite antichains

Minors are the basic substructures of matroids. In this section, we consider a problem for matroid minors that is motivated by a recent celebrated theorem for graphs.

A graph H is a *minor* of a graph G if H is isomorphic to a graph obtained from G by a sequence of single-edge deletions, single-edge contractions, and deletions of isolated vertices. Evidently, if a graph G_1 is a minor of a graph G_2, then $M(G_1)$ is a minor of $M(G_2)$. We know from considering the graphs in Figs 7.1 and 7.2 that the converse of this fails in general. However, it follows from Whitney's 2-isomorphism theorem (see [21]) that the converse holds if G_1 is loopless and 3-connected.

One common feature of the lists of excluded minors in Theorems 7.17–7.19 is that all of them are finite—in each case, there are only finitely many obstructions to the specified matroid property. Wagner conjectured that every minor-closed class of graphs has a finite list of excluded minors (see, for example, [13, p. 155]). We call a set of graphs or matroids an *antichain* if no member of the set is isomorphic to a minor of another member of the set.

As the culmination of a long series of difficult papers, Robertson and Seymour [14] proved Wagner's conjecture by establishing the following result.

Theorem 7.20 *There is no infinite antichain of graphs.*

Given the intimate links between graphs and matroids, it is natural to ask whether this result extends to matroids. But, even before Theorem 7.20 was proved, it was known that infinite antichains of matroids exist (see, for example, [1, p. 155]). The set of cycles $\{C_3, C_4, C_5, \ldots\}$ is clearly infinite, and no member of this set is isomorphic to a subgraph of another. We can use this set of graphs, or, indeed, any infinite set of simple graphs with the last property, to build an infinite antichain of matroids as follows. Embed each C_n in the plane, viewing its vertices as points of a matroid and its edges as lines of the matroid as in Example 7.2; then add one extra point on each line to obtain a matroid M_n that consists of a ring of n 3-point lines. It is not difficult to check that $\{M_3, M_4, M_5, \ldots\}$ is an infinite antichain of matroids.

We now know that Theorem 7.20 fails for the class of all matroids. A problem that is currently attracting much research attention is whether Theorem 7.20 can be generalized to the class of binary matroids: *Is there an infinite antichain of binary matroids?*

The level of difficulty of the graph result suggests that proving the non-existence of such an infinite antichain will be very hard indeed. The only apparent way that this problem could be easily resolved is by producing such an infinite antichain. In [4], there is a discussion of some of the classes of matroids that have infinite antichains and some that do not.

7.9 Conclusion

This chapter has touched on just a small part of the widespread interaction between graphs and matroids, with the emphasis being on how each subject has enriched the other. From its inception, matroid theory has looked to graph theory as a rich source of examples and conjectures. While many graph results fail for matroids in general, others have analogues for certain restricted classes, such as binary matroids. But graph theory has also benefited from this relationship. Matroid results can produce new graph results when applied to both graphic matroids and their duals; they can provide new insights into graph theory; and they can suggest new and interesting conjectures for graphs.

Many fruitful aspects of this interaction between graphs and matroids have not been mentioned here, including, for example, those relating to the Tutte polynomial, colouring and flow problems, the dual relationship between Eulerian and bipartite graphs, and the existence of disjoint spanning trees. It is not feasible to provide a list of all such results here. Instead, we refer the reader to the extensive bibliographies in [2], [12] and [20].

The links between graphs and matroids are already numerous. We close with the challenging words of Tutte [19, p. 497] who, in a commentary on [18], wrote: 'if a theorem about graphs can be expressed in terms of edges and circuits only it probably exemplifies a more general theorem about matroids'.

References

1. T. Brylawski, Constructions, *Theory of Matroids* (ed. N. White), Cambridge University Press, 1986, pp. 127–223.
2. T. Brylawski and J. Oxley, The Tutte polynomial and its applications, *Matroid Applications* (ed. N. White), Cambridge University Press, 1993, pp. 123–225; *MR* **93k**: 05060.
3. W. H. Cunningham, On matroid connectivity, *J. Combin. Theory (B)* **30** (1981), 94–99; *MR* **83f**: 05015.
4. G. Ding, B. Oporowski and J. Oxley, On infinite antichains of matroids, *J. Combin. Theory (B)* **63** (1995), 21–40; *MR* **95j**: 05063.
5. R. Halin, Zur Theorie der n-fach zusammenhängenden Graphen, *Abh. Math. Sem. Univ. Hamburg* **33** (1969), 133–164; *MR* **41**#3310.
6. R. Halin, Untersuchungen über minimale n-fach zusammenhängende Graphen, *Math. Ann.* **182** (1969), 175–188; *MR* **43**#4705.
7. M. Lemos, On 3-connected matroids, *Discrete Math.* **73** (1989), 273–283; *MR* **89m**: 05035.

8. W. Mader, Connectivity and edge-connectivity in finite graphs, *Surveys in Combinatorics* (ed. B. Bollobás), London Math. Soc. Lecture Notes **38**, Cambridge University Press, 1979, pp. 66–95; *MR* **81d**: 05044.
9. W. Mader, On vertices of degree n in minimally n-connected graphs and digraphs, *Combinatorics, Paul Erdős is Eighty* (ed. D. Miklós, V. T. Sós and T. Szönyi), Bolyai Math. Studies, to appear.
10. J. G. Oxley, On matroid connectivity, *Quart. J. Math. Oxford (2)* **32** (1981), 193–208; *MR* **82g**: 05040.
11. J. G. Oxley, On connectivity in matroids and graphs, *Trans. Amer. Math. Soc.* **265** (1981), 47–58; *MR* **82e**: 05052.
12. J. G. Oxley, *Matroid Theory*, Oxford University Press, 1992; *MR* **94d**: 05033.
13. N. Robertson and P. D. Seymour, Graph minors—a survey, *Surveys in Combinatorics 1985* (ed. I. Anderson), London Math. Soc. Lecture Notes **103**, Cambridge University Press, 1985, pp. 153–171; *MR* **87e**: 05130.
14. N. Robertson and P. D. Seymour, Graph minors. XX. Wagner's conjecture, to appear.
15. P. D. Seymour, Decomposition of regular matroids, *J. Combin. Theory (B)* **28** (1980), 305–359; *MR* **82j**: 05046.
16. W. T. Tutte, Matroids and graphs, *Trans. Amer. Math. Soc.* **90** (1959), 527–552; *MR* **21**#337.
17. W. T. Tutte, A theory of 3-connected graphs, *Indag. Math.* **23** (1961), 441–455; *MR* **25**#3517.
18. W. T. Tutte, Connectivity in matroids, *Canad. J. Math.* **18** (1966), 1301–1324; *MR* **34**#5706.
19. W. T. Tutte, *Selected Papers of W. T. Tutte*, Vol. II (ed. D. McCarthy and R. G. Stanton), Charles Babbage Research Centre, Winnipeg, 1979; *MR* **81k**: 01058b.
20. D. J. A. Welsh, *Matroid Theory*, London Math. Soc. Monographs 8, Academic Press, 1976; *MR* **55**#148.
21. H. Whitney, 2-isomorphic graphs, *Amer. J. Math.* **55** (1933), 245–254.
22. H. Whitney, On the abstract properties of linear dependence, *Amer. J. Math.* **57** (1935), 509–533.

8
Codes

ROBERT T. CURTIS AND TONY R. MORRIS

In this chapter adjacency matrices of graphs are used to generate binary codes. By placing suitable conditions on the graph, we can ensure that the resulting code has certain desirable properties. In particular, we use the method to construct explicitly all binary self-dual doubly-even codes of length less than or equal to 24.

8.1 Introduction

In a recent paper [4], it was shown how the binary Golay code can be obtained from the graph of the icosahedron. In this chapter we exploit the technique of generating codes from graphs and show how, by imposing various conditions on the graph, we can ensure that the resulting code has certain desirable properties. In particular, we are able to produce all self-dual doubly-even binary codes of lengths 8, 12 and 24, simply by writing down a collection of graphs satisfying the conditions.

It is worth pointing out that this method has a number of important benefits. We obtain not only a natural basis for the code, together with a mnemonic for remembering a generating matrix, but also a simple description of all of the codewords. This enables us to read off the minimum weight directly from the graph, and immediately write down all short codewords. In addition, as symmetries of the graph are symmetries of the code, the task of calculating the full group of the code is simplified.

8.2 Self-dual doubly-even codes

Let F be a finite field, and let F^n denote the vector space of n-tuples of elements in F. A *linear code* C is simply a subspace of F^n, and its vectors are known as *codewords*. If the code C has dimension k, it is said to be of *type* $[n, k]$ and n is called the *length* of the code. If $\mathbf{a} = (a_1, a_2, \ldots, a_n)$ and $\mathbf{b} = (b_1, b_2, \ldots, b_n)$ are codewords of C, then the *Hamming distance* $d(\mathbf{a}, \mathbf{b})$ between $\mathbf{a}$ and $\mathbf{b}$ is the number of places where they differ. The *Hamming weight* $wt(\mathbf{a})$ of a codeword $\mathbf{a}$ is the number of non-zero entries in $\mathbf{a}$. These last two notions are clearly linked, since

$$d(\mathbf{a}, \mathbf{b}) = wt(\mathbf{a} - \mathbf{b}).$$

Perhaps the most important parameter of an error-correcting code is its *minimum Hamming distance*, defined by

$$\min_{\mathbf{a} \ne \mathbf{b}} wt(\mathbf{a} - \mathbf{b}),$$

since it is this parameter that controls the number of errors that can be corrected (see, for example, MacWilliams and Sloane [7]). A code of length n, dimension k and minimum distance d is said to be of *type* $[n, k, d]$. Since C is closed under differences, we see that the minimum distance of a linear code C equals the minimum weight of its non-zero codewords.

We define the *natural inner product* on the vector space F^n by

$$\mathbf{a} \cdot \mathbf{b} = \sum_i a_i b_i;$$

if $\mathbf{a} \cdot \mathbf{b} = 0$, then $\mathbf{a}$ and $\mathbf{b}$ are *orthogonal*. If C is a linear code, then the *dual code* $C^\perp$ is the set of vectors that are orthogonal to all codewords of C—that is,

$$C^\perp = \{\mathbf{u} \mid \mathbf{u} \cdot \mathbf{v} = 0, \text{ for all } \mathbf{v} \in C\}.$$

Clearly, $C^\perp$ is a linear code, and if C is of type $[n, k]$, then $C^\perp$ is of type $[n, n-k]$. If $C = C^\perp$, then C is called *self-dual*, in which case $n = 2k$.

If A_i is the number of codewords in C of weight i, then the polynomial

$$W_C(x, y) = \sum_i A_i x^{n-i} y^i$$

is the *Hamming weight enumerator* of C. The famous *MacWilliams identities* (see [7, p. 127]) give us the following relationship between the weight enumerators of C and $C^\perp$:

$$W_{C^\perp}(x, y) = \frac{1}{|C|} W_C(x + (q-1)y, x - y),$$

where q is the order of the field F. For self-dual binary codes, where $q = 2$, this reduces to

$$W_C(x, y) = \frac{1}{|C|} W_C(x + y, x - y).$$

8.3 Invariant theory

A code C is *doubly-even* if $wt(\mathbf{a})$ is divisible by 4 for all $\mathbf{a}$ in C. If C is a self-dual doubly-even binary code of length n, then it follows from the MacWilliams identities that

$$W_C(x, y) = \frac{1}{2^{n/2}} W_C(x + y, x - y) = W_C\left(\frac{x+y}{\sqrt{2}}, \frac{x-y}{\sqrt{2}}\right).$$

Furthermore, since $W_C(x, y)$ involves only powers of y^4, we have

$$W_C(x, y) = W_C(x, iy).$$

Thus $W_C(x,y)$ is preserved by the operations

$$(x,y) \to \left(\frac{x+y}{\sqrt{2}}, \frac{x-y}{\sqrt{2}}\right) \quad \text{and} \quad (x,y) \to (x, iy).$$

It follows that the polynomial $W_C(x,y)$ is *invariant* under the group

$$L = \left\langle \frac{1}{\sqrt{2}}\begin{pmatrix} 1 & 1 \\ 1 & -1 \end{pmatrix}, \begin{pmatrix} 1 & 0 \\ 0 & i \end{pmatrix} \right\rangle,$$

whose order is 192.

We now define a_j to be the maximum number of linearly independent homogeneous polynomials of degree j that are invariant under L, and write

$$\Phi(\lambda) = \sum_j a_j \lambda^j.$$

In 1897, Molien [8] proved that, for any finite group Γ of $m \times m$ complex matrices,

$$\Phi(\lambda) = \frac{1}{|\Gamma|} \sum_{A \in \Gamma} \frac{1}{\det(\mathbf{I}_m - \lambda \mathbf{A})}.$$

Calculating this *Molien series* for our group L, we obtain the simple form

$$\Phi(\lambda) = \frac{1}{(1-\lambda^8)(1-\lambda^{24})},$$

and so $a_j = 0$ unless j is divisible by 8; in particular, this yields the important result that

a self-dual doubly-even binary code is of type $[8r, 4r, d]$.

Expanding the above series, we obtain

$$\Phi(\lambda) = 1 + \lambda^8 + \lambda^{16} + 2\lambda^{24} + 2\lambda^{32} + 2\lambda^{40} + 3\lambda^{48} + \cdots.$$

The requisite number of linearly independent homogeneous polynomials of each degree is obtained by taking combinations of just two polynomials—one of degree 8 and the other of degree 24. Two such are the *extended Hamming code* $\mathcal{H}$ of type $[8,4,4]$, with weight enumerator

$$W_{\mathcal{H}}(x,y) = x^8 + 14x^4y^4 + y^8,$$

and the *Golay code* $\mathcal{G}$ of type $[24, 12, 8]$, with weight enumerator

$$W_{\mathcal{G}}(x,y) = x^{24} + 759x^{16}y^8 + 2576x^{12}y^{12} + 759x^8y^{16} + y^{24}.$$

We describe these two codes later, using the graph construction. Their significance is demonstrated by a theorem of Gleason [6], which states that *the weight enumerator of any self-dual doubly-even binary code is a polynomial in $W_{\mathcal{H}}(x,y)$ and $W_{\mathcal{G}}(x,y)$*.

For a fuller treatment of the relevant invariant theory, see MacWilliams and Sloane [7].

8.4 Constructing binary codes from graphs

Let G be a graph with no multiple edges (but possibly loops) and vertex set $S = \{1,2,3,\ldots,n\}$, and let $\mathbf{C} = (c_{ij})$ be its adjacency matrix. We further let $V(S)$ denote the n-dimensional vector space over $GF(2)$ with basis $B = \{v_i : i \in S\}$, and we interpret $\mathbf{C}$ as the matrix of a linear transformation from V to V with respect to this basis. We identify each vector in V with the subset of the vertices of G to which it corresponds.

We are particularly interested in the eigenspaces $V^{(\lambda)}$ of $\mathbf{C}$, for $\lambda \in GF(2)$; note that $V^{(0)}$ consists of those sets of vertices X for which any vertex of G is adjacent to an even number of vertices in X, and $V^{(1)}$ consists of those sets X for which every vertex not in X is adjacent to an even number of vertices in X and every vertex in X is joined to an odd number of vertices in X.

We now define a bipartite graph $\widehat{G}$ with twice as many vertices as G, as follows:

(1) the vertices are $1,2,\ldots,n,\overline{1},\overline{2},\ldots,\overline{n}$;
(2) i is joined to $\overline{j}$ if and only if i is adjacent to j in G, and these are the only adjacencies.

Thus $\widehat{\mathbf{C}}$, the adjacency matrix of $\widehat{G}$, takes the form

$$\widehat{\mathbf{C}} = \begin{pmatrix} 0 & \mathbf{C} \\ \mathbf{C} & 0 \end{pmatrix}.$$

The codes in which we are particularly interested are the eigenspaces of $\widehat{\mathbf{C}}$ for the eigenvalue 1. That is, we seek the kernels of matrices of the form

$$\widehat{\mathbf{C}} + \mathbf{I} = \begin{pmatrix} \mathbf{I} & \mathbf{C} \\ \mathbf{C} & \mathbf{I} \end{pmatrix};$$

such a matrix has rank at least n. Since $\mathbf{C}$ is symmetric, we have, with $\sim$ denoting row equivalence,

$$\begin{aligned}
\operatorname{rank}(\widehat{\mathbf{C}} + \mathbf{I}) = n &\iff (\mathbf{I} \mid \mathbf{C}) \sim (\mathbf{C} \mid \mathbf{I}) \\
&\iff (\mathbf{C}^{-1} \mid \mathbf{I}) \sim (\mathbf{C} \mid \mathbf{I}) \\
&\iff \mathbf{C}^{-1} = \mathbf{C} \\
&\iff \mathbf{C}^2 = \mathbf{I} \\
&\iff \mathbf{C}\mathbf{C}^T = \mathbf{I} \\
&\iff (\widehat{\mathbf{C}} + \mathbf{I})(\widehat{\mathbf{C}} + \mathbf{I})^T = 0 \\
&\iff \text{the rows of } \widehat{\mathbf{C}} + \mathbf{I} \text{ span a self-dual code.}
\end{aligned}$$

Graph-theoretically, the condition $\mathbf{C}^2 = \mathbf{I}$ means that

(P1) *for each pair of vertices of G, the number of vertices joined to both is even, and each vertex has odd degree.*

Thus every graph with this property gives rise to a self-dual binary code. To ensure that the resulting code is also doubly-even, we require

(P2) *each vertex of G has degree congruent to 3 (modulo 4).*

Example 8.1 Consider $G = K_4$. Each vertex has degree 3, and each pair of vertices has just two common neighbours, so properties (P1) and (P2) hold. The dimension is 4, and the codewords are readily seen to be

$$0, \quad \{1,2,3,4,\bar{1},\bar{2},\bar{3},\bar{4}\}, \quad \{i,\bar{j},\bar{k},\bar{l}\} \text{ (4 such)}, \quad \{\bar{i},j,k,l\} \text{ (4 such)},$$
$$\text{and} \quad \{i,j,\bar{i},\bar{j}\} \text{ (6 such)}, \quad \text{where } \{i,j,k,l\} = \{1,2,3,4\}.$$

These correspond to subgraphs of $\widehat{G}$ isomorphic to the empty graph, $\widehat{G}$, $K_{1,3}$ and $2K_2$. Thus $\mathcal{C}$ is an $[8,4,4]$ code with fourteen codewords of weight 4, and is plainly the Hamming code (see MacWilliams and Sloane [7, p. 30]).

Example 8.2 Now let G be the graph of the icosahedron (shown in Fig. 8.1), which is regular of degree 5.

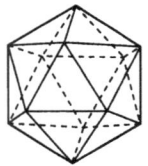 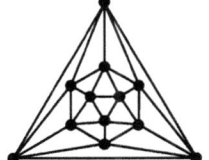

FIG. 8.1

It clearly does not satisfy property (P2). However, it is easily checked that for each pair of vertices, there are either no vertices joined to both, or exactly two. Thus property (P1) holds, and the binary code $\mathcal{C}$ obtained is self-dual but not doubly-even. Note that each codeword has the form $(X \mid \overline{Y})$, where X is any subset of S and Y consists of all vertices of S that are joined to an odd number of vertices of X.

The code $\mathcal{C}$ has 2×12 codewords of weight 6, consisting of a vertex of S (or $\overline{S}$) and the five vertices of $\overline{S}$ (or S) joined to it in $\widehat{G}$. Moreover, there are a further 20×2 codewords of weight 6, consisting of three vertices forming a triangle T on one side and the three vertices joined to just one vertex of T on the other. It is easily seen that these 64 codewords are the only ones of weight 6, and that none has smaller weight. This graph fails to give a doubly-even code, since the vertices have degree 5. However, the complement of G (see the graph G_{28} on page 125) satisfies both (P1) and (P2), and the new code is therefore doubly-even and self-dual. It is in fact the *binary Golay code* $\mathcal{G}$ (see [4] and [1, p. 359]). For a treatment of the above over an arbitrary field, see Curtis [4].

8.5 How many codes arise like this?

From the above discussion, it is apparent that some of the most interesting binary self-dual doubly-even codes arise in this manner, and that, when they do, there is a very convenient way of working with the code. However, it is natural to ask whether this is in some sense freakish behaviour, or, at the other extreme, whether most such codes arise in this way. The answers to these questions are contained in the following theorem, whose generalization to arbitrary finite fields, and to the Hermitian inner product when $|F| = q^2$, is given in [5, Theorem 2.1]; by a *fixed-point-free involution*, we mean a permutation of order 2 which moves every vertex.

Theorem 8.1 *A binary self-dual code C is equivalent (under permutations of the basis vectors) to one obtained from a graph in the above manner if and only if its group of automorphisms contains a fixed-point-free involution.*

Proof Suppose that C has a generating matrix of the form $(\mathbf{I} \mid \mathbf{C})$, where $\mathbf{C}$ is symmetric and $\mathbf{C}^2 = \mathbf{I}$. Then we have

$$(\mathbf{I} \mid \mathbf{C}) \begin{pmatrix} \mathbf{0} & \mathbf{I} \\ \mathbf{I} & \mathbf{0} \end{pmatrix} = (\mathbf{C} \mid \mathbf{I}) \sim (\mathbf{I} \mid \mathbf{C}^{-1}) \sim (\mathbf{I} \mid \mathbf{C}).$$

Thus the matrix preserves the code and corresponds to a fixed-point-free permutation of the basis vectors.

Conversely, let C be a self-dual binary code of dimension n and length $2n$, with basis $\{\mathbf{v}_1, \mathbf{v}_2, \ldots, \mathbf{v}_{2n}\}$. Further, let α be the linear transformation of the vector space induced by the permutation $\pi = (1,2)(3,4)\ldots(2n-1,2n)$ acting on the basis vectors, and suppose that C is preserved by α. We must prove that it is possible to produce a set of vectors $\{\mathbf{u}_1, \mathbf{u}_2, \ldots, \mathbf{u}_n\}$, with $\mathbf{u}_i \in \{\mathbf{v}_{2i-1}, \mathbf{v}_{2i}\}$, such that

$$\langle \mathbf{u}_1, \mathbf{u}_2, \ldots, \mathbf{u}_n \rangle \cap C = \mathbf{0}.$$

Certainly $\mathbf{v}_1 \notin C$, since $\mathbf{v}_1 \cdot \mathbf{v}_1 = 1 \neq 0$, so we choose $\mathbf{u}_1 = \mathbf{v}_1$. Now suppose that we have chosen $\{\mathbf{u}_1, \mathbf{u}_2, \ldots, \mathbf{u}_m\}$ such that

$$\langle \mathbf{u}_1, \mathbf{u}_2, \ldots, \mathbf{u}_m \rangle \cap C = \mathbf{0},$$

but that

$$\langle \mathbf{u}_1, \mathbf{u}_2, \ldots, \mathbf{u}_m, \mathbf{v}_{2m+1} \rangle \cap C \neq \mathbf{0} \neq \langle \mathbf{u}_1, \mathbf{u}_2, \ldots, \mathbf{u}_m, \mathbf{v}_{2m+2} \rangle \cap C.$$

Let

$$\mathbf{u} = \sum_{i=1}^{m} \lambda_i \mathbf{u}_i + \mathbf{v}_{2m+1} \in C \quad \text{and} \quad \mathbf{v} = \sum_{j=1}^{m} \mu_j \mathbf{u}_j + \mathbf{v}_{2m+2} \in C,$$

where $\lambda_i, \mu_j \in F$ and $\alpha(\mathbf{v}_{2m+1}) = \mathbf{v}_{2m+2}$. Then

$$\alpha(\mathbf{u}) \cdot \mathbf{v} = \mathbf{v}_{2m+2} \cdot \mathbf{v}_{2m+2} = 1 \neq 0.$$

This contradicts the fact that $\alpha(\mathbf{u})$ and $\mathbf{v}$ are both vectors in the self-dual code $\mathcal{C}$. Thus the required linearly independent set of vectors can be found, and as basis for the space we take

$$\{\mathbf{u}_1, \mathbf{u}_2, \ldots, \mathbf{u}_n, \alpha(\mathbf{u}_1), \alpha(\mathbf{u}_2), \ldots, \alpha(\mathbf{u}_n)\},$$

which is simply a permutation of its original basis. We now take a generating matrix for the code, written with respect to this basis, and row-reduce it to obtain

$$(\mathbf{I} \mid \mathbf{C}) \sim \alpha(\mathbf{I} \mid \mathbf{C}) \sim (\mathbf{C} \mid \mathbf{I}) \sim (\mathbf{C}^{-1} \mid \mathbf{I}).$$

But the self-duality of $\mathcal{C}$ implies that $\mathbf{CC}^T = 1$, and so $\mathbf{C} = \mathbf{C}^{-1} = \mathbf{C}^T$. Thus $\mathbf{C}$ is symmetric, as required. □

Self-dual binary codes are highly symmetric objects, particularly when they are doubly-even, and it turns out that each such code of length not exceeding 24 possesses such an automorphism. Indeed, the work of Conway and Pless [3] shows that this is also close to being true for the eighty distinct doubly-even self-dual binary codes of length 32. In his PhD thesis [9], the second author has systematically enumerated all graphs on 4, 8 or 12 vertices with properties (P1) and (P2); these twenty-eight graphs G_1 to G_{28} are presented in the next section, and are tabulated with their properties in Table 8.1 on page 126. Of course, the group of symmetries of each graph is a subgroup of the full group of symmetries of the code. Thus, to obtain the latter, we may (for instance) seek fusion among the orbits on minimal codewords of the former.

The structure of a given code is often best understood through invariant subcodes, known as *components*. We can construct a larger code by taking a direct sum of familiar components and adjoining additional codewords known as *glue*. The third column of Table 8.1 indicates the *component-type* of the code obtained from each graph G_1 to G_{28}. (To see precisely how these components are stuck together using glue, see [3].) The symbol u_n denotes the number of codewords of weight n, and the fourth column of the table gives u_4, the number of *tetrads* in the code. Otherwise, standard ATLAS notation for split extensions and wreath products is used to indicate the structure of groups of automorphisms (see [2]).

8.6 Graphs with the required properties

In the diagrams on pages 123–125, the larger circles represent orbits under a group of automorphisms of the graph, labelled according to how many vertices they contain; lines and smaller circles represent edges between these suborbits.

A diagram for the graph of the icosahedron (Fig. 8.1) is shown in Fig. 8.2.

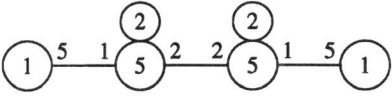

FIG. 8.2

Similarly, diagram G_3 represents the graph $3K_2 * K_2$, with 6 vertices joined in pairs, each of which is joined to a further 2 vertices that are themselves joined to one another (see Fig. 8.3).

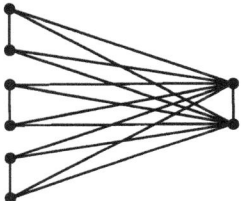

FIG. 8.3

Note that diagrams G_{25} and G_{28} involve bold lines that represent loops in the graph, so that in these two cases each vertex is joined to itself. In diagram G_{16} the subgraph on the two 3-orbits is a hexagon, as indicated, and in diagram G_{24} the subgraph on the first two 4-orbits is drawn explicitly with the black vertices and the white vertices representing the two suborbits. In each of the other cases, the diagram is self-explanatory.

Graphs on 4 and 8 vertices, leading to self-dual doubly-even binary codes of lengths 8 and 16

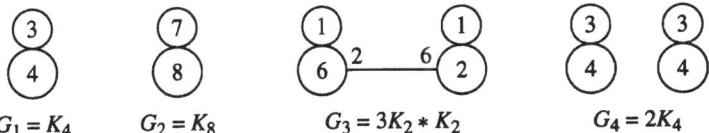

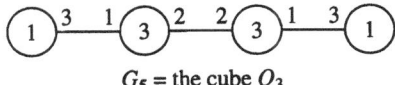

Graphs on 12 vertices,

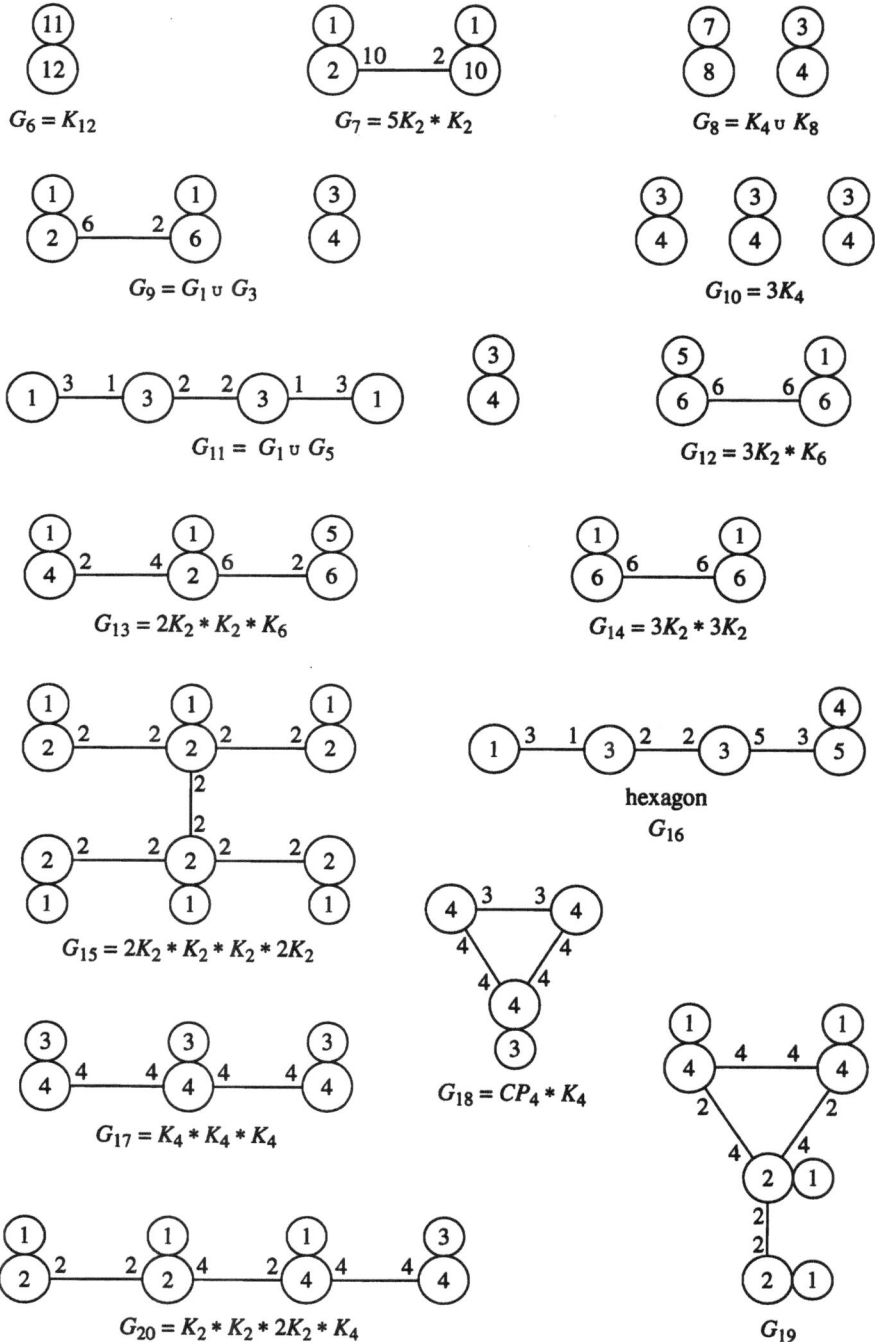

leading to self-dual doubly-even binary codes of length 24

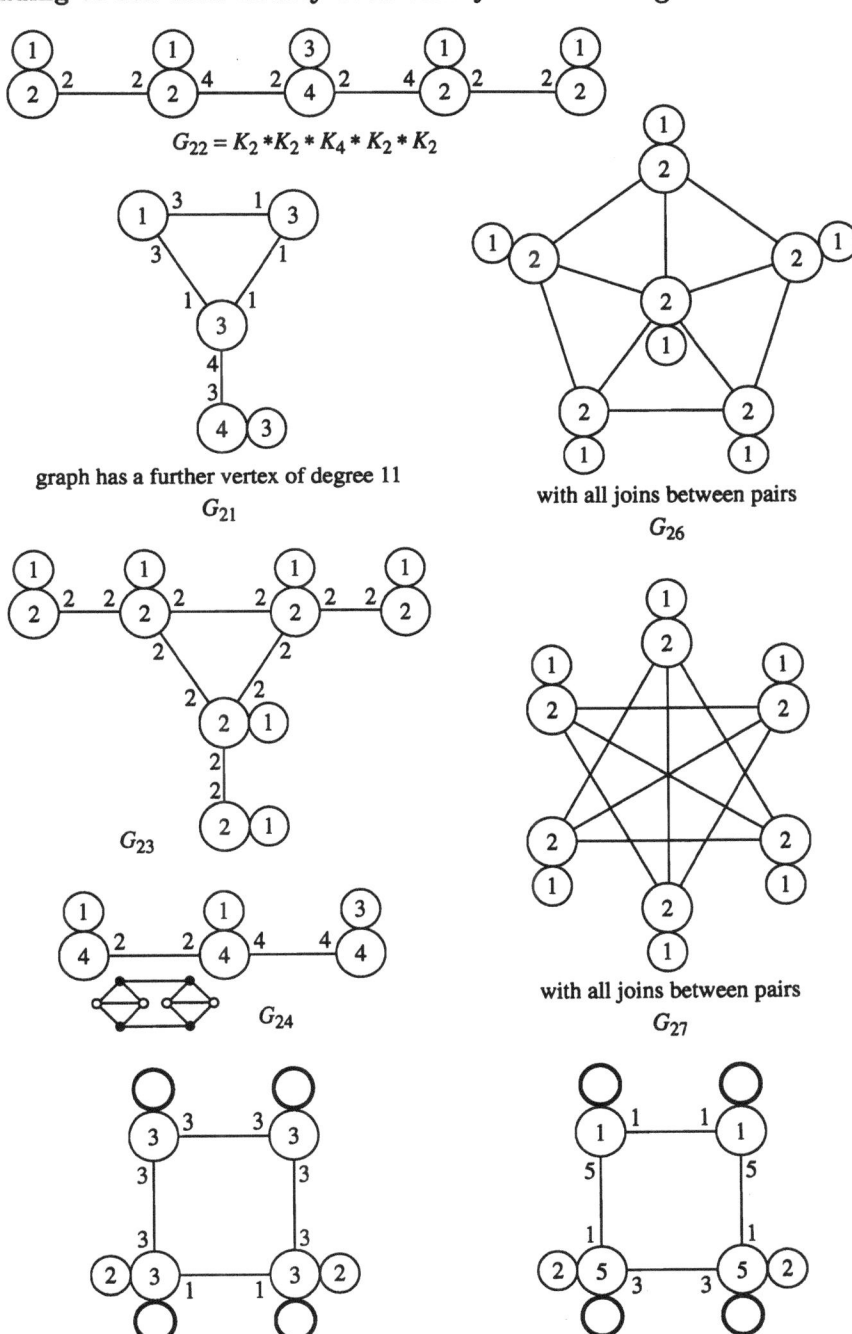

$G_{22} = K_2 * K_2 * K_4 * K_2 * K_2$

graph has a further vertex of degree 11
G_{21}

G_{23}

G_{24}

G_{25}

with all joins between pairs
G_{26}

with all joins between pairs
G_{27}

G_{28}

Table 8.1 *The self-dual doubly-even binary codes of length ≤ 24*

graph	length	components	u_4	graph group	code group
G_1	8	$\mathbf{e}_8$	14	S_4	$2^3 : L_2(7)$
G_2	16	$\mathbf{d}_{16}$	28	S_8	$2^7 : S_8$
G_3	16	$\mathbf{d}_{16}$	28	$2 \times (2 \wr S_3)$	$2^7 : S_8$
G_4	16	$2\mathbf{e}_8$	28	$S_4 \wr 2$	$(2^3 : L_2(7)) \wr 2$
G_5	16	$2\mathbf{e}_8$	28	$S_4 \times 2$	$(2^3 : L_2(7)) \wr 2$
G_6	24	$\mathbf{d}_{24}$	66	S_{12}	$2^{11} : S_{12}$
G_7	24	$\mathbf{d}_{28}$	66	$(2 \wr S_5) \times 2$	$2^{11} : S_{12}$
G_8	24	$\mathbf{d}_{16} + \mathbf{e}_8$	42	$S_8 \times S_4$	$(2^3 : L_2(7)) \times (2^7 : S_8)$
G_9	24	$\mathbf{d}_{16} + \mathbf{e}_8$	42	$(2 \times (2 \wr S_3)) \times S_4$	$(2^3 : L_2(7)) \times (2^7 : S_8)$
G_{10}	24	$3\mathbf{e}_8$	42	$S_4 \wr S_3$	$2^3 : L_2(7) \wr S_3$
G_{11}	24	$3\mathbf{e}_8$	42	$(S_4 \times 2) \times S_4$	$2^3 : L_2(7) \wr S_3$
G_{12}	24	$2\mathbf{d}_{12}$	30	$S_6 \times (2 \wr S_3)$	$(2^5 : S_6) \wr 2$
G_{13}	24	$2\mathbf{d}_{12}$	30	$D_8 \times 2 \times S_6$	$(2^5 : S_6) \wr 2$
G_{14}	24	$2\mathbf{d}_{12}$	30	$(2 \wr S_3) \wr 2$	$(2^5 : S_6) \wr 2$
G_{15}	24	$2\mathbf{d}_{12}$	30	$2^6 : D_8$	$(2^5 : S_6) \wr 2$
G_{16}	24	$\mathbf{d}_{10} + 2\mathbf{e}_7$	24	$S_3 \times S_5$	$2^4 : S_5 \times (L_2(7) \wr 2)$
G_{17}	24	$3\mathbf{d}_8$	18	$(S_4 \wr 2) \times S_4$	$2^3 : S_4 \wr S_3$
G_{18}	24	$3\mathbf{d}_8$	18	$S_4 \times S_4 \times 2$	$2^3 : S_4 \wr S_3$
G_{19}	24	$3\mathbf{d}_8$	18	$(D_8 \wr 2) \times 2^2$	$2^3 : S_4 \wr S_3$
G_{20}	24	$3\mathbf{d}_8$	18	$2^2 \times D_8 \times S_4$	$2^3 : S_4 \wr S_3$
G_{21}	24	$3\mathbf{d}_8$	18	$S_4 \times S_3$	$2^3 : S_4 \wr S_3$
G_{22}	24	$3\mathbf{d}_8$	18	$S_4 \times 2^4 : 2$	$2^3 : S_4 \wr S_3$
G_{23}	24	$3\mathbf{d}_8$	18	$2^6 : S_3$	$2^3 : S_4 \wr S_3$
G_{24}	24	$3\mathbf{d}_8$	18	$D_8 \times S_4$	$2^3 : S_4 \wr S_3$
G_{25}	24	$4\mathbf{d}_6$	12	$S_3^3 : 2$	$(2^2 : S_3) \wr S_4$
G_{26}	24	$6\mathbf{d}_4$	6	$2^6 : D_{10}$	$2^{12} : 3 \cdot S_6$
G_{27}	24	$6\mathbf{d}_4$	6	$2^6 : D_{12}$	$2^{12} : 3 \cdot S_6$
G_{28}	24	$\mathbf{g}_{24}$	0	$A_5 \times 2$	M_{24}

8.7 Counting tetrads

Finally, we demonstrate how easily the short codewords may be read from the graph. As we mentioned in Section 8.5, all codewords consist of any subset X of the vertices of the graph G on one side, together with those joined to an odd number of vertices of X on the other. Thus in particular, tetrads (codewords of weight 4) must either have one on one side and three on the other, in which case they correspond to vertices of degree 3 in G, or they have two on each side. If we look at diagram G_6, the complete graph K_{12}, we see that there are no tetrads of the first type, since each vertex has degree 11. Moreover, the group of the graph (the symmetric group S_{12}) is doubly transitive, and so there is only one type of pair to consider. Since each vertex of this pair is joined to its mate, and not to itself, these two points occur on the other side. However, all other vertices are joined to both of them. Thus there are $(12 \times 11)/2 = 66$ tetrads in the code.

Consider now diagram G_{12}, which consists of 6 vertices joined in pairs, and a further 6 vertices of degree 11. Again there are no vertices of degree 3, and so there are no tetrads of the first type. However, two vertices of degree 11 are adjacent, and all other vertices are adjacent to both. Similarly, any pair of vertices of degree 7 can easily be seen to give rise to a tetrad. However, a vertex of degree 11 with one of degree 7 gives rise to a codeword of weight 8. Thus this code possesses $\binom{6}{2} + 3 + (6 \times 4)/2 = 30$ tetrads.

References

1. A. E. Brouwer, A. M. Cohen and A. Neumaier, *Distance-Regular Graphs*, Springer-Verlag, 1989; *MR* **90e**: 05001.
2. J. H. Conway, R. T. Curtis, S. P. Norton, R. A. Parker and R. A. Wilson, *An ATLAS of Finite Groups*, Clarendon Press, 1985.
3. J. H. Conway and V. Pless, On the enumeration of self-dual codes, *J. Combin. Theory (A)* **28** (1980), 26–53; *MR* **81a**: 94032.
4. R. T. Curtis, The regular dodecahedron and the binary Golay code, *Ars Combin.* **29B** (1990), 55–64.
5. R. T. Curtis, On graphs and codes, *Geom. Dedicata* **41** (1992), 127–134; *MR* **92m**: 94014.
6. A. M. Gleason, Weight polynomials of self-dual codes and the MacWilliams identities, *Actes Congrès Intern. de Math.*, Vol. 3 (1970), 211–215; *MR* 54#12354.
7. F. J. MacWilliams and N. J. A. Sloane, *The Theory of Error-Correcting Codes*, North-Holland, 1977; *MR* **57**#5408.
8. T. Molien, Über die Invarianten der linearen Substitutionsgruppe, *Sitzungsber. König Preuss. Akad. Wiss.* (1897), 1152–1156.
9. A. R. Morris, *Self-Dual Codes Generated by Graphs*, PhD Thesis, University of Birmingham, 1993.
10. V. Pless, A classification of self-orthogonal codes over GF(2), *Discrete Math.* **3** (1972), 209–246; *MR* 46#3200.
11. V. Pless and N. J. A. Sloane, On the classification and enumeration of self-dual codes, *J. Combin. Theory (A)* **18** (1975), 313–335; *MR* **51**#1242.

9
Groups

PETER J. CAMERON

In this chapter, the connections between a graph and its automorphism group are described. The main themes are that most graphs have very little symmetry; abstract automorphism groups can be prescribed independently of most graph-theoretic properties; vertex-transitivity, however, entails various structural properties; and still higher degrees of symmetry can be expected to lead to a complete classification.

9.1 Introduction

The most important connection between graphs and groups is the fact that every graph G has an *automorphism group*, consisting of all permutations of the vertex set which map edges to edges and non-edges to non-edges. For most graphs, the automorphism group is trivial, consisting of only the identity permutation. However, every group is the automorphism group of some graph, and highly symmetric graphs have a number of special properties not shared by arbitrary graphs.

This chapter is an account of these matters. It is not a complete survey; in particular, there is very little discussion of several ways in which graph theory has been used to study finite groups. The work of Fischer on 3-transposition groups is briefly mentioned in Section 9.9. Fischer's groups are not the only sporadic simple groups first constructed as automorphism groups of graphs; see Higman and Sims [18] for an elegant example.

Rather than give the most general theorems available, we have tried to give the flavour by means of motivation and a few sample results. For more detailed surveys, see Babai [2], Cameron [11], Babai and Goodman [6]. Only finite graphs and groups are discussed; the infinite is very different territory! We refer to Wielandt [28] for undefined terms from group theory.

9.2 An example: the Petersen graph

The Petersen graph, shown in Fig. 9.1, provides a good introduction to the concept of symmetry of graphs. We can see from the picture that it has at least the symmetry of a pentagon (a dihedral group of order 10). A little imagination suggests that the outer pentagon and inner pentagram can be interchanged by an automorphism (although this cannot be represented by a Euclidean transformation), giving rise to a group of order 20 which acts transitively on the vertices.

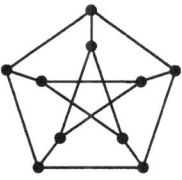
FIG. 9.1

FIG. 9.2

In fact, the full automorphism group is six times as large as this. There are various ways to see this. Figure 9.2 shows a drawing of the Petersen graph exhibiting a symmetry of order 3; this shows at least that the full group has order divisible by 3. Alternatively, we may argue abstractly. The vertices can be labelled with the unordered pairs from $\{1, 2, 3, 4, 5\}$ in such a way that two vertices are adjacent if and only if their labels are disjoint, as in Fig. 9.3. Now any permutation of $\{1, \ldots, 5\}$, acting on the labels, is an automorphism; so the symmetric group S_5 on five symbols is a group of automorphisms.

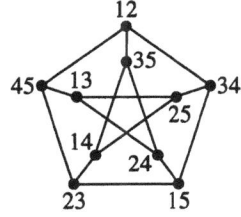
FIG. 9.3

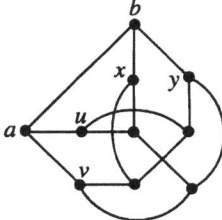
FIG. 9.4

Now S_5 is the full automorphism group. For let ab be an edge, and let u and v be the other two neighbours of a, and x and y those of b. Now any further vertex is joined to exactly one of u and v, and to exactly one of x and y, and any choice is realized by just one vertex. (This follows from the special properties of the Petersen graph—it has diameter 2 and girth 5; see Fig. 9.4.) So only the identity automorphism can fix these six vertices. It follows that different automorphisms have different effects on (a, b, u, v, x, y); so the number of automorphisms does not exceed the number of configurations like this one, which is easily seen to be 120. Since we have already found a group of order 120 acting on the graph, it must be the full automorphism group.

9.3 Three kinds of groups

To motivate some well-studied questions on graph automorphisms, we take a quick (and very much simplified) trip through the history of group theory, looking at the changing concept of a group.

A layman's conception of a group is as the symmetry group or *automorphism group* of some mathematical object, consisting of all permutations which preserve the object. In the case of a graph, this means the set of all permutations of the vertices that preserve the adjacency relation; in other words, those that map edges to edges and non-edges to non-edges.

By the mid-19th century, a group was taken to mean a *permutation group*: any set of permutations that is closed under composition and inversion and contains the identity permutation. It is clear that every automorphism group is a permutation group. Conversely, it can be shown that every finite permutation group is the automorphism group of some relational structure, though the arity of the relation might have to be as large as one less than the total number of points. (The *arity* of a relation is the number of arguments in an instance of the relation; thus a graph can be described by the binary relation of adjacency. For infinite groups, relational structures do not always suffice, and topological concepts are necessary.)

This raises the question: *which permutation groups are automorphism groups of graphs?* (Here the set of points permuted by the group must be taken as the vertex set of the graph.) It is not expected that this question will have a simple solution; indeed, the answer depends on the kind of graph considered. For example, the identity group acting on two points is not the automorphism group of a simple graph, although it is an automorphism group if either loops or directed arcs are allowed. The alternating group on four points is not the automorphism group of any kind of graph. Research in this area has concentrated on the concept of a *2-closed permutation group*; this can be defined as a group which is the full automorphism group of some edge-coloured digraph, preserving the colours of the edges. The automorphism group of a simple graph is 2-closed, but the converse is false.

Following the trend to abstraction and axiomatization in the second half of the 19th century, a group was redefined as a set with an operation satisfying the familiar axioms. We temporarily call such a structure an *abstract group*. Then any permutation group is an abstract group and Cayley's Theorem shows that every abstract group is isomorphic to a permutation group. Cayley's well-known proof constructs the *regular representation*, in which a group acts on itself by right multiplication.

As before, we may ask: *which abstract groups are automorphism groups of graphs?*

9.4 Universal classes

The above question was answered by Frucht [14].

Theorem 9.1 *Every group is the automorphism group of a graph.*

The proof of this impressive result is revealing, and perhaps a little disappointing. It is trivial to find a graph whose automorphism group contains a given group—a sufficiently large null graph will do; the problem is to restrict the symmetry of the graph. We begin with Cayley's construction, translated into graphical terms. Given a group Γ with a generating set S, we construct a *Cayley colour digraph* with vertex set Γ, in which (g, sg) is an arc of colour s for each $s \in S$, $g \in \Gamma$. The group of colour-preserving automorphisms is precisely Γ, acting by right multiplication as in Cayley's proof. Now we replace the coloured arcs by 'gadgets', one for each colour. Figure 9.5 illustrates this with a graph whose automorphism group is cyclic of order 3, constructed by this method. The exact details of the gadgets are not important; we require only that

the original vertices can be distinguished (in the example, they have degree greater than 4);

different colours (that is, different generators) are replaced by non-isomorphic gadgets;

no automorphism of a gadget fixes the two vertices of attachment;

there is an automorphism interchanging the vertices of attachment if and only if the corresponding generator has order 2.

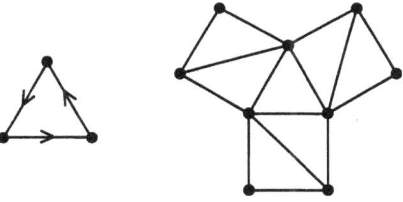

FIG. 9.5

Subsequently, Frucht [15] showed that *every group is the automorphism group of a cubic graph*. This has inspired a large number of similar results. We call a class $\mathcal{C}$ of structures *universal* if every finite group is the automorphism group of a structure in $\mathcal{C}$. The next result gives a sample of results on universality, due to Frucht, Sabidussi, Mendelsohn, Babai, Kantor, and others. A few results about structures other than graphs are included. In almost all cases, these are proved by 'encoding' a graph into a structure of the type considered. The structures are finite, except in case (i).

Theorem 9.2 *The following classes are universal:*
 (a) *graphs of degree k, for any fixed $k \geq 3$;*
 (b) *bipartite graphs;*
 (c) *strongly regular graphs;*
 (d) *Hamiltonian graphs;*
 (e) *k-connected graphs, for $k \geq 1$;*
 (f) *k-chromatic graphs, for $k \geq 2$;*
 (g) *switching classes of graphs;*
 (h) *lattices;*
 (i) *projective planes (possibly infinite);*
 (j) *Steiner triple systems;*
 (k) *symmetric designs.*

Another class of results involves structures for which there is some obvious restriction on the automorphism group. For example, a tournament cannot admit an automorphism of order 2—such an automorphism would necessarily interchange the ends of some arc—and so the automorphism group has odd order. This is the only restriction, as Moon [22] showed.

Theorem 9.3 *Every group of odd order is the automorphism group of a tournament.*

In view of these results, it becomes interesting to find a class in which there is some non-obvious restriction on the automorphism groups. It was observed by Pólya that not every group is the automorphism group of a tree. Indeed, we have the following.

Theorem 9.4 *The class of automorphism groups of trees is the smallest class containing the trivial group and closed under direct product and the operations 'wreath product with the symmetric group S_n of degree n', for each $n > 1$.*

(Rather than attempt a formal definition of the wreath product, we note that if Γ is the automorphism group of a connected graph G, then the wreath product of Γ with S_n is the automorphism group of the disjoint union of n copies of G. An automorphism of this structure is obtained by applying independently-chosen automorphisms to the n components, and following this with an arbitrary permutation of the components.)

Usually such a precise description does not exist. For example, every composition factor of the automorphism group of a planar graph is either a cyclic group or an alternating group. More generally, Babai showed the following.

Theorem 9.5 *If Γ is the automorphism group of a graph embeddable in the orientable surface of genus k, then there is a number $f(k)$ such that any composition factor of Γ is cyclic, alternating, or of order at most $f(k)$.*

Still more generally, Babai [2] showed that no class of finite graphs which is closed under subgraphs and contractions can be universal, except for the class of all graphs.

9.5 Bounds

Frucht's graphs are much larger than their automorphism groups, because of the addition of many asymmetric gadgets in the construction. So it is natural to ask how small a graph with given automorphism group can be. These matters are discussed further in the survey by Babai and Goodman [6], to which we refer for details and references.

Babai showed that, with just three exceptions, any group Γ is the automorphism group of a graph on which Γ has just two regular vertex orbits and hence $2|\Gamma|$ vertices.

For most groups, we can do better. A *Cayley graph* for Γ is obtained from a Cayley colour digraph, with respect to a given generating set, by ignoring colours and directions. If Γ acts regularly on the vertex set of G, then G is necessarily a Cayley graph for Γ. Hetzel and Godsil (see Babai [3]) showed the following.

Theorem 9.6 *If Γ is not an abelian group of exponent greater than 2, a generalized dicyclic group, or one of finitely many others, then there is a Cayley graph for Γ whose automorphism group is Γ.*

Can anything be said about groups Γ which are automorphism groups of graphs with $o(|\Gamma|)$ vertices? The symmetric group S_n is the extreme example; its order is $n!$ but it is the automorphism group of the n-vertex null graph. More surprisingly, Wells and Liebeck showed that the alternating group A_n is the automorphism group of a graph with roughly 2^n vertices. On the other hand, a cyclic group of prime-power order q cannot be the automorphism group of a graph with fewer than q vertices, since it is not isomorphic to a permutation group on fewer than q points. Recent work by Babai, Goodman and Pyber suggests that the presence of large subgroups of this form in a group Γ may be the only obstruction to the existence of a small graph with automorphism group Γ.

The situation is surprisingly different for edges. Babai and Goodman [5] and Goodman [16] showed that the number of edges can be bounded linearly, but the number of edge orbits is not bounded by a constant.

Theorem 9.7

(a) *Every group Γ is the automorphism group of a graph with at most $c|\Gamma|$ edges, for some constant c;*

(b) *there exist infinitely many groups Γ for which any graph with automorphism group Γ has at least $\sqrt{(c \log |\Gamma|)}$ edge orbits.*

Despite the negative result of part (b), a bounded number of edge orbits does suffice for many special classes of groups (abelian groups, simple groups, etc.).

9.6 Random graphs

A well-known fact, perhaps attributable to Erdős and Rényi [12], is that almost every graph has trivial automorphism group. This means that the proportion of such graphs among all graphs on n vertices, tends to 1 as $n \to \infty$. The observation can be refined in various ways. The convergence is very rapid; for almost all graphs, the asymmetry can be recognized by very fast algorithms; and, for almost all graphs, the 'nearest' graph which has a non-trivial automorphism is obtained by inserting and deleting edges so that two vertices have the same neighbours.

However, as a cautionary tale, note that, on average, a tree has exponentially many automorphisms.

Faced with this situation, we can restrict the type of graphs considered. It is known that almost all regular graphs of given degree k (> 2) are asymmetric. No such result is known for strongly regular graphs. However, two types of strongly regular graph (those derived from Steiner triple systems and Latin squares) are very common, and may predominate in the enumeration of all strongly regular graphs. For these types, almost all are asymmetric.

Another approach was taken by Cameron [9]. Given a group Γ, consider those n-vertex graphs whose automorphism group contains Γ. The proportion of such graphs whose automorphism group is precisely Γ tends to a limit as $n \to \infty$, but this limit is not necessarily 1. For example, if Γ is the dihedral group of order 10, then the limit is $\frac{1}{3}$. This is because, of those graphs which admit Γ, almost all have Γ acting on a set of five vertices and fixing the rest of the graph. A random graph admitting Γ consists of a random graph on $n - 5$ vertices, and five 'special' vertices on which Γ acts as the symmetry group of a pentagon, all joined to a random subset of the first $n - 5$ vertices. The induced subgraph on the five special vertices may be a complete graph, a null graph, or a 5-cycle; only in the third case is Γ the full group (almost surely).

It would be interesting to know whether similar results hold under hypotheses tending to work against such very local symmetries. For example, does a similar result hold for regular graphs?

Yet another approach is to start with a group acting in a more global fashion. There has recently been considerable interest in *random Cayley graphs* for groups. As might be expected, for most groups Γ, the full automorphism group of a random Cayley graph, defined by choosing a random generating set S for Γ, is precisely Γ (see Babai and Godsil [4]). Random Cayley graphs tend to have other interesting properties. For example, Alon and Roichman [1] showed that they are almost always good expanders. This notion, of some importance in computer science, means that any set of fewer than half of the vertices has a relatively large number of neighbours—a desirable property for a communication network!

9.7 Vertex-transitive graphs

We now turn to the 'other side' of the subject, which is more interesting to algebraists: *which graphs are highly symmetric?* We would expect that graphs with a high degree of symmetry should be completely classifiable; for those with less symmetry, we shall be content with establishing general properties.

In this section, we consider *vertex-transitive graphs*, those whose automorphism group acts transitively on the vertices. (A group acts *transitively* on a set if it contains an element mapping any point of the set to any other. We shall apply this concept to vertices, edges, and various other configurations in a graph.)

Any Cayley graph is vertex-transitive. The Petersen graph is an example of a vertex-transitive graph which is not a Cayley graph. On the other hand, any vertex-transitive graph on a prime number of vertices is a Cayley graph for the cyclic group of prime order. Recently, McKay and Praeger [21] have begun investigating the set of numbers n for which vertex-transitive non-Cayley graphs on n vertices exist, and the proportion of vertex-transitive graphs which are not Cayley graphs.

Clearly, a vertex-transitive graph is regular but vertex-transitive graphs have many properties not shared by all regular graphs. The next result, due to Mader, Watkins, Little, Grant, Holton, Babai and others, gives some examples. A graph is *vertex-primitive* if there is no equivalence relation on the vertex set preserved by all automorphisms, apart from the trivial relations of equality and the 'universal' equivalence.

Theorem 9.8 *Let G be a connected k-regular vertex-transitive graph of order n. Then*

(a) *G is $\lfloor \frac{2}{3}(k+1) \rfloor$-connected (even k-connected if it is vertex-primitive), and k-edge-connected;*

(b) *G has a 1-factor if n is even;*

(c) *G has a cycle of length at least $\sqrt{6n}$;*

(d) *the product of the clique number and the independence number of G is at most n.*

It has been conjectured that with finitely many exceptions, a connected vertex-transitive graph is Hamiltonian. The Petersen graph, of course, is one of these exceptions.

The *Hadwiger number* of a graph is the smallest number k for which some component of the graph can be contracted to the complete graph K_k. Recently, Babai and Thomassen have obtained structure theorems for connected vertex-transitive graphs with prescribed Hadwiger number. They showed that a sufficiently large vertex-transitive graph with Hadwiger number k is either toroidal or 'ring-like'—that is, the vertices can be divided into sets $S_0, \ldots, S_{n-1}$, with cardinalities bounded by a function of the Hadwiger number, such that all edges join vertices in the same set or in consecutive sets (mod n); the automorphism group induces a cyclic or dihedral group on this family of sets. Clearly, arbitrarily large toroidal vertex-transitive graphs can be obtained as quotients of plane lattices—for example, rectangular grids with opposite sides identified. The proof of this substantial result involves many geometrical ideas, including isoperimetric inequalities for the hyperbolic plane.

9.8 Distance-transitive graphs

A connected graph G is *distance-transitive* if, given two pairs (x_1, y_1) and (x_2, y_2) of vertices with $d(x_1, y_1) = d(x_2, y_2)$, where d is the graph distance, there is an automorphism of G carrying the first pair to the second. Again, the Petersen graph provides an example.

This is such a strong symmetry condition that we expect a complete classification. At present, such a classification has not been achieved, but a number of mathematicians are working towards it. We briefly outline the methods being used. For a more detailed account, with a description of the state of the art in 1988 and references for most of the results in this section, see van Bon and Cohen [7]. Throughout this section, G is a distance-transitive graph of degree greater than 2.

Smith [24] has shown that if G is not vertex-primitive, then either it is bipartite, or it is *antipodal*; that is, the relation of being either equal or at maximum distance is an equivalence relation on the vertex set. In each case, we can 'collapse' G to a smaller distance-transitive graph. In the bipartite case, take a bipartite block, and join two vertices if their distance in G is 2. The antipodal reduction identifies antipodal vertices, leading, for example, from the dodecahedron graph to the Petersen graph.

So the classification falls into two parts: first classify the vertex-primitive distance-transitive graphs, and then find all the 'bipartite doubles' and 'antipodal covers'. The first part of this programme is further advanced than the second, which is so far somewhat *ad hoc*.

A *Hamming graph* $H(d,q)$ has as vertex set the d-tuples from an alphabet of q symbols, two vertices being adjacent if they differ in just one coordinate. The name arises from the fact that the graph distance is exactly the Hamming distance familiar from coding theory. Hamming graphs are distance-transitive. A group is *affine* if it acts on the points of an affine space and contains the translations of the space. A group Γ is *almost simple* if it has a simple normal subgroup N and is contained in the automorphism group of N. Praeger, Saxl and Yokoyama [23] have shown that if G is a primitive distance-transitive graph, then either G is a Hamming graph (or the complement of one, in the case $d = 2$), or its automorphism group is affine or almost simple.

In order to find the primitive distance-transitive graphs with affine or almost simple automorphism groups, a very powerful tool, the classification of finite simple groups, must be used. In the almost simple case, this tells us the possible groups, and we have to decide in each case how the group can act distance-transitively on a graph; this has been done for most simple groups. In the affine case, further group-theoretic analysis is required; much of the work on this case has been done by van Bon.

Several other symmetry conditions have been considered. In some cases, such as *rank 3 graphs* or *t-tuple transitive graphs*, the classification has been completed using group theory, or is a special case of the distance-transitive graph classification. In other cases, such as *s-arc transitive graphs*, no detailed classification is expected, and we are satisfied with information about the local structure of the graph.

As an example, we discuss t-tuple transitive graphs. We say that two t-tuples $(x_1, \ldots, x_t)$ and $(y_1, \ldots, y_t)$ of vertices of a graph G are *isomorphic* if

$$x_i = x_j \iff y_i = y_j \quad \text{and} \quad x_i \sim x_j \iff y_i \sim y_j.$$

The graph G is called *t-tuple transitive* if, whenever two t-tuples are isomorphic, there is an automorphism of G carrying the first t-tuple to the second. Now all 5-tuple transitive graphs can be determined by purely combinatorial arguments. Cameron [10] has shown that such a graph must be a disjoint union of complete graphs of the same size, or the complement of such a graph (a regular complete multipartite graph), or the pentagon, or the line graph of $K_{3,3}$. All these graphs are t-tuple transitive for all t. (In the terminology of Chapter 5, they are *homogeneous*.) Subsequently, the t-tuple transitive graphs were classified successively for $t = 4, 3, 2$; increasing amounts of group theory, all depending on the classification of finite simple groups, were used in the proofs. (A connected graph is 2-tuple transitive if and only if it is distance-transitive with diameter 2; these graphs are also called *rank 3 graphs*.) Of course, 1-tuple transitivity is the same as vertex-transitivity; as we have seen, there is no hope of a classification in this case.

9.9 Local structure

The topic of this section does not explicitly involve groups, but draws most of its motivation from group theory. Let H be a graph. A graph G is *locally H* if, for each vertex v of G, the induced subgraph on the set of neighbours of v is isomorphic to H. A graph is *locally uniform* if it is locally H for some H. Any vertex-transitive graph is locally uniform, and any locally uniform graph is regular, but neither of these implications can be reversed; the class of locally uniform graphs lies strictly between the classes of regular and vertex-transitive graphs.

The connection with group theory is best seen in the work of Fischer [13] on 3-transposition groups. The idea can be described as follows. Let Γ be a finite group of even order, and let T be a set of involutions (elements of order 2) in Γ which is closed under conjugation (for example, a conjugacy class of involutions). The *commuting graph* $K(T)$ of T is obtained by taking T as vertex set, with two involutions being adjacent if and only if they commute. Now, if T is a conjugacy class, then Γ acts vertex-transitively on $K(T)$ (by conjugation), and $K(T)$ is locally $K(T')$, where $T' = T \cap C_\Gamma(\tau) \setminus \{\tau\}$, for $\tau \in T$, and $C_\Gamma(\tau)$ is the centralizer of τ in Γ. Under sufficiently strong hypotheses, T' is a conjugacy class in $C_\Gamma(\tau)$. Then a characterization of $K(T)$ by its local structure allows an inductive approach to determining groups satisfying the hypotheses.

For example, let Γ be the symmetric group S_n, and let T be the conjugacy class of transpositions. Two transpositions commute if and only if their supports are disjoint; so the commuting graph G_n is the complement of the line graph of K_n. (Figure 9.3 exhibits the Petersen graph as G_5.) Furthermore, we see that G_n is locally G_{n-2}, and so, for example, G_7 is locally Petersen.

This method was used by Fischer to determine groups generated by a conjugacy class of *3-transpositions*; these are involutions with the property that the product of any two has order at most 3. This property is satisfied by the transpositions in S_n; indeed, this was Fischer's motivating example. But several other classes of groups arise, including some geometrically defined (symplectic and orthogonal) groups, and three sporadic groups which now bear Fischer's name.

To return to the general problem: Bulitko [8] and Winkler [29] showed that, given H, the question of whether a locally H graph exists is recursively insoluble. Attention has focussed on finding graphs which are locally H for some fixed graph H. For example, Hall [17] found that there are exactly three locally Petersen graphs, on 21, 63 and 65 vertices. All three of these graphs are commuting graphs of involutions in certain groups; the smallest comes from the group S_7, in accordance with the comments above.

A related question is whether, given a finite graph H, there exists an infinite graph which is locally H. Weetman [26, 27] gave two general results on this question. He showed that if H is regular and has girth at least 6, then there exists an infinite graph that is locally H; in fact, unless H is very highly symmetric, there are $2^{\aleph_0}$ non-isomorphic infinite 'extensions'. On the other hand, he gave sufficient conditions on a graph H of diameter 2 for all graphs which are locally H to be finite of bounded order. These conditions are sufficiently general that, for example, no strongly regular graph is known not to satisfy them.

9.10 Computational aspects

Two natural computational problems in this area are:

given a graph, find (generating permutations for) its automorphism group;

given two graphs, decide whether they are isomorphic.

These two problems are closely related. Certainly, the second reduces to the first, since two connected graphs are isomorphic if and only if their disjoint union has an automorphism interchanging them. Moreover, a graph has an automorphism carrying x to y if and only if the graphs obtained by adding a suitable 'tail' to x and to y are isomorphic.

These problems are important in computer science, because they belong to a select class of problems which have neither been shown to have polynomial-time solutions nor been proved NP-complete (or harder). They may lie strictly between these two extremes. Their relation with random computation is also curious. As noted above, the algorithm which says 'the automorphism group of G is trivial' is almost always correct. It is possible to be just a little more subtle; there are efficient algorithms which, for almost all graphs, *prove* that the automorphism group really is trivial.

For some special classes, the isomorphism problem is known to have a polynomial algorithm. The best result of this kind is for graphs of bounded degree, a result of Luks [19]. It involves some careful group-theoretic analysis.

On the practical side, if a graph has a reasonably large automorphism group, we can use this group to help us, for example, in simplifying hard searching problems on the graph. Informally, we often start an argument with the phrase, 'By symmetry, we may assume ... '. The computer system GRAPE developed by Soicher [25] is based on this philosophy. It uses efficient algorithms for permutation groups behind the scenes. It also uses McKay's program **nauty** (McKay [20]), which quickly calculates the automorphism groups of large graphs.

References

1. N. Alon and Y. Roichman, Random Cayley graphs and expanders, *Random Structures Appl.* **5** (1994), 271–284.
2. L. Babai, Automorphism groups of graphs and edge-contraction, *Discrete Math.* **8** (1974), 13–20; *MR* **48**#10881.
3. L. Babai, On the abstract group of automorphisms, *Combinatorics* (ed. H. N. V. Temperley), London Math. Soc. Lecture Notes **52**, Cambridge University Press, 1981, pp. 1–40; *MR* **83a**: 05064.
4. L. Babai and C. D. Godsil, On the automorphism groups of almost all Cayley graphs, *European J. Combin.* **3** (1982), 9–15; *MR* **84d**: 05092.
5. L. Babai and A. J. Goodman, Subdirectly reducible groups and edge-minimal graphs with given automorphism group, *J. London Math. Soc.* **47** (1993), 417–432; *MR* **94d**: 20031.
6. L. Babai and A. J. Goodman, On the abstract group of automorphisms, *Coding Theory, Design Theory, Group Theory* (ed. D. Jungnickel and S. A. Vanstone), Wiley, 1993, pp. 121–143; *MR* **94e**: 05122.
7. J. T. M. van Bon and A. M. Cohen, Prospective classification of distance-transitive graphs, *Combinatorics '88, Vol. I* (ed. A. Barlotti *et al.*), Mediterranean Press, 1992, pp. 25–38; *MR* **94g**: 05038.
8. V. K. Bulitko, Graphs with prescribed environments of the vertices (Russian) *Trudy Mat. Inst. Steklov* **133** (1973), 78–94; *MR* **55**#7846.
9. P. J. Cameron, On graphs with given automorphism group, *European J. Combin.* **1** (1980), 91–96; *MR* **82f**: 05051.
10. P. J. Cameron, 6-transitive graphs, *J. Combin. Theory (B)* **28** (1980), 168–179; *MR* **81g**: 05076.
11. P. J. Cameron, Automorphism groups of graphs, *Selected Topics in Graph Theory 2* (ed. L. W. Beineke and R. J. Wilson), Academic Press, 1983, pp. 89–127.
12. P. Erdős and A. Rényi, Asymmetric graphs, *Acta Math. Acad. Sci. Hungar.* **14** (1963), 295–315; *MR* **27**#6258.
13. B. Fischer, Finite groups generated by 3-transpositions I, *Invent. Math.* **13** (1971), 232–246; *MR* **45**#3557.
14. R. Frucht, Herstellung von Graphen mit vorgegebener abstrakter Gruppe, *Compositio Math.* **6** (1938), 239–250.
15. R. Frucht, Graphs of degree three with a given abstract group, *Canad. J. Math.* **1** (1949), 365–378; *MR* **13**–857b.
16. A. J. Goodman, The edge-orbit conjecture of Babai, *J. Combin. Theory (B)* **57** (1993), 26–35; *MR* **94h**: 05038.
17. J. I. Hall, Locally Petersen graphs, *J. Graph Theory* **4** (1980), 173–187; *MR* **81h**: 05076.
18. D. G. Higman and C. C. Sims, A simple group of order 44,352,000, *Math. Z.* **105** (1968), 110–113; *MR* **37**#2854.
19. E. M. Luks, Isomorphism of graphs of bounded valence can be tested in polynomial time, *J. Comput. System Sci.* **25** (1982), 42–65; *MR* **84a**: 68063.

20. B. D. McKay, nauty user's guide (version 1.5), Technical report TR-CS-90-02, Computer Science Department, Australian National University, 1990.
21. B. D. McKay and C. E. Praeger, Vertex-transitive graphs which are not Cayley graphs, I, *J. Austral. Math. Soc. (A)* **56** (1994), 53–63; *MR* **95b**: 05098.
22. J. W. Moon, Tournaments with a given automorphism group, *Canad. J. Math.* **16** (1964), 485–489; *MR* **29**#603.
23. C. E. Praeger, J. Saxl and K. Yokoyama, Distance transitive graphs and finite simple groups, *Proc. London Math. Soc. (3)* **55** (1987), 1–21; *MR* **88c**: 05062.
24. D. H. Smith, Primitive and imprimitive graphs, *Quart. J. Math. Oxford (2)* **22** (1971), 551–557; *MR* **48**#5926.
25. L. H. Soicher, GRAPE: a system for computing with graphs and groups, *Groups and Computation* (ed. L. Finkelstein and W. M. Kantor), DIMACS Series in Discrete Mathematics and Theoretical Computer Science **11**, Amer. Math. Soc., 1993, pp. 287–291; *MR* **94c**: 20003.
26. G. M. Weetman, A construction of locally homogeneous graphs, *J. London Math. Soc. (2)* **50** (1994), 68–86; *MR* **95d**: 05113.
27. G. M. Weetman, Diameter bounds for graph extensions, *J. London Math. Soc. (2)* **50** (1994), 209–221; *MR* **95h**: 05064.
28. H. Wielandt, *Finite Permutation Groups*, Academic Press, 1964; *MR* **32**#1252.
29. P. M. Winkler, Existence of graphs with a given set of r-neighborhoods, *J. Combin. Theory (B)* **34** (1983), 165–176; *MR* **85e**: 05135.

10
Geometry

EDWARD R. SCHEINERMAN

There are many ways to connect graph theory with geometry. In this chapter we forge a link by means of intersection graphs, whose vertices are geometrical objects that are adjacent if and only if they have non-empty intersection.

10.1 Introduction

There are a great many connections between graph theory and geometry. For example, we can consider the plane to be a graph: the points of the plane are the vertices and two vertices are adjacent if they are exactly distance 1 apart. A famous problem is to determine the chromatic number of this graph; it is known to lie in the range 4 to 7.

In this chapter, we take a different and rather specialized approach to linking geometry and graph theory, by exploring the families of graphs that arise as the intersection graphs of geometrical objects.

What is an intersection graph? Suppose that we are given a finite collection of sets $S_1, S_2, \ldots, S_n$. Usually these sets are geometrical objects, such as intervals on a line or curves in the plane. We can think of the sets as being vertices of a graph, and we draw an edge between two vertices precisely when the corresponding sets have non-empty intersection. The resulting graph is an *intersection graph*.

For example, consider Fig. 10.1. On the left is a family of sets, and on the right is the corresponding intersection graph of these sets. Notice that the vertices x, y and z form a triangle K_3. Indeed, the three sets X, Y and Z have non-empty intersection, but this fact cannot be deduced just from the graph. All that is required is that the sets X, Y and Z have non-empty pairwise intersections.

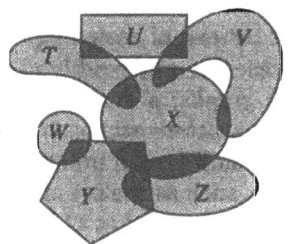
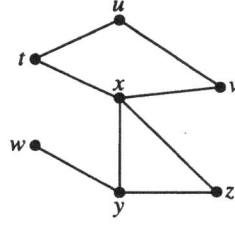

FIG. 10.1

More formally, let Σ denote a family of sets. Typically, Σ represents a family of geometrical objects, such as discs in the plane or curves in space. A Σ-*representation* of G is an assignment of some member of Σ to each vertex of G in such a way that two vertices are adjacent exactly when their assigned sets intersect. A graph which has a Σ-representation is called a Σ-*graph*, and the set of all Σ-graphs is denoted by $\Omega(\Sigma)$. A graph has an *intersection representation* if it has a Σ-representation for some family Σ.

For example, suppose that Σ is the above set of compact connected subsets of the plane. Then the graph in Fig. 10.1 is a Σ-graph, and a representation is the mapping $u \mapsto U, \ldots, z \mapsto Z$. It is easy to check that K_n, $K_{r,s}$ and the Petersen graph are all intersection graphs of compact connected subsets of the plane. An example of a graph that is *not* such an intersection graph is given in Section 10.5.

This notion of intersection representation is our link between graph theory and geometry. We consider various families of geometrical sets Σ and explore the properties of the corresponding Σ-graphs. Our exploration follows (more or less) the dimension of the geometrical objects in Σ. Arguably, the most important family of sets that we consider is that of the intervals on the real line, which give rise to the *interval graphs* (see Section 10.3). All the other graphs that we discuss may be thought of as generalizations of these.

Given a collection Σ of geometrical objects, our goal is to describe, as completely as possible, the class $\Omega(\Sigma)$ of Σ-graphs. Note that $\Omega(\Sigma)$ is a *hereditary class* of graphs, in the sense that if G is a Σ-graph, then so is each of its induced subgraphs. Thus one way to describe $\Omega(\Sigma)$ is to find $\mathcal{F}(\Sigma)$, the set of *minimal non-Σ-graphs*. Because $\Omega(\Sigma)$ is hereditary, a graph is a Σ-graph if and only if it contains no graph in $\mathcal{F}(\Sigma)$ as an induced subgraph. Another way to study $\Omega(\Sigma)$ is to prove theorems that describe the structure of Σ-graphs. Additionally, where possible, we seek efficient algorithms for recognizing Σ-graphs.

10.2 Dimension 0: discrete sets

The simplest geometrical object is a point. Let S be a set containing just one point, and let $\Sigma = \{S\}$. Then the Σ-graphs are just the complete graphs, and $\mathcal{F}(\Sigma) = \{2K_1\}$.

Let us cast our net a little wider. Instead of assigning the *same* one-point set to all the vertices of a graph, we now assign to each vertex any singleton point set. Now the Σ-graphs that we can form are exactly those in which each component is complete, and so $\mathcal{F}(\Sigma) = \{P_3\}$.

The situation becomes more interesting if instead of assigning singleton sets to vertices, we assign sets of size 2. In this case, we can readily check that the class of Σ-graphs is exactly the class of *line graphs*; a survey on line graphs, including structural and forbidden subgraph characterizations, is given in [18].

The next step is to assign to each vertex *any* finite set of points. Let Σ denote the family of all finite subsets of any infinite set, such as the set of integers. Then every graph is a Σ-graph (see [23]). The proof is simple. Let G be a graph, and suppose that the edges of G are labelled with distinct integers. For each vertex v

of G, let $E(v)$ denote the set of edges that are incident with v. Then $vw \in E(G)$ if and only if $E(v) \cap E(w) \neq \varnothing$. Thus the mapping $v \mapsto E(v)$ is an intersection representation of G.

To make this more interesting, we can seek to minimize the number of different points used in the representation. Let G be a graph with $V(G) = \{v_1, \ldots, v_n\}$. Suppose that $f : V(G) \to \Sigma$ is an intersection representation, and define the *size* $|f|$ of f to be
$$|f| = |f(v_1) \cup f(v_2) \cup \cdots \cup f(v_n)|.$$
The *intersection number* $i(G)$ of a graph G is the minimum value of $|f|$, taken over all representations f of G.

For example, consider the graph in Fig. 10.2. A representation of size 3 is shown, and hence $i(G) \leq 3$. Since this representation is as small as possible, $i(G) = 3$. Note that we have shaded three cliques of G; all vertices in the first clique have a 1 in their set, and likewise for the second and third cliques.

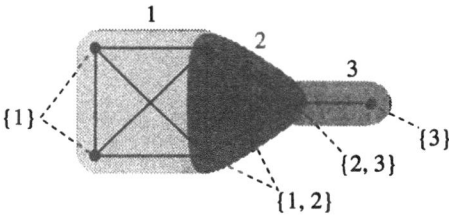

FIG. 10.2

This works in general. Suppose that we have a family $\{X_1, \ldots, X_t\}$ of cliques of G, with the property that each pair of adjacent vertices of G occurs in one of the cliques X_i. Such a family is an *edge covering by cliques* of G. We can now construct an intersection representation of G by assigning to each vertex v the set $\{i : v \in X_i\}$. Conversely, given an intersection representation f of G, we can construct an edge cover by cliques of G by letting $X_i = \{v : i \in f(v)\}$. This yields our first result.

Theorem 10.1 *The intersection number $i(G)$ of a graph G is equal to the minimum number of cliques that cover all the edges of G.*

Since the set of edges of G forms a clique cover of G, it follows that $i(G) \leq |E(G)|$. If G is triangle-free, then we have equality, but if G has a triangle, then we have strict inequality. *Turán's Theorem* tells us that the maximum number of edges in a triangle-free graph of order n is $\lfloor n^2/4 \rfloor$, and this is achieved only by the complete bipartite graph with part sizes as close as possible. Likewise, among all graphs on n vertices, the one with highest intersection number is the same complete bipartite graph (see [9]).

Theorem 10.2 *If G is a graph on n vertices, then $i(G) \leq \lfloor n^2/4 \rfloor$. Furthermore, equality holds if and only if $G = K_{r,s}$, with $r = \lfloor n/2 \rfloor$ and $s = \lceil n/2 \rceil$.*

10.3 Dimension 1: interval graphs

We now come to the star of our show: the family of interval graphs. An *interval graph* is an intersection graph of closed intervals on the real line. In fact, it does not matter whether we use closed intervals, open intervals, or any sort of intervals on the real line; we always arrive at the same class of graphs. Note that the graph in Fig. 10.2 is an interval graph. To see why, replace $\{1\}$ by the interval $[0,1]$, $\{1,2\}$ by $[1,2]$, $\{2,3\}$ by $[2,3]$, and $\{3\}$ by $[3,4]$. Similarly, any complete graph is an interval graph: we simply assign to all vertices the same interval.

There are a number of ways to characterize the family of interval graphs, and we present some here; for more details, see [10] or [14].

Chordal graphs The 4-cycle is not an interval graph, as a moment's doodling will show. Indeed, except for the triangle, no cycle is an interval graph. Moreover, no graph containing a chordless cycle can be an interval graph. Thus every interval graph must be a *chordal graph*—that is, it contains no induced cycle of length greater than 3 (see Section 10.4).

Not all chordal graphs are interval graphs. For example, consider the chordal graph G in Fig. 10.3.

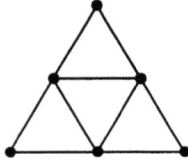

FIG. 10.3

Suppose, for a contradiction, that G is an interval graph, and fix a representation. Consider a pair of intervals that do not intersect. One of these must lie entirely to the left of the other. Indeed, these intervals induce a partial order on the vertices. We can therefore give an orientation to the edges of the complement $\overline{G}$, with $x \to y$ whenever the interval for x lies entirely to the left of the interval for y. Such an orientation must be *transitive*: if the edges xy and yz are oriented as $x \to y$ and $y \to z$, then the graph must include the edge xz and its orientation must be $x \to z$.

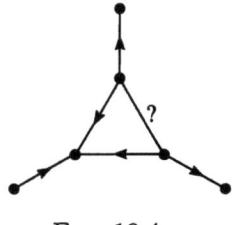

FIG. 10.4

Now consider the graph in Fig. 10.4, which is the complement of that in Fig. 10.3. We claim it has no transitive orientation. For, if it did, then, without loss of generality, we could orient the lower left edge as shown. Transitivity forces us to orient the remaining marked edges as shown. However, no matter how we orient the edge marked with a question mark, we violate transitivity. Thus, $\overline{G}$ does not have a transitive orientation, and so G is not an interval graph.

Thus all interval graphs must be chordal, and their complements must be transitively orientable. It is a lovely fact that the converse of this statement also holds (see [13]).

Theorem 10.3 *A graph is an interval graph if and only if it is chordal and its complement has a transitive orientation.*

Chordal graphs (and, in particular, interval graphs) are *perfect*: for each induced subgraph, the clique number equals the chromatic number.

Asteroidal triples Now let G be the graph in Fig. 10.4. It is interesting to note that G is also chordal but is not an interval graph. We could check this by observing that its complement (in Fig. 10.3) has no transitive orientation. Instead, we prove this in a different manner.

Let a, b and c be the vertices of degree 1 in G, and let P_{ab}, P_{bc} and P_{ac} be the shortest paths between them; then vertex a is not adjacent to any vertex on P_{bc}. Suppose, for a contradiction, that G is an interval graph, and fix a representation. Notice that the intervals for the path P_{bc} must be disjoint from the interval for a. Thus the interval for a must lie entirely to the left, or entirely to the right, of all the intervals for P_{bc}. A similar statement holds for b and P_{ac}, and also for c and P_{ab}. However, one of the intervals for a, b and c must lie between the other two, giving the required contradiction.

In general, an *asteroidal triple* is a collection of three vertices, each pair of which is connected by a path that does not meet the third vertex or any of its neighbours. It follows that a graph with an asteroidal triple cannot be an interval graph. This gives us a second characterization of interval graphs (see [22]).

Theorem 10.4 *A graph is an interval graph if and only if it is chordal and contains no asteroidal triples.*

Consecutive 1s It is well known that intervals on the real line enjoy the Helly property (see [17]): *if a collection of intervals are pairwise non-disjoint, then there is a point lying in all the intervals*; in symbols, if $I_1, I_2, \ldots, I_n$ are intervals for which $I_i \cap I_j \neq \varnothing$ for all i and j, then $\bigcap I_i \neq \varnothing$. This is easy to prove: we just consider the right-most left end-point in the collection.

Let G be an interval graph, and let $X_1, X_2, \ldots, X_t$ be a list of its maximal cliques. Fix an interval representation for G. Since the vertices in each clique are pairwise adjacent, their intervals intersect pairwise. By the Helly property, there must be a point common to all of these intervals. Let x_i be a point common to the intervals from the vertices in the clique X_i and, without loss of generality, suppose that $x_1 < x_2 < \cdots < x_n$. Now consider any vertex v. Its interval contains some (and perhaps all) of the points x_i, and the points x_i in the interval must be consecutive.

We can express this fact in matrix terms. Let $\mathbf{M} = (m_{ij})$ be the *maximal-clique incidence matrix*, whose rows are indexed by the vertices of G, and whose columns are indexed by the maximal cliques. The entries of $\mathbf{M}$ are 0s and 1s, with $m_{ij} = 1$ when $v_i \in X_j$. We say that $\mathbf{M}$ has the *consecutive 1s property* if we can reorder its columns so that the 1s in each row appear consecutively. The maximal-clique incidence matrix of an interval graph must have the consecutive 1s property, and conversely (see [11]).

Theorem 10.5 *Let G be a graph, and let $\mathbf{M}$ be its maximal-clique incidence matrix. Then G is an interval graph if and only if $\mathbf{M}$ has the consecutive 1s property.*

Forbidden subgraphs The set of minimal non-interval graphs gives us another way to characterize interval graphs. Since paths are interval graphs, but cycles (other than C_3) are not, such cycles are minimal non-interval graphs. The graphs in Figs 10.3 and 10.4 are also minimal non-interval graphs.

The remaining minimal non-interval graphs are shown in Fig. 10.5; the left two diagrams represent infinite families of graphs, and in each case the path of similar vertices may be as long as we wish; indeed, the graphs in Figs 10.3 and 10.4 may be considered as the first two graphs in these families. Thus the complete forbidden subgraph characterization consists of three infinite families of graphs and two sporadic examples (see [22]).

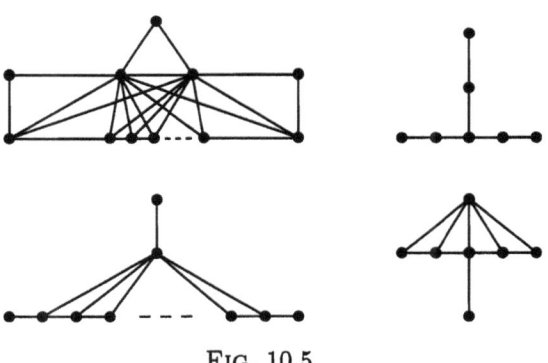

Fig. 10.5

Theorem 10.6 *A graph is an interval graph if and only if it does not contain as an induced subgraph a cycle of length 4 or more, or any of the graphs in Fig. 10.5.*

Algorithms Efficient algorithms exist for recognizing interval graphs and constructing an interval representation (see, for example, [3] and [4]). From the interval representation, one can readily compute such hard parameters as independence number, clique number, chromatic number, etc.

Variations A number of variations of interval graphs have been studied; see [14] for a good overview. *Unit interval graphs* are the intersection graphs of intervals all of the same length. *Tolerance graphs* are similar to interval graphs, in that each vertex is assigned an interval, but in addition, each vertex gets a tolerance and two vertices are adjacent when the *length* of the intersection exceeds the tolerance of either. In a *multiple-interval graph*, each vertex is represented by the union of a fixed number of intervals.

There are extraordinarily many variations of the interval graph concept. The exploration which we consider here is to increase the dimension beyond that of the real line and examine the intersection graphs of interval-analogues in higher dimensions.

10.4 Dimension 1 and a little: trees and circles

How do intervals generalize to higher dimensions? As we travel from the line to the plane and beyond, there are two intermediate stops: trees and circles. We can branch out by replacing the real line by a tree, or we can double back and replace the line by a circle. In each case, we have a generalization of the concept of interval graph.

Subtrees of a tree Where a geometer sees intervals on a line, a graph theorist might see subpaths of a path graph. A natural generalization of a path is a tree. Let $\Sigma = \{T_1, T_2, \ldots, T_n\}$ be a collection of subtrees of a fixed tree, and let G be the intersection graph of Σ—thus we draw an edge between v_i and v_j if and only if T_i and T_j have a common vertex. Which graphs arise in this manner? A bit of doodling will convince you that cycles with four or more vertices cannot be so represented, and hence only chordal graphs admit such a representation. The surprise is that *all* chordal graphs are achievable (see [12] and [38]).

Theorem 10.7 *A graph is the intersection graph of subtrees of a tree if and only if it is chordal.*

Indeed, chordal graphs can be seen as a generalization of trees. A *simplicial vertex* is a vertex whose neighbours form a clique. Thus an end-vertex is a special sort of simplicial vertex. Every chordal graph has a simplicial vertex (see [7] and [14]).

Many theorems about chordal graphs, such as Theorem 10.7, are proved by induction, by deleting a simplicial vertex. Let G be a chordal graph, and let v_1 be a simplicial vertex. Since $G - v_1$ is chordal, it also has a simplicial vertex v_2. Let v_3 be simplicial in $G - v_1 - v_2$, and so on. We can therefore decompose G by iteratively removing simplicial vertices. Such a decomposition of G is called a *perfect elimination order*.

Theorem 10.8 *A graph is chordal if and only if it has a perfect elimination order.*

Arcs of a circle Another slightly higher-dimensional analogue of an interval on a line is an arc of a circle. The characterization and recognition problems for circular arc graphs are more difficult, partly because circular arcs do not satisfy the Helly property: three arcs can intersect pairwise but have no point in common.

However, there is a *cyclic neighbourhood-inclusion property* that serves to characterize circular arc graphs and can be used to recognize them in polynomial time.

Given a collection Σ of arcs of a circle, we label them $A_1, A_2, \ldots, A_n$ as they are first encountered in moving anticlockwise around the circle (starting from an arbitrary point); Fig. 10.6 shows an example of such a labelling. It is not difficult to see that if A_i and A_j overlap (for example, A_1 and A_4 in Fig. 10.6), then in going anticlockwise either from A_i to A_j or from A_j to A_i, one must encounter all of the intermediate arcs. Hence the vertices of the intersection graph G can be labelled $v_1, v_2, \ldots, v_n$ so that if $v_i v_j \in E(G)$, then either v_i is adjacent to $v_{i+1}, v_{i+2}, \ldots, v_j$, or v_j is adjacent to $v_{j+1}, v_{j+2}, \ldots, v_i$ (with subscripts considered modulo n). It turns out that any graph that can be so labelled is a circular arc graph.

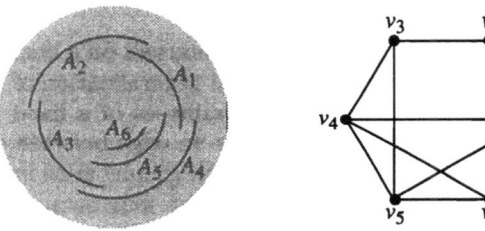

FIG. 10.6

Theorem 10.9 *A graph is a circular arc graph if and only if there exists a cyclic neighbourhood-inclusion labelling of its vertices.*

10.5 Dimensions 2 and higher

As we move into the plane and beyond, there are many ways to generalize intervals. Intervals on the line are curves, compact connected sets, discs, boxes, convex sets, and line segments. All of these notions are equivalent on the line, but are different in $\mathbf{R}^2$ and beyond, and each yields an interesting generalization of interval graphs.

Strings A *string graph* is the intersection graph of a set of curves in the plane. Although not all compact connected subsets of the plane are curves, it turns out that the intersection graph of compact connected sets is always a string graph. (It makes no difference if the curves are simple or not.) It is easy to check that complete graphs, complete multipartite graphs, circular arc graphs and planar graphs are all string graphs. Also included are *split graphs*, which are those graphs whose vertex set can be partitioned into an independent set and a clique. It can also be shown that a graph G is a split graph if and only if both G and $\overline{G}$ are chordal.

Doodling suggests that *all* graphs are string graphs, but this is not the case, as is shown by the following example (see [8] and [30]). Let G be the graph obtained from K_5 by replacing each edge by a path of length 2. Label the five original vertices $v_1, \ldots, v_5$ and the ten new vertices w_{ij}, where each w_{ij} is adjacent only to v_i and v_j. Suppose, for a contradiction, that G is a string graph, and draw the fifteen curves representing its vertices. Then the five curves corresponding to the vertices v_i are pairwise disjoint, as are the ten w_{ij} curves. Shrink the v_i curves to a point, dragging the w_{ij} curves with them. This gives a crossing-free drawing of K_5 in the plane. As this is impossible, G is not a string graph.

It is NP-hard to recognize string graphs (see [21] and [24]), and it is unknown if the string graph recognition problem is in NP. In three dimensions and higher, every graph is the intersection graph of a set of curves.

Convex sets Closed intervals are the *convex* compact subsets of the line. The intersection graphs of convex compact subsets of the plane include all planar graphs, but not all graphs, and such graphs have not yet been characterized. However, it is known that every graph is an intersection graph of convex compact subsets of $\mathbf{R}^3$ (see [40]).

Balls and coins A *ball* is a subset of $\mathbf{R}^n$ of the form

$$B(\mathbf{x}, r) = \{\mathbf{y} \in \mathbf{R}^n : |\mathbf{x} - \mathbf{y}| \leq r\}.$$

Closed intervals are 1-dimensional balls, and so a natural generalization of an interval graph is the intersection graph of balls in some dimension. The least integer d for which G is the intersection graph of balls in $\mathbf{R}^d$ is called the *sphericity* of G (see [27]). Spherical representations of graphs arise in chemistry, where the atoms of a molecule are the vertices of a graph. Each atom is modelled as a sphere, and an intersection of spheres corresponds to a chemical bond. Every graph is the intersection graph of spheres in some dimension, and there are graphs with arbitrarily high sphericity; a proof of this is given in Section 10.6.

Returning to the plane, a *coin representation* of a graph G is an intersection representation of G by discs (2-dimensional balls) in the plane, with the added condition that when discs intersect, they do so in exactly one point—that is, they are externally tangent. Every graph with a coin representation must be planar: to see this, embed each vertex at the centre of its representing disc, and join adjacent vertices by straight-line segments. A beautiful theorem is that all planar graphs arise in this fashion (see [5] and [32]).

Theorem 10.10 *A graph is a coin graph if and only if it is planar.*

Boxes A *box* is a Cartesian product of closed intervals. Each box in $\mathbf{R}^n$ is determined by a pair of points $(a_1, a_2, \ldots, a_n)$ and $(b_1, b_2, \ldots, b_n)$:

$$[a_1, b_1] \times [a_2, b_2] \times \cdots \times [a_n, b_n] = \{\mathbf{x} \in \mathbf{R}^n : a_i \leq x_i \leq b_i \text{ for each } i\}.$$

Thus 1-dimensional boxes are closed intervals, 2-dimensional boxes are axis-parallel rectangles with their interiors, and so on.

For each graph G, there is a least integer d so that G is the intersection graph of boxes in $\mathbf{R}^d$ (see [6] or [27]); this integer is called the *boxicity* of the graph. The generalized octahedron $K_{2,2,...,2}$ has boxicity equal to the number of 2s. In particular, the octahedron graph $K_{2,2,2}$ cannot be represented as the intersection graph of boxes in $\mathbf{R}^2$, but (see [31]) every planar graph is the intersection graph of boxes in $\mathbf{R}^3$.

Line segments A line segment is another higher-dimensional analogue of a closed interval. In the plane, the set of line segment intersection graphs is a proper subset of the set of string graphs; see [8] for a construction, and Section 10.6 below for another proof. The intersection graphs of line segments in $\mathbf{R}^3$ form a proper superset of those arising from planar line segments. However, no new graphs are created by considering line segments in higher dimensions (see [28]).

One of my favourite questions is: *is every planar graph the intersection graph of line segments in the plane?* The answer is unknown, but a partial result is the following (see [16]).

Theorem 10.11 *Every bipartite planar graph is the intersection graph of line segments in the plane, using only vertical and horizontal line segments.*

The above open question can be made even more enticing: *is every planar graph the intersection graph of line segments in the plane which have only one of four slopes (say, $+1$, -1, 0 and ∞)?*

We now consider a special family of line segment graphs for which there is a complete characterization. Draw two parallel lines in the plane, and consider a collection of line segments that have one end-point on each of these lines. The intersection graphs of such line segments are called *permutation graphs* and have the following characterizations (see [14]).

Theorem 10.12 *Let G be a graph with vertex set $\{1, 2, \ldots, n\}$. The following are equivalent:*

(a) G is a permutation graph;

(b) there is a permutation $\pi \in S_n$ with the property that $ij \in E(G)$ if and only if $(i-j)(\pi(i) - \pi(j)) < 0$;

(c) both G and $\overline{G}$ have transitive orientations;

(d) G is the comparability graph of a partially ordered set of dimension at most 2.

10.6 Counting methods via real algebraic geometry

Given two graph properties $\mathcal{P}$ and $\mathcal{Q}$, one often wishes to prove theorems of the form $\mathcal{P} \Rightarrow \mathcal{Q}$—that is, if G has property $\mathcal{P}$, then G also has property $\mathcal{Q}$. In many cases, $\mathcal{Q}$ is a family of intersection graphs. For example, our open question from the previous section has

$\mathcal{P} = \{\text{planar graphs}\}$ and $\mathcal{Q} = \Omega(\Sigma)$, where $\Sigma = \{\text{line segments in the plane}\}$,
and we want to prove that $\mathcal{P} \Rightarrow \mathcal{Q}$.

To prove such an assertion, one must show that all graphs with property $\mathcal{P}$ have a Σ-representation, and typical proofs are constructive. However, to *disprove* an assertion of the form $\mathcal{P} \Rightarrow \Omega(\Sigma)$, we need to show that some graph in $\mathcal{P}$ does *not* have a Σ-representation. Such irrepresentability results can be tricky to prove (see [29]). Another technique is to show that there are 'more' graphs in $\mathcal{P}$ than there are in $\Omega(\Sigma)$. Typically, these sets are both infinite, and so we need to be a little more careful.

Let $\mathcal{P}(n)$ be the number of labelled graphs on n vertices with property $\mathcal{P}$. For example, if $\mathcal{P}$ is the property 'is a tree', then Cayley's Theorem states that $\mathcal{P}(n) = n^{n-2}$. If $\mathcal{P}$ is the property 'is a cycle', then $\mathcal{P}(n) = n!/(2n)$, for $n \geq 3$.

Generally, we do not need to compute $\mathcal{P}(n)$ exactly—indeed, this is usually quite difficult—but need only find a general approximation. We use the big-oh order notation: if g is a positive function, then $f(n) = O(g(n))$ means that $f(n) \leq Cg(n)$ for some constant C and all sufficiently large n, and $f(n) = LO(g(n))$ means that $\log f(n) = O(\log g(n))$; 'O' stands for 'order' and 'LO' stands for 'logarithmic order'.

Now let $\mathcal{P}$ be the property 'is a string graph' and let $\mathcal{Q}$ be the property 'is an intersection graph of line segments in the plane'; clearly, $\mathcal{Q} \Rightarrow \mathcal{P}$. To prove that $\mathcal{P} \not\Rightarrow \mathcal{Q}$, we recall that all the split graphs have property $\mathcal{P}$. There are at least $2^{n^2/4}$ split graphs on n labelled vertices, and hence $\mathcal{P}(n) \geq 2^{n^2/4}$. If every string graph were a line segment graph, then we should also have $\mathcal{Q}(n) \geq 2^{n^2/4}$. We show, however, that $\mathcal{Q}(n) = LO(n^{4n})$, and therefore $\mathcal{Q}(n) < \mathcal{P}(n)$ for large n; this is a contradiction.

The problem is: how do we bound $\mathcal{Q}(n)$? A line segment intersection representation for a graph G with property $\mathcal{Q}$ can be thought of as a list of n line segments. Each line segment can be specified by the coordinates of its end-points (a, b) and (c, d). The entire representation can thus be encoded as a vector of length $4n$, such as $\mathbf{x} = (a_1, b_1, c_1, d_1, \ldots, a_n, b_n, c_n, d_n)$.

Given such a vector $\mathbf{x}$, we can recover the graph G. To check whether ij is in $E(G)$, we look at the numbers $a_i, b_i, c_i, d_i, a_j, b_j, c_j, d_j$. To determine whether the line segments cross, we define

$$p_{ij}(\mathbf{x}) = \det\left(\begin{pmatrix} 1 & a_i & b_i \\ 1 & c_i & d_i \\ 1 & a_j & b_j \end{pmatrix} \begin{pmatrix} 1 & a_i & b_i \\ 1 & c_i & d_i \\ 1 & c_j & d_j \end{pmatrix} \right).$$

Then the line segments assigned to the vertices i and j cross if and only if $p_{ij}(\mathbf{x}) \leq 0$ and $p_{ji}(\mathbf{x}) \leq 0$, since each determinant tells us that the end-points of one segment lie on opposite sides of the line determined by the other segment. Thus, given the vector $\mathbf{x}$, we can reconstruct G by computing several polynomial functions and checking their signs. If two vectors $\mathbf{x}$ and $\mathbf{y}$ give the *same* sign patterns when we evaluate these polynomials, then they must encode the same graph.

Thus an upper bound on the number of sign patterns for this family of polynomials yields an upper bound on the number of graphs with property $\mathcal{Q}$. More formally, let $q_1, q_2, \ldots, q_r$ be polynomial functions from $\mathbf{R}^\ell$ to $\mathbf{R}$, and let

$\sigma(\mathbf{x}) = (\text{sgn}(q_1(\mathbf{x})), \ldots, \text{sgn}(q_r(\mathbf{x})))$. Let N denote the number of different sign patterns for the q_i—that is,

$$N = \left|\{\sigma(\mathbf{x}) : \mathbf{x} \in \mathbf{R}^\ell\}\right|.$$

Since each entry of the r-vector $\sigma(\mathbf{x})$ can have one of three possible values ($+1$, -1 or 0), we have $N \leq 3^n$. This, however, does not take advantage of the fact that the q_i are *polynomials*. Consequently, we can obtain a much tighter bound (see [39] and [1]).

Theorem 10.13 *Let $q_1, \ldots, q_r : \mathbf{R}^\ell \to \mathbf{R}$ be polynomials of degree at most d. If $r \geq \ell$, then the number N of different sign patterns for these polynomials is bounded by*

$$N \leq \left(\frac{8edr}{\ell}\right)^\ell. \tag{*}$$

Let us apply this to the class $\Omega(\{\text{line segments}\})$ to obtain an upper bound on $\mathcal{Q}(n)$. The graphs with property $\mathcal{Q}$ are represented as $4n$-vectors. We take $\ell = 4n$, and examine the signs of $r = 2\binom{n}{2}$ polynomials, each of degree 4. By Theorem 10.13, we have

$$\mathcal{Q}(n) \leq N \leq \left(\frac{8e \cdot 4 \cdot 2\binom{n}{2}}{4n}\right)^{4n} = (O(1)n)^{4n} = LO(n^{4n}),$$

as required.

More generally, let Σ be a family of geometrical sets, and let k be a positive integer. We say (see [1]) that the sets in Σ admit k *degrees of freedom* provided that each set in Σ can be uniquely specified using k real numbers, and that we can tell if two sets in Σ intersect just by checking the signs of polynomials computed from their $2k$ parameters. Arguing as above, we have the following result.

Theorem 10.14 *Let Σ be a family of sets that admits k degrees of freedom, and let $\mathcal{P} = \Omega(\Sigma)$. Then $\mathcal{P}(n) = LO(n^{kn})$.*

Earlier, we noted that there are graphs with unbounded sphericity. To prove this by contradiction, suppose that all graphs have sphericity at most d; in other words, let $\Sigma = \{\text{balls in } \mathbf{R}^d\}$ and suppose that $\mathcal{P} = \Omega(\Sigma)$ contains all graphs; then $\mathcal{P}(n) = 2^{\binom{n}{2}}$. Now a ball in $\mathbf{R}^d$ can be described by $d+1$ parameters, corresponding to its centre $\mathbf{x}$ and radius r, and the balls $B(\mathbf{x}, r)$ and $B(\mathbf{y}, s)$ intersect if and only if

$$(\mathbf{x} - \mathbf{y}) \cdot (\mathbf{x} - \mathbf{y}) - (s - r)^2 \leq 0;$$

we simply check the sign of a polynomial. Thus, by Theorem 10.14, we have

$$2^{\binom{n}{2}} = \mathcal{P}(n) = LO(n^{(d+1)n}).$$

This contradiction establishes the result.

References

1. N. Alon and E. R. Scheinerman, Degrees of freedom versus dimension for containment orders, *Order* **5** (1988), 11–16; *MR* **89g**: 06002.
2. S. Benzer, On the topology of the genetic fine structure, *Proc. Nat. Acad. Sci. U.S.A.* **45** (1959), 1607–1620.
3. K. S. Booth and G. S. Lueker, Linear algorithms to recognize interval graphs and test for the consecutive ones property, *Proc. Seventh ACM Sympos. Theory of Computing*, ACM Press, 1975, pp. 255–265; *MR* **55**#1777.
4. K. S. Booth and G. S. Lueker, Testing for the consecutive ones property, interval graphs, and graph planarity using PQ-tree algorithms, *J. Comput. System Sci.* **13** (1976), 335–379; *MR* **55**#6932.
5. G. R. Brightwell and E. R. Scheinerman, Representations of planar graphs, *SIAM J. Discrete Math.* **6** (1993), 214–229; *MR* **95d**: 05043.
6. M.B. Cozzens, Higher and multi-dimensional analogues of interval graphs, PhD Thesis, Rutgers University, 1981.
7. G. A. Dirac, On rigid circuit graphs, *Abh. Math. Sem. Univ. Hamburg* **25** (1961), 71–76; *MR* **24**#A57.
8. G. Ehrlich, S. Even and R. E. Tarjan, Intersection graphs of curves in the plane, *J. Combin. Theory (B)* **21** (1976), 8–20; *MR* **58**#21842.
9. P. Erdős, A. Goodman and L. Pósa, The representation of a graph by set intersections, *Canad. J. Math.* **18** (1966), 106–112; *MR* **32**#4034.
10. P. C. Fishburn, *Interval Orders and Interval Graphs: A Study of Partially Ordered Sets*, Wiley, 1985; *MR* **86m**: 06001.
11. D. R. Fulkerson and O. A. Gross, Incidence matrices and interval graphs, *Pacific J. Math.* **15** (1965), 835–855; *MR* **32**#3881.
12. F. Gavril, The intersection graphs of subtrees in trees are exactly the chordal graphs, *J. Combin. Theory (B)* **16** (1974), 47–56; *MR* **48**#10868.
13. P. C. Gilmore and A. J. Hoffman, A characterization of comparability graphs and of interval graphs, *Canad. J. Math.* **16** (1964), 539–548; *MR* **31**#87.
14. M. C. Golumbic, *Algorithmic Graph Theory and Perfect Graphs*, Academic Press, 1980; *MR* **81e**: 68081.
15. G. Hajós, Über eine Art von Graphen, *Internat. Math. Nachr.* **11** (1957), 65.
16. I. Ben-Arroyo Hartman, I. Newman and R. Ziv, On grid intersection graphs, *Discrete Math.* **87** (1991), 41–52; *MR* **92a**: 5042.
17. E. Helly, Über Mengen Körper mit gemeinschaftlichen Punkten, *Jahresber. Deutsch. Math.-Verein* **32** (1923), 175–176.
18. R. L. Hemminger and L. W. Beineke, Line graphs and line digraphs, *Selected Topics in Graph Theory* (ed. L. W. Beineke and R. J. Wilson), Academic Press, 1978, pp. 271–305.
19. M. Katchalski and A. C. F. Liu, Intersection patterns of families of convex sets, *Canad. J. Math.* **34** (1982), 921–931; *MR* **84i**: 52011.
20. V. Klee, What are the intersection graphs of arcs in a circle?, *Amer. Math. Monthly* **76** (1969), 810–813.

21. J. Kratochvíl and J. Matoušek, String graphs requiring exponential representation, *J. Combin. Theory (B)* **53** (1991), 1–4; *MR* **92g**: 05080.
22. C. G. Lekkerkerker and J. Ch. Boland, Representation of a finite graph by a set of intervals on the real line, *Fund. Math.* **51** (1974), 45–64; *MR* **25**#2596.
23. E. Marczewski, Sur deux propriétés des classes d'ensembles, *Fund. Math.* **33** (1945), 303–307; *MR* 7-420.
24. M. Middendorf and F. Pfeiffer, Weakly transitive orientations, Hasse diagrams and string graphs, *Discrete Math.* **111** (1993), 393–400; *MR* **93j**: 05071.
25. W. F. Ogden and F. S. Roberts, Intersection graphs of families of convex sets with distinguished points, *Combinatorial Structures and Their Applications* (ed. R. Guy et al.), Gordon and Breach, 1970, pp. 311–313.
26. P. L. Renz, Intersection representation of graphs by arcs, *Pacific J. Math.* **34** (1970), 501–510; *MR* **42**#5839.
27. F. S. Roberts, On the boxicity and cubicity of a graph, *Recent Progress in Combinatorics* (ed. W. T. Tutte), Academic Press, 1969, pp. 301–310; *MR* **40**#5489.
28. E. R. Scheinerman, Intersection classes and multiple intersection parameters of graphs, PhD Thesis, Princeton University, 1984.
29. E. R. Scheinerman, Irrepresentability by multiple intersection, or why the interval number is unbounded, *Discrete Math.* **55** (1985), 195–211; *MR* **86k**: 05100.
30. F. W. Sinden, Topology of thin film RC circuits, *Bell Sys. Tech. J.* (1966), 1639–1662.
31. C. Thomassen, Interval representations of planar graphs, *J. Combin. Theory (B)* **40** (1986), 9–20; *MR* **87d**: 05071.
32. W. Thurston, *The Geometry and Topology of Three-Manifolds*, manuscript.
33. A. C. Tucker, Characterizing circular-arc graphs, *Bull. Amer. Math. Soc.* **76** (1970), 1257–1260; *MR* **43**#1877.
34. A. C. Tucker, Matrix characterizations of circular-arc graphs, *Pacific J. Math.* **39** (1971), 535–545; *MR* **46**#8915.
35. A. C. Tucker, Structure theorems for some circular-arc graphs, *Discrete Math.* **7** (1974), 167–195; *MR* **52**#203.
36. A. C. Tucker, An efficient test for circular-arc graphs, *SIAM J. Comput.* **9** (1980), 1–24; *MR* **81a**: 68074.
37. J. Urrutia, Intersection graphs of some families of plane curves, PhD Thesis, University of Waterloo, 1980.
38. J. R. Walter, Representations of chordal graphs as subtrees of a tree, *J. Graph Theory* **2** (1978), 265–267; *MR* **58**#21868.
39. H. E. Warren, Lower bounds for approximation by nonlinear manifolds, *Trans. Amer. Math. Soc.* **133** (1968), 167–178; *MR* **37**#1871.
40. G. Wegner, Eigenschaften der Nervan homologische-einfacher Familiern im R^n, PhD Thesis, University of Göttingen, 1967.

11
Topology

LOWELL W. BEINEKE

The primary focus of this chapter involves putting graphs on surfaces. Questions that we discuss include the following. Which graphs are planar; that is, which graphs can be drawn in the plane without any edges crossing? If a graph is not planar, what is the smallest number of crossings in any drawing of it? How many planar graphs are needed to form a given graph? In what surfaces can a non-planar graph be embedded?

11.1 Introduction

There are many links between graph theory and topology, but the strongest is that of drawings and embeddings of graphs on surfaces. Our survey of this area of mathematics is divided into two parts, the first (Sections 11.2–11.4) on graphs in the plane, including the topics of crossing number and thickness, and the second (Sections 11.5–11.8) on embeddings in other surfaces, both orientable and non-orientable.

In Section 11.2, we concentrate on basic results on planarity, including *Euler's Polyhedron Formula* and *Kuratowski's Theorem*. This is followed by discussions of the *thickness* of a graph (the minimum number of planar graphs into which it can be decomposed) and the *crossing number* (the number of crossings it must have when drawn in the plane).

The *genus* of a graph is the number of handles that must be added to a sphere in order for there to be an embedding of the graph. In Sections 11.5 and 11.6 we explore this concept, along with its connection with map-colouring problems. Embeddings on non-orientable surfaces are discussed in Section 11.7. Because general results other than bounds are rare in this area, we focus considerable attention on complete graphs and complete bipartite graphs in Sections 11.3–11.7. In the final section, we return to the ideas involved in Kuratowski's Theorem and discuss its analogues for other surfaces.

In order to keep the number of references down, we frequently give a secondary source rather than the original. In particular, the excellent book *Graph Theory 1736-1936* by Biggs, Lloyd and Wilson [6] is given as the reference for early work on topological graph theory.

11.2 Planar graphs

When one is drawing a graph, it is natural to ask whether it can be drawn in the plane with no two of the edges crossing each other. It is not difficult to convince oneself that the complete graph K_5 and the complete bipartite graph $K_{3,3}$ in Fig. 11.1 cannot be drawn in this way. Surprisingly, these constitute the only real obstacles to crossing-free drawings, a fact first proved in 1930 by Kuratowski [16]. Interestingly, this theorem was announced independently by Frink and Smith [8] in the same year.

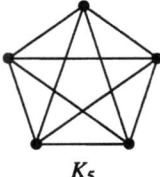 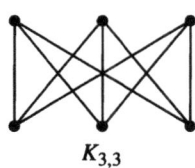

K_5 $K_{3,3}$

FIG. 11.1

In a *drawing* of a graph G in the plane, each vertex becomes a point in the plane and each edge becomes a simple curve joining the two points corresponding to its vertices. If the only points at which two edges meet is an end-vertex of both, then the drawing is an *embedding*. A graph that can be embedded in the plane is *planar*.

Kuratowski's Theorem gives an elegant theoretical criterion for determining whether a graph is planar. In order to state it formally, we need another definition. Two graphs are *homeomorphic* if there is some graph from which both can be obtained by subdividing edges (see Chapter 1). Informally, this means that one can get from one graph to the other by suppressing or inserting vertices of degree 2 in the edges. Two homeomorphic graphs are shown in Fig. 11.2.

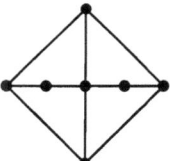 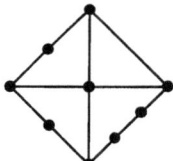

FIG. 11.2

Theorem 11.1 (Kuratowski's Theorem) *A graph is planar if and only if it has no subgraph homeomorphic to K_5 or $K_{3,3}$.*

Several other criteria for planarity are known, but we mention only one, a dual result of Theorem 11.1 due first to Wagner [25]. Again, we need a definition for its statement. One graph is *contractible* to another if the second can be obtained from the first through edge-contractions. For example, Fig. 11.3 shows that the Petersen graph is contractible to K_5, although it has no subgraph homeomorphic to K_5.

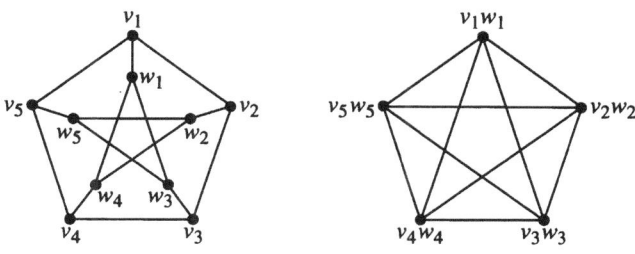

FIG. 11.3

Theorem 11.2 *A graph is planar if and only if it has no subgraph contractible to K_5 or $K_{3,3}$.*

When the vertices and edges of a graph embedded in the plane or on a sphere are removed, the maximal connected sets that remain are the *regions* (or *faces*) of the embedded graph. The number of regions is related to the numbers of vertices and edges of the graph by Euler's Polyhedron Formula. Euler presented his result in a letter to Goldbach in 1750, but he never succeeded in proving it. Early proofs of this result, which might well be called the *Fundamental Theorem of Topological Graph Theory*, were given by Legendre in 1794, l'Huilier in 1811 and 1812, and Cauchy in 1813 (see [6]).

Theorem 11.3 (Euler's Polyhedron Formula) *If a plane embedding of a connected graph has n vertices, m edges and r regions, then $n - m + r = 2$.*

For later reference we note a couple of elementary consequences of this formula. The first tells us that a planar graph has 'not too many' edges, and the second that there is a vertex of low degree.

Corollary 11.4 *If G is a planar graph with n (≥ 3) vertices, then*

(a) G has at most $3n - 6$ edges;
(b) G has at most $2n - 4$ edges, if it is bipartite;
(c) G has at most $g(n - 2)/(g - 2)$ edges, if it has girth g.

Corollary 11.5 *In every planar graph, some vertex has degree 5 or less.*

Testing a graph for planarity is not difficult. In fact, Hopcroft and Tarjan [14] gave an algorithm that is linear in the order of the graph, not only for determining whether a graph is planar, but for finding an embedding if it is. For more on other aspects of planar graphs, such as duality and other tests for planarity, see [27]. To conclude this section, we note that in 1922 E. Steinitz answered the question of when a graph is the skeleton of some polyhedron.

Theorem 11.6 (Steinitz' Theorem) *A graph is polyhedral if and only if it is planar and 3-connected.*

11.3 Thickness

Planarity is important if shorting is to be avoided on a circuit board. An interesting related question is to determine how many different circuit boards with crossing-free drawings are needed to print a given non-planar network. For example, as shown in Fig. 11.4, the complete graph K_8 would require only two boards.

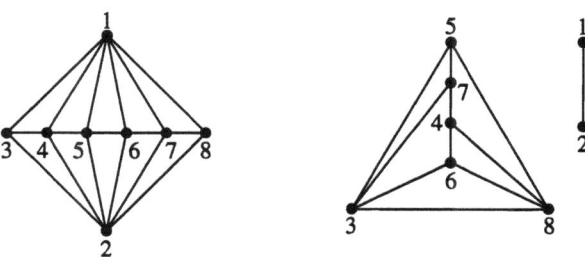

FIG. 11.4

The *thickness* $t(G)$ of a graph G is the minimum number of planar graphs whose union is G. We note that although planarity is invariant under homeomorphism of graphs, thickness is not. In fact, if each edge of a graph is replaced by a path of length 3, the result is the union of two forests, one with the middle edges of the paths, the other a collection of stars. Thus every non-planar graph is homeomorphic to a graph of thickness 2.

Mansfield [17] showed that determining thickness is NP-hard, and so our discussion will focus on general bounds and the thickness of specific graphs. We begin with two sharp lower bounds that follow from Corollary 11.4(a) and (b).

Theorem 11.7 *If G is a graph with n (≥ 3) vertices and m edges, then*

$$t(G) \geq \left\lceil \frac{m}{3n-6} \right\rceil. \tag{11.1}$$

Furthermore, if G is bipartite, then

$$t(G) \geq \left\lceil \frac{m}{2n-4} \right\rceil. \tag{11.2}$$

Complete graphs For the complete graph K_n, inequality (11.1) can be rewritten in the form $t(K_n) \geq \lfloor (n+7)/6 \rfloor$, since $\lceil p/q \rceil = \lfloor (p+q-1)/q \rfloor$ for any positive integers p and q. Now K_5 is non-planar and K_8 is the union of two planar graphs, and so equality holds for all $n \leq 8$. However, Battle, Harary and Kodama showed in 1962 that the inequality is strict for $n = 9$ and 10. It is somewhat surprising, then, that equality does hold for all other values of n. Except for $n \equiv 4 \pmod{6}$, this was shown by Beineke and Harary in 1965 (see [2]); the remaining case was established in 1976, independently by V. B. Alekseev and V. S. Gonchakov and by J. M. Vasak (see [27] for references).

Theorem 11.8 *The thickness of the complete graph is*

$$t(K_n) = \left\lfloor \frac{n+7}{6} \right\rfloor, \quad \text{except that } t(K_9) = t(K_{10}) = 3.$$

The complete proof of this result is too lengthy to present here, but includes an interesting decomposition of the generalized octahedron, the complete graph K_{6r} with a perfect matching removed. It consists of r isomorphic triangulations of the plane based on the octahedron; Fig. 11.5 shows the triangulation for $r = 3$. The decomposition of K_{6r+4} into $r + 1$ planar graphs is then obtained by appropriately adding four more vertices to each of these r graphs, making a few adjustments, and then constructing one more planar graph.

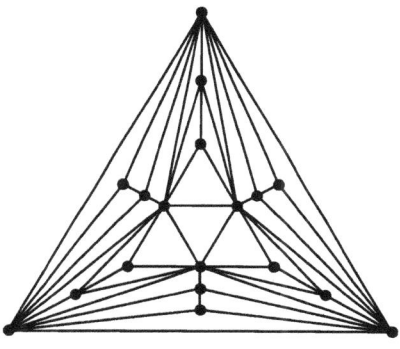

FIG. 11.5

For complete bipartite graphs, there are no known exceptions to the bound given by inequality (11.2), but the problem has not been entirely solved. Beineke, Harary and Moon [3] found the answer for 'most' graphs in the family.

Theorem 11.9 *For $r \leq s$, the thickness of the complete bipartite graph is*

$$t(K_{r,s}) = \left\lceil \frac{rs}{2(r+s-2)} \right\rceil,$$

except possibly when r and s are both odd and there exists an integer k satisfying $(r+5)/4 \leq k \leq (r-3)/2$ for which $s = \lfloor 2k(r-2)/(r-2k) \rfloor$.

The unknown cases are relatively rare: if r is even, there are none, and if r is odd, there are at most $(r-7)/4$. For $r \leq 30$, there are just six graphs $K_{r,s}$ whose thickness remains unknown: $K_{19,29}$, $K_{19,47}$, $K_{23,27}$, $K_{25,59}$, $K_{27,71}$ and $K_{29,129}$. No progress has been made on this problem in the past thirty years.

A colouring problem Much of the interest in topological graph theory was inspired by the four colour problem and its extension to other surfaces (see later sections of this chapter). The *Earth–Moon map-colouring problem* is one that is related to graph thickness. It can be stated as follows. *Each country in some set* on the Earth can have a lunar colony. If, on maps of the two surfaces, each colony is to be coloured the same as its parent country, and if neighbouring colonies as well as neighbouring countries must have different colours, how many colours are needed?

Given such a pair of maps, we consider the graph whose vertices correspond to the countries, with two vertices being adjacent if the corresponding countries are neighbours on the Earth or their colonies are neighbours on the Moon. Thus we are looking at the chromatic number of graphs of thickness at most 2, which we call *biplanar graphs*. By Corollary 11.4(a), a biplanar graph of order n has at most $6n - 12$ edges. Consequently, some vertex has degree at most 11, and so it follows by induction that the chromatic number of such a graph never exceeds 12. On the other hand, T. Sulanke discovered that $K_{11} - C_5$ is biplanar and has chromatic number 9. Since K_9 itself is not biplanar, this is a rarity in map-colouring problems—a situation in which a complete graph is not optimal.

It is interesting to note that if there are more than two satellites—that is, when each country can have a region on each of k spheres—then the complete graph thickness theorem (together with the Euler-type upper bound) cuts down the number of possibilities for the chromatic number by 1.

Theorem 11.10 *The maximum chromatic number among all graphs of thickness k is*

$$\begin{array}{ll} 4, & \text{if } k = 1; \\ 9, 10, 11 \text{ or } 12, & \text{if } k = 2; \\ 6k-2, 6k-1 \text{ or } 6k, & \text{if } k \geq 3. \end{array}$$

11.4 Crossing numbers

In this section we consider the problem of determining the minimum number of crossings of edges in any drawing of a given graph G, called its *crossing number* $\nu(G)$. For planar graphs, this number is of course 0, and it is easy to draw K_5 and $K_{3,3}$ with one crossing each. The drawings in Fig. 10.6 show that $\nu(K_6) \leq 3$ and $\nu(K_{4,4}) \leq 4$. Equality actually holds in both cases, although the reverse inequalities are more difficult to establish. This situation is typical of crossing number problems—the greatest difficulty consists in showing that one can do no better than a particular known construction. It was shown by Garey and Johnson [9] that determining the crossing number is an NP-complete problem.

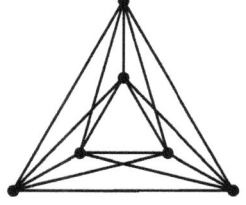
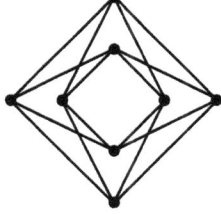

Fig. 11.6

We note a couple of points on conventions in drawings. For obvious reasons, an edge is not allowed to pass through a vertex. Normally, we also do not allow more than two edges to pass through a given point, but in describing a drawing it may be convenient to permit this. It presents no intrinsic difficulty provided that we count each pair of edges through the point as contributing a different crossing.

Complete bipartite graphs The study of the crossing number of the complete bipartite graph goes back to *Turán's brick factory problem*. There are a set of ovens for baking bricks and a set of storage sites where bricks are kept, and there is a track from each oven to each store. The problem is to minimize the number of times that tracks cross or, in terms of graphs, to find the crossing number of the corresponding complete bipartite graph.

Except when one of the partite sets is small, this 50-year-old problem remains open. The best known upper bound for $\nu(K_{r,s})$ is achieved in a simple drawing. In the Cartesian plane, place r vertices on the x-axis so that, as nearly as possible, half are on the negative side and half on the positive side, and then do the same for s vertices on the y-axis. Now join each pair of vertices on different axes by a line segment (see Fig. 11.7 for the case $r = 5$ and $s = 7$). This gives a total of

$$\left\lfloor \frac{r}{2} \right\rfloor \left\lfloor \frac{s}{2} \right\rfloor + \left\lfloor \frac{r}{2} \right\rfloor \left\lceil \frac{s}{2} \right\rceil + \left\lceil \frac{r}{2} \right\rceil \left\lfloor \frac{s}{2} \right\rfloor + \left\lceil \frac{r}{2} \right\rceil \left\lceil \frac{s}{2} \right\rceil$$

crossings, which can be simplified to the expression given in the next theorem.

Theorem 11.11 *The crossing number of the complete bipartite graph satisfies the inequality*

$$\nu(K_{r,s}) \leq \left\lfloor \frac{r}{2} \right\rfloor \left\lfloor \frac{r-1}{2} \right\rfloor \left\lfloor \frac{s}{2} \right\rfloor \left\lfloor \frac{s-1}{2} \right\rfloor.$$

Kleitman [15] proved that equality holds when the smaller set has at most six vertices. Thus the smallest open case is $K_{7,7}$ and its crossing number is known to be 77, 79 or 81.

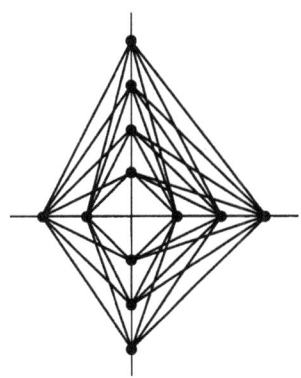

Fig. 11.7

Complete graphs Here too, the problem remains open except for a few cases, and again the main result is an upper bound obtained by a construction. The one that we give is due to Blažek and Koman [7]; we describe it in terms of a drawing on a tin can, which of course has its equivalent on a sphere or in the plane.

Assume first that n is even, say $n = 2k$, and place k vertices equally spaced on each rim of a can. If two vertices are on the same end, join them by a line segment across that end, while if they are on different ends, join them by a shortest helical curve around the side. The construction for K_8, where $k = 4$, is shown in flattened-out form in Fig. 11.8.

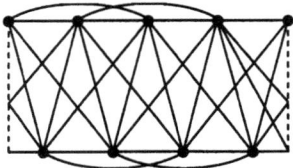

FIG. 11.8

On each end, there is precisely one crossing for each set of four vertices, giving $2\binom{k}{4}$ crossings on the two ends. It is somewhat more difficult to count the lateral crossings, of which there are $k\binom{k}{3}$, and we omit the details of that argument. The total number of crossings in this drawing of K_{2k} is thus equal to $\binom{k}{2}\binom{k-1}{2}$.

For the odd case, say $n = 2k - 1$, we delete one vertex from the case $n = 2k$. The resulting drawing has $\binom{k-1}{2}^2$ crossings. The two expressions can be combined into the form given in the following theorem.

Theorem 11.12 *The crossing number of the complete graph satisfies the inequality*

$$\nu(K_n) \leq \frac{1}{4} \left\lfloor \frac{n}{2} \right\rfloor \left\lfloor \frac{n-1}{2} \right\rfloor \left\lfloor \frac{n-2}{2} \right\rfloor \left\lfloor \frac{n-3}{2} \right\rfloor,$$

with equality if $n \leq 10$.

The following table gives the known values of $\nu(K_n)$.

n	4	5	6	7	8	9	10
$\nu(K_n)$	0	1	3	9	18	36	60

Guy [12] has noted that $\lim_{n \to \infty} \nu(K_n)/n^4$ exists and lies between $\frac{1}{80}$ and $\frac{1}{64}$, and consequently the bound in the theorem is of the right order of magnitude.

Products of cycles Interest in the crossing numbers of the products of cycles originated with the observation of Harary, Kainen and Schwenk [13] that these graphs can all be embedded on a torus, but their crossing numbers are unbounded. They conjectured that if $r \leq s$, then $\nu(C_r \times C_s) = (r-2)s$. It is easy to see that this number is an upper bound (see Fig. 11.9 for the case $C_4 \times C_5$), but equality has been established only for $r \leq 5$. For $r = 3$ and 4, the results were obtained by Beineke and Ringeisen [4] using the fact, first established by R. B. Eggleton, that $C_4 \times C_4$ (the 4-dimensional cube) has crossing number 8. The crossing number of $C_5 \times C_5$ was recently found by Richter and Thomassen [18], and this has been extended by Klešč, Richter and Stobert to arbitrary $C_5 \times C_s$. Apart from variations of the graphs discussed here, we are not aware of any large crossing numbers that are known exactly.

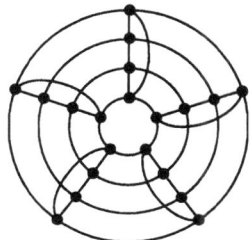

FIG. 11.9

Theorem 11.13 *For $r \leq s$, the crossing number of the product of two cycles satisfies the inequality*

$$\nu(C_r \times C_s) \leq (r-2)s,$$

with equality if $r \leq 5$.

Linear drawings In 1936, K. Wagner proved that every planar graph can be embedded in the plane with each edge being a straight line segment; this result frequently goes by the name of *Fáry's Theorem* because of an independent discovery.

We say that a drawing of a graph G in the plane is *linear* if all of the edges are segments, and we define the *linear crossing number* $\overline{\nu}(G)$ to be the minimum number of crossings in such a drawing. The drawing of $K_{r,s}$ in Fig. 11.7 is linear, while that of K_n in Fig. 11.8 is not. In fact, $\overline{\nu}(K_n) = \nu(K_n)$ for $n \leq 7$, but $\overline{\nu}(K_8) = \nu(K_8) + 1$. Curiously, however, $\overline{\nu}(K_9) = \nu(K_9)$, but that is the last time equality holds: $\nu(K_{10}) = 60$, while $\overline{\nu}(K_{10}) = 61$ or 62, and for each $n \geq 11$ it is known that $\overline{\nu}(K_n) > \nu(K_n)$, although none of the values is known exactly (see Guy [12]).

For graphs in general, Bienstock and Dean [5] extended the Wagner–Fáry Theorem about as far as it can go. It is surprising that there is so great a difference between the behaviour of the linear crossing numbers of the graphs with $\nu(G) = 3$ and those with $\nu(G) = 4$.

Theorem 11.14 *If G is a graph for which $\nu(G) \leq 3$, then $\overline{\nu}(G) = \nu(G)$, but for $k \geq 4$, there are graphs G with $\nu(G) = k$ and $\overline{\nu}(G)$ arbitrarily large.*

11.5 Orientable surfaces and rotation systems

If a graph is not planar, it is natural to ask about its embeddability on other surfaces, and that is the main subject of the second half of this chapter. We begin by considering orientable compact 2-manifolds. *Compact 2-manifolds* are surfaces with the following two properties:

each point has a neighbourhood homeomorphic to an open disc;

every covering of the surface with open discs has a finite subcovering.

Such a surface is called *orientable* if the following property also holds:

a clockwise sense of rotation can be assigned consistently around all points.

The simplest examples of such surfaces are the sphere and the torus; the plane itself does not qualify because it is not compact. A less formal description is that these are the surfaces obtainable by adding handles to a sphere. Brahana's classic theorem (see [27]) implies that when handles are added to the sphere, the result in effect depends only on the number added and not on their placement. Consequently, we speak of *the orientable surface S_h with h handles*; the number h is called the *genus* of the surface. Figure 11.10 depicts the double-torus S_2, both as customarily drawn and as a sphere with two handles.

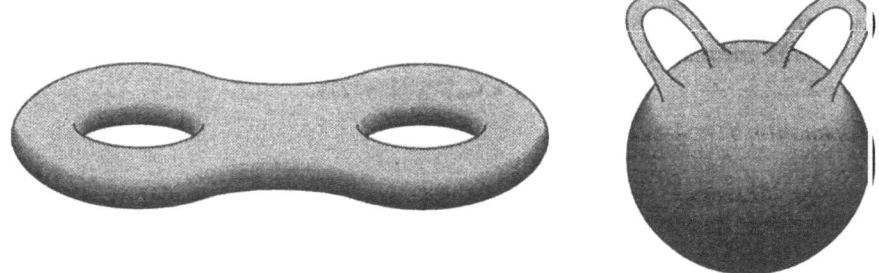

FIG. 11.10

In working with these surfaces, we frequently use a diagrammatic representation of a $4h$-gon in which the sides are identified according to the pattern

$$\alpha_1, \beta_1, \alpha_1^{-1}, \beta_1^{-1}, \ldots, \alpha_h, \beta_h, \alpha_h^{-1}, \beta_h^{-1}.$$

Figure 11.11 shows the torus and double-torus represented in this way.

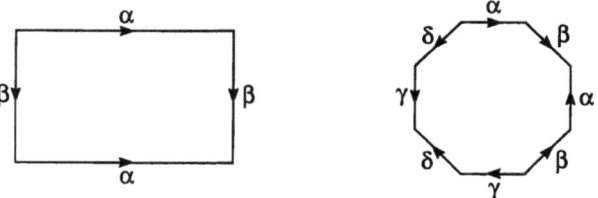

FIG. 11.11

As in the case of the plane, in an *embedding* of a graph G in a surface S, each vertex of G becomes a point of S and each edge becomes a simple curve joining the corresponding points, and the curves do not meet except at their ends. It is not difficult to see that any graph can be embedded in a surface with enough handles—simply add a handle wherever two edges in a drawing cross.

The *regions* of an embedding of a graph are also defined as for the plane; they are the components that remain if the vertices and edges of the embedded graph are deleted from the surface. An embedding is called *cellular* (or a *2-cell embedding*) if each region is homeomorphic to an open disc. In Fig. 11.12 we show three embeddings of K_4 in the torus, each represented in two ways; only the last of the three is cellular.

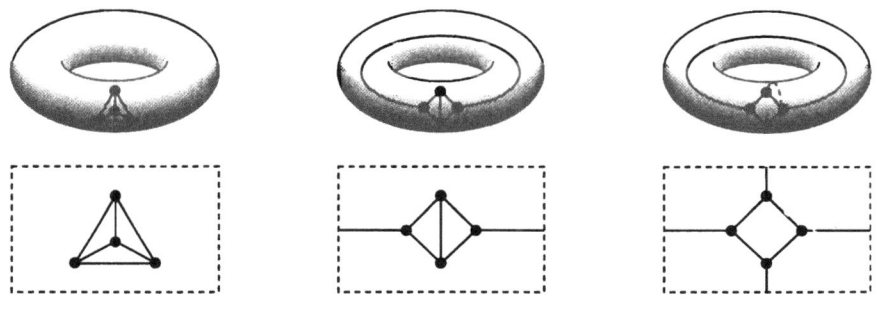

FIG. 11.12

It is interesting that each orientable surface has its own polyhedron formula, a result first published by l'Huilier in 1812/13 (see [6]).

Theorem 11.15 *If an embedding of a graph G in the orientable surface S_h is cellular and has n vertices, m edges and r regions, then*

$$n - m + r = 2 - 2h.$$

Consider now a cellular embedding of a graph G in an orientable surface. We can describe the embedding by listing the edges encountered in going around each region in a clockwise direction, keeping the region on the right.

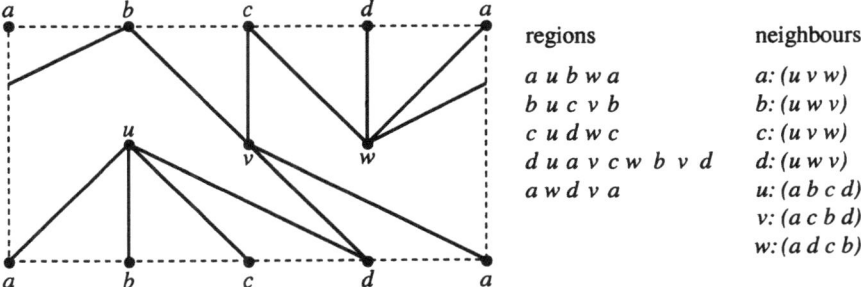

FIG. 11.13

For example, the embedding of $K_{3,4}$ on the torus depicted in Fig. 11.13 has the regions as listed. Note that each edge appears twice, once in each direction.

The embedding can also be described by listing the neighbours of each vertex in the order that the edges to them are encountered while going anticlockwise around the vertices. The regions can then be recovered from these neighbour lists, treated as cyclic permutations, by applying the following *rotation rule*: after the edge xy, take the edge yz, where z is the successor to x in the permutation of y's neighbours.

If in our example we begin by going from a to u, we must then go from u to b since b follows a in u's permutation. Then, in b's permutation, w follows u, so we take the edge from b to w. Continuing in this way, we get the boundary of the first region, and eventually we obtain the entire set. Note that vertices can occur more than once on a region, and so it may be necessary to duplicate an edge in order to guarantee that one has finished the boundary of a region.

The gist of the next theorem is that, by means of the rotation rule, each set of cyclic permutations of the neighbours of the vertices of a given connected graph yields a cellular embedding of the graph, and all cellular embeddings can be obtained in this way. Such rotation schemes originated in the work of Heffter (see [7]).

Theorem 11.16 (Rotation Scheme Theorem) *Let G be a connected graph with n vertices and m edges, let $\Pi = \{\pi_1, \pi_2, \ldots, \pi_n\}$ be a set of cyclic permutations of the neighbours of the vertices of G, and let $W_1, W_2, \ldots, W_r$ be the closed walks obtained by applying the rotation rule to Π. Then the walks are the boundaries of the regions of a cellular embedding of G in the orientable surface S_h, where $h = (2 - n + m - r)/2$.*

Rotation schemes thus provide all possible embeddings of a graph. However, the work required rapidly becomes prohibitive for large graphs, since just a single vertex of degree d has $(d-1)!$ cyclic permutations of its neighbours. In the next section, we discuss the minimum genus among the surfaces in which a given graph is embeddable.

11.6 Genus and chromatic numbers

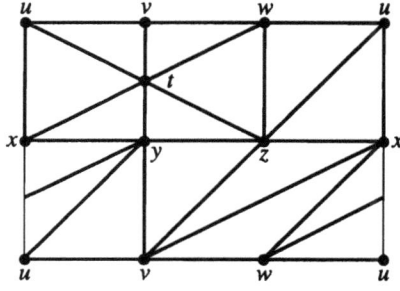

FIG. 11.14

The *genus* $\gamma(G)$ of a graph G is defined to be the minimum genus of any orientable surface in which G is embeddable. For example, since K_7 is not planar but is embeddable in the torus (as shown in Fig. 11.14), its genus is 1. Planar graphs are precisely the graphs of genus 0.

Although there are linear algorithms for planarity, such as that of Hopcroft and Tarjan [14], Thomassen [24] showed that determining the genus of a graph is NP-complete. Therefore, we are again restricted to considering bounds and specific families of graphs. The lower bounds given in the next result follow from Theorem 11.15; they are an important tool in determining the genus of graphs.

Theorem 11.17 *If G is a graph with n vertices and m edges, then*
(a) $\gamma(G) \geq \frac{1}{6}m - \frac{1}{2}n + 1$;
(b) $\gamma(G) \geq \frac{1}{4}m - \frac{1}{2}n + 1$, if G is bipartite;
(c) $\gamma(G) \geq \lceil (g-2)/2g \rceil m - \frac{1}{2}n + 1$, if G has girth g.

Complete graphs The question of the genus of the complete graph has its roots in the famous four colour problem. After demonstrating the flaw in Kempe's 'solution' to that problem, Heawood (see [6]) posed the corresponding problem for other surfaces—namely, how many colours are needed to colour all maps on S_h, given that neighbouring countries must have different colours? We shall return to this question shortly, but first we look at the genus of the complete graph.

It follows from Theorem 11.17(a) that $\gamma(K_n) \geq (n-3)(n-4)/12$ for $n \geq 3$. The form of this expression suggests that considering congruence classes modulo 12 might prove fruitful, and that is indeed what was done. The form also suggests that the cases when the fraction is an integer (that is, when $n \equiv 0, 3, 4$ or 7) might be the easiest to handle, since then the embeddings are triangulations. Curiously, however, the irregular case $n \equiv 5$ was the first to be settled; this occurred in 1954 and the solution was due to Gerhard Ringel. By 1968, all twelve cases had fallen. Various people were involved at different stages, but the bulk of the work was done by Ringel and Youngs [20].

Theorem 11.18 (Ringel–Youngs Theorem) *The genus of the complete graph K_n ($n \geq 3$) is*

$$\gamma(K_n) = \left\lceil \frac{(n-3)(n-4)}{12} \right\rceil.$$

The full proof appears in Ringel's book [19], and a summary of some of the main features has been given by White [26]. One of the primary techniques was the use of 'current graphs', so called because the arcs are labelled to satisfy Kirchhoff's Current Law. A dual approach, using 'voltage graphs', has since been developed and seems to be more useful in other situations. This theory is due to Gross and Tucker (see [11]).

As we have been following the trail of complete bipartite graphs as well as complete graphs, we note that Ringel proved the following result in 1965.

Theorem 11.19 *The genus of the complete bipartite graph is*

$$\gamma(K_{r,s}) = \left\lceil \frac{(r-2)(s-2)}{4} \right\rceil.$$

Chromatic numbers We now return to Heawood's map-colouring problem. The *chromatic number* $\chi(S)$ of a surface S is the minimum number of colours that suffice to colour all maps on S; equivalently, it is the maximum chromatic number of any graph embeddable in S. Although Heawood himself (see [6]) stated the value of $\chi(S_h)$, he actually proved it just for the torus. In general, he established it only as an upper bound, the argument running as follows. For $h > 0$, let

$$f(h) = \left\lfloor \frac{7 + \sqrt{1 + 48h}}{2} \right\rfloor.$$

Theorem 11.15 implies that every graph embeddable in S_h has a vertex of degree less than $f(h)$. That $\chi(S_h) \leq f(h)$ can then be deduced by induction on the order of S_h-embeddable graphs. The Ringel–Youngs Theorem provides what is needed for the reverse inequality. If K_n is embeddable in S_h, then clearly $\chi(S_h) \geq n$. By taking $n = f(h)$, we obtain the desired result. Thus, except for the sphere, the chromatic number of every orientable surface was known by 1968, eight years before the Appel–Haken proof of the Four Colour Theorem. Note that the same formula holds also for that case.

Theorem 11.20 *The chromatic number of the orientable surface S_h is*

$$\chi(S_h) = \left\lfloor \frac{7 + \sqrt{1 + 48h}}{2} \right\rfloor.$$

11.7 Non-orientable surfaces

We next consider embeddings of graphs in surfaces (compact 2-manifolds) that do not have the property of orientability. The simplest such surface is the projective plane, and its most convenient representation is as a closed disc with each pair of diametrically opposite points identified (see Fig. 11.15).

FIG. 11.15

The same idea is used in adding a *crosscap* to a surface: remove an open disc and identify the pairs of opposite points on the boundary. The Klein bottle in Fig. 11.16(a) is equivalent to the sphere with two crosscaps added.

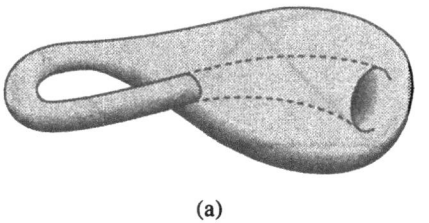
(a)

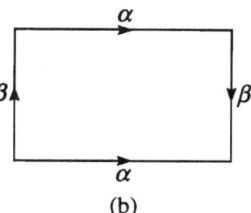
(b)

FIG. 11.16

There are many parallels in the theories of the two types of surfaces, as well as some important differences that we note later. Brahana's surface classification theorem includes the fact that every non-orientable surface can be obtained by adding some number $k \geq 1$ of crosscaps to the sphere; this is called the *non-orientable surface of genus k*, denoted by $\tilde{S}_k$.

No surface with a crosscap is orientable. In fact, if h handles and k crosscaps are added to the sphere (with $k > 0$), the result is the same as if $2h + k$ crosscaps were added. This has an interesting corollary: the addition of two crosscaps to a non-orientable surface is equivalent to adding one handle, but for orientable surfaces this is not so.

Non-orientable surfaces, like their orientable relatives, have convenient depictions as polygons: $\tilde{S}_k$ can be represented as a $2k$-gon with its sides labelled $\alpha_1, \alpha_1, \alpha_2, \alpha_2, \ldots, \alpha_k, \alpha_k$. Figure 11.16(b) shows the Klein bottle $\tilde{S}_2$ represented in the more traditional way of $\alpha\beta\alpha^{-1}\beta$.

We return now to the topic of graph embeddings, beginning with the non-orientable version of Euler's Polyhedron Formula, first published by Tietze in 1910 (see [6]).

Theorem 11.21 *If an embedding of a graph G in the non-orientable surface $\tilde{S}_k$ is cellular and has n vertices, m edges and r regions, then*

$$n - m + r = 2 - k.$$

Interestingly, even though the Klein bottle and the torus have the same polyhedron formula, K_7 is embeddable in the latter but not the former. This was first shown by P. Franklin in 1934, and it is the one exception to a general result that we present later.

The *crosscap number* (or *non-orientable genus*) $\tilde{\gamma}(G)$ of a graph G is defined to be the minimum value of k for which G is embeddable in $\tilde{S}_k$, with $\tilde{\gamma}(G)$ defined to be 0 if G is planar. Figure 11.17 shows K_6 on the projective plane and $K_{4,4}$ on the Klein bottle. Since K_6 is non-planar and $K_{4,4}$ can be shown not to be projective-planar, it follows that $\tilde{\gamma}(K_6) = 1$ and $\tilde{\gamma}(K_{4,4}) = 2$.

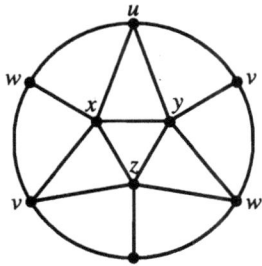 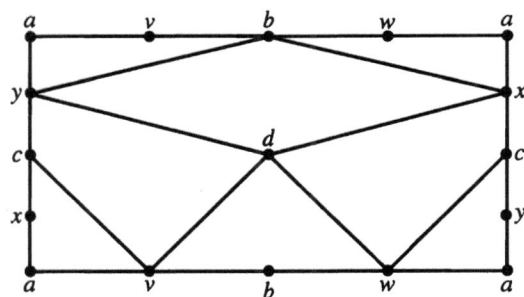

Fig. 11.17

The following result is the non-orientable analogue of Theorem 11.17.

Theorem 11.22 *If G is a graph with n vertices and m edges, then*
(a) $\tilde{\gamma}(G) \geq \frac{1}{3}m - n + 2$;
(b) $\tilde{\gamma}(G) \geq \frac{1}{2}m - n + 2$, *if G is bipartite;*
(c) $\tilde{\gamma}(G) \geq [(g-2)/g]m - n + 2$, *if G has girth g.*

Our next two theorems were both found by Ringel, in 1954 and 1965, respectively (see [19]).

Theorem 11.23 *The crosscap number of the complete graph K_n ($n \geq 3$) is*

$$\tilde{\gamma}(K_n) = \left\lceil \frac{(n-3)(n-4)}{6} \right\rceil,$$

except that $\tilde{\gamma}(K_7) = 3$.

Theorem 11.24 *The crosscap number of the complete bipartite graph $K_{r,s}$ ($r, s \geq 2$) is*

$$\tilde{\gamma}(K_{r,s}) = \left\lceil \frac{(r-2)(s-2)}{2} \right\rceil.$$

Just as the genus does for orientable surfaces, the crosscap number provides what is needed to find the chromatic number of non-orientable surfaces. We note that this time there is one exceptional surface—the Klein bottle.

Theorem 11.25 *For $k > 0$, the chromatic number of the non-orientable surface $\tilde{S}_k$ is*

$$\chi(\tilde{S}_k) = \left\lfloor \frac{7 + \sqrt{1+24k}}{2} \right\rfloor,$$

except that $\chi(\tilde{S}_2) = 6$.

There are some further interesting comparisons to be made between the parameters $\gamma(G)$ and $\tilde{\gamma}(G)$. As noted earlier, any surface with a crosscap is non-orientable. It follows that for any graph G, $\tilde{\gamma}(G) \leq 2\gamma(G) + 1$. There is, however, no bound in the other direction, as there are graphs of arbitrarily large orientable genus that are projective-planar.

Other differences appear in cellular embeddings of graphs (see [27] for more details and references). If $\gamma(G) = h$, then every embedding of G on S_h is cellular (a result of J. W. T. Youngs), but the corresponding statement for the crosscap number does not hold. In particular, although $\tilde{\gamma}(K_7) = 3$, not every embedding of K_7 in $\tilde{S}_3$ is cellular. Furthermore, while the orientable genus is additive over the blocks of a graph (a theorem of Battle, Harary, Kodama and Youngs), the crosscap number is not—the graph consisting of two copies of K_7 is a counter-example.

11.8 Kuratowski-type theorems

Each surface has a family of graphs that play the roles that K_5 and $K_{3,3}$ play in Kuratowski's Theorem. In this section, we look at these families. In particular, we note that the family is always finite.

Given a surface S, we define the *minimal forbidden family* $\mathcal{F}(S)$ to be the set of graphs with the following three properties:

none of the graphs in $\mathcal{F}(S)$ is embeddable in S;

each graph that is not embeddable in S contains a subgraph homeomorphic to some graph in $\mathcal{F}(S)$;

no graph in $\mathcal{F}(S)$ is homeomorphic to a subgraph of another graph in the family.

The only surface other than the sphere for which the minimal forbidden family has been determined is the projective plane: $\mathcal{F}(\tilde{S}_1)$ consists of 103 graphs (see Glover, Huneke and Wang [10]). It is known that there are more than 800 minimal forbidden subgraphs for the torus.

Recall that a graph F is a *minor* of a graph G if F can be obtained from a subgraph of G by a sequence of deletions and contractions of edges. It is clear that if an edge of a planar graph is deleted or contracted, then the resulting graph is also planar, so that *every minor of a planar graph is planar*. Kuratowski's Theorem thus has the following 'minor' version: *a graph is planar if and only if it has neither K_5 nor $K_{3,3}$ as a minor*.

The finiteness of the minimal forbidden family for every non-orientable surface was established by Archdeacon and Huneke in 1980 (see [1]). In contrast, the orientable case was not settled until 1984, when Robertson and Seymour [21] proved their spectacular result on graph minors.

Theorem 11.26 (Robertson–Seymour Theorem) *Every infinite collection of graphs contains at least one graph that is a minor of another.*

The *minor-minimal family* $\mathcal{M}(S)$ of forbidden subgraphs of a surface S is defined analogously to the family $\mathcal{F}(S)$:

none of the graphs in $\mathcal{M}(S)$ is embeddable in S;

every graph that is not embeddable in S has some graph in $\mathcal{M}(S)$ as a minor;

no graph in $\mathcal{M}(S)$ is a minor of another.

Although the two families $\mathcal{M}(S)$ and $\mathcal{F}(S)$ are the same for the plane, this is not generally the case. In fact, in contrast to the 103 homeomorphically minimal forbidden graphs for the projective plane, the minor-minimal family has just 35 members. The following theorems are consequences of the Robertson–Seymour Theorem.

Theorem 11.27 *For any surface S, the family of minor-minimal graphs that are not embeddable in S is finite.*

Theorem 11.28 *For any surface S, the family of homeomorphically minimal graphs that are not embeddable in S is finite.*

We note that there are classes of 'surfaces' for which these theorems do not hold. The simplest example, found by Širáň and Gvozdjak [23], is obtained by taking two spheres and identifying two points on one with two points on the other.

We conclude the chapter with an interesting result on embedding graphs in 3-dimensional space. It is easy to see that every graph has a crossing-free embedding in 3-space—simply place the vertices along a line and draw each edge in a different plane containing that line. It was observed by H.-G. Bothe that every embedding of K_6 has two 'linked cycles' (see Fig. 11.18), and that K_6 is minimal with respect to this property.

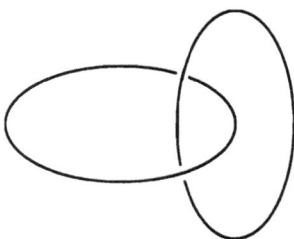

FIG. 11.18

Eventually, H. Sachs found seven such graphs. They are shown in Fig. 11.19, and curiously, each of them has just fifteen edges. In addition to K_6, the list includes the Petersen graph and $K_{4,4}$ with one edge removed. Recently, Robertson, Seymour and Thomas [22] verified that Sachs' list is complete.

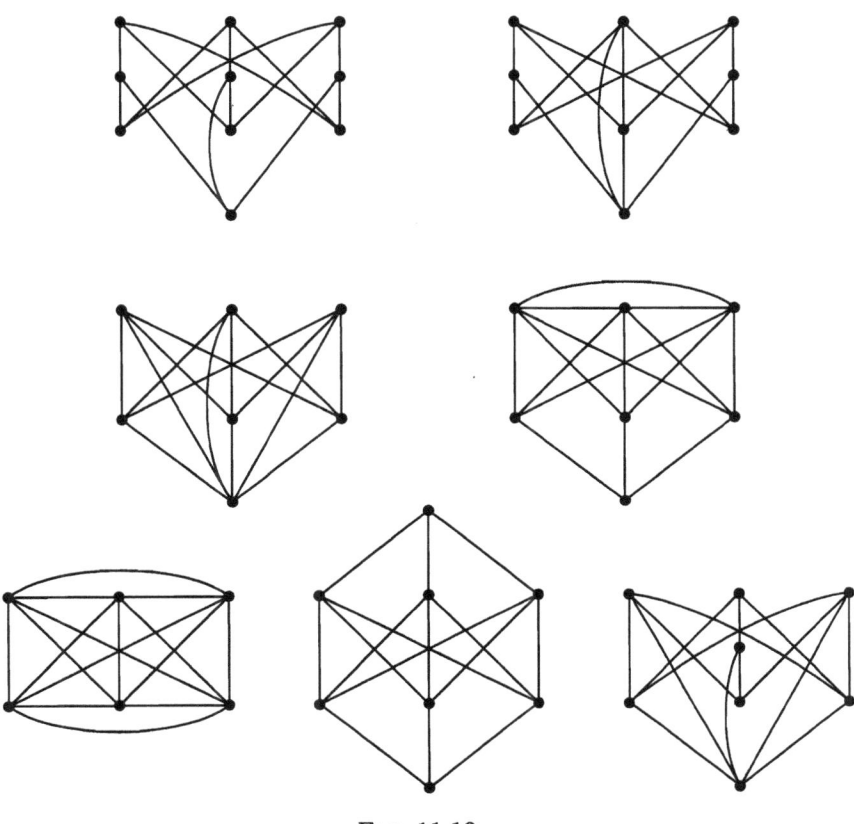

Fig. 11.19

Theorem 11.29 *The minor-minimal family of forbidden subgraphs for graphs having three-dimensional embeddings with no linked cycles consists of the seven graphs in Fig. 11.19.*

References

1. D. Archdeacon and P. Huneke, A Kuratowski theorem for nonorientable surfaces, *J. Combin. Theory (B)* **46** (1989), 173–231; *MR* **90f**: 05049.
2. L. W. Beineke and F. Harary, The thickness of the complete graph, *Canad. J. Math.* **21** (1969), 850–859; *MR* **32**#4032.
3. L. W. Beineke, F. Harary and J. W. Moon, On the thickness of the complete bipartite graph, *Proc. Cambridge Philos. Soc.* **60** (1964), 1–6; *MR* **28**#1611.
4. L. W. Beineke and R. D. Ringeisen, On the crossing numbers of products of cycles and graphs of order four, *J. Graph Theory* **4** (1980), 145–155; *MR* **81i**: 05059.
5. D. Bienstock and N. Dean, Bounds for rectilinear crossing numbers, *J. Graph Theory* **17** (1993), 333–348; *MR* **94e**: 05134.

6. N. L. Biggs, E. K. Lloyd and R. J. Wilson, *Graph Theory 1736-1936* (paperback edn), Clarendon Press, 1986; *MR* **56**#2771.
7. J. Blažek and N. Koman, A minimal problem concerning complete plane graphs, *Theory of Graphs and its Applications* (ed. M. Fiedler), Czechoslovak Academy of Sciences, 1964, pp. 113–117; *MR* **30**#4249.
8. O. Frink and P. A. Smith, Irreducible non-planar graphs [Abstract 179], *Bull. Amer. Math. Soc.* **36** (1930), 214.
9. M. R. Garey and D. S. Johnson, Crossing number is NP-complete, *SIAM J. Algebraic Discrete Methods* **4** (1983), 312–316; *MR* **85a**: 68070.
10. H. H. Glover, J. P. Huneke and C. S. Wang, 103 graphs that are irreducible for the projective plane, *J. Combin. Theory (B)* **27** (1979), 332–370; *MR* **81h**: 05060.
11. J. L. Gross and T. W. Tucker, *Topological Graph Theory*, Wiley-Interscience, 1987; *MR* **88h**: 05034.
12. R. K. Guy, Crossing numbers of graphs, *Graph Theory and Applications* (ed. Y. Alavi *et al.*), Lecture Notes in Math. **303**, Springer-Verlag, 1972, pp. 111–124; *MR* **49**#2442.
13. F. Harary, P. C. Kainen and A. J. Schwenk, Toroidal graphs with arbitrarily high crossing numbers, *Nanta Math.* **6** (1973), 58–67; *MR* **49**#103.
14. J. Hopcroft and R. E. Tarjan, Efficient planarity testing, *J. Assoc. Comput. Mach.* **21** (1974), 549–568; *MR* **50**#11841.
15. D. J. Kleitman, The crossing number of $K_{5,n}$, *J. Combin. Theory* **9** (1970), 315–323; *MR* **43**#6123.
16. K. Kuratowski, Sur le problème des courbes gauches en topologie, *Fund. Math.* **15** (1930), 271–283; English translation by J. Jaworowski, *Graph Theory, Lagow 1981* (ed. M. Borowiecki *et al.*), Lecture Notes in Math. **1018**, Springer-Verlag, 1983, pp. 1–13.
17. A. Mansfield, Determining the thickness of graphs is NP-hard, *Math. Proc. Cambridge Philos. Soc.* **93** (1983), 9–23; *MR* **84c**: 68032.
18. R. B. Richter and C. Thomassen, Intersection of curve systems and the crossing number of $C_5 \times C_5$, *Discrete Comput. Geom.* **13** (1995), 149–159; *MR* **95j**: 05081.
19. G. Ringel, *Map Color Theorem*, Springer-Verlag, 1974; *MR* **50**#1955.
20. G. Ringel and J. W. T. Youngs, Solution of the Heawood map-coloring problem, *Proc. Nat. Acad. Sci. U.S.A.* **60** (1968), 438–445; *MR* **37**#3959.
21. N. Robertson and P. D. Seymour, Generalizing Kuratowski's theorem, *Congr. Numer.* **45** (1984), 129–138; *MR* **86f**: 05058.
22. N. Robertson, P. D. Seymour and R. Thomas, A survey of linkless embeddings, *Graph Structure Theory*, Contemp. Math. **147**, Amer. Math. Soc., 1993.
23. J. Širáň and P. Gvozdjak, Kuratowski-type theorems do not extend to pseudosurfaces, *J. Combin. Theory (B)* **54** (1992), 209–212; *MR* **93a**: 05056.
24. C. Thomassen, The graph genus problem is NP-complete, *J. Algorithms* **10** (1989), 568–576; *MR* **91d**: 68054.

25. K. Wagner, Über eine Eigenschaft der ebenen Komplexe, *Math. Ann.* **114** (1937), 570–590.
26. A. T. White, The proof of the Heawood conjecture, *Selected Topics in Graph Theory* (ed. L. W. Beineke and R. J. Wilson), Academic Press, 1978, pp. 51–81.
27. A. T. White and L. W. Beineke, Topological graph theory, *Selected Topics in Graph Theory* (ed. L. W. Beineke and R. J. Wilson), Academic Press, 1978, pp. 15–49.

12
Knots

DOMINIC WELSH

This chapter illustrates some of the uses of graph theory in the study of knots. These include using the intimate relationship between some knot polynomials and the Tutte polynomial of a graph to prove a long-standing Tait conjecture, and also the use of graphs to provide bounds in the knot enumeration problem.

12.1 Introduction

Early work on knot theory was motivated by its applications. For example, Gauss [6] was concerned with the magnetic field which a knotted wire would produce if it carried an electric current. Kelvin and others tried to understand atoms as knots, hoping that the table of knots would reproduce the regularities of the table of elements.

In the last ten years or so there has been an upsurge of interest, due in some part to applications of knots in such subjects as biology (enzyme recognition), chemistry (polymer theory) and physics (statistical mechanics). An interesting recent application of knot theory is what has been described as 'the topological approach to enzymology'. The object is to unravel the secrets of enzyme action. As there is no known direct method of observing enzymes in action, one approach suggested is to observe the change that an enzyme causes in the topological state of the molecule on which the enzyme is acting.

The DNA molecule can be regarded as a polymer which is long and threadlike, and one of the reasons why knot theory is useful in the study of DNA is that various naturally occurring enzymes (called topoisomerases) alter the way in which the DNA is embedded in three-dimensional space. These alterations include increasing the 'coiling', passing one strand through another by an enzyme-induced break in the molecule, breaking strands and rejoining to different ends (recombinant enzymes). Linear DNA is a natural substrate for enzyme action, although this is of little help to an experimentalist since there can be no interesting observable topological change in an unconstrained linear piece of string. The idea is therefore to get the enzyme to operate on *circular* DNA molecules. This can be done, and then the reaction product should give (by its knot/link structure) a description of the operating enzyme. In other words, the aim is to be able to characterize enzymes by the family of knots that they produce when reacted with unknotted circular substrate.

Knots 177

In his very interesting survey, Sumners [19] reports how other enzymes can be characterized by the family of knots (or links) that they produce. As a test of the model, the effect was observed of allowing successive encounters of unknotted circular DNA with Tn3 resolvase. The experiment achieved four successive strand exchanges on eleven occasions, and in each case the knot type produced was the 6_2 knot shown in Fig. 12.1. This is exactly the outcome that is predicted by the model. Since there are eight distinct 6-crossing links, it was argued that the probability of the outcome achieved in the experiment being by chance is $(1/8)^{11} \sim 1.2 \times 10^{-10}$. Since their model explained all the observations, the authors concluded that it is correct.

FIG. 12.1

12.2 Basic concepts

We now give an introduction to combinatorial knot theory, with the emphasis on classification and the complexity of recognition problems.

Classical knot theory is concerned with embeddings of a circle (1-sphere) S^1 into Euclidean three-dimensional space $\mathbf{R}^3$ or the 3-sphere S^3. A *knot* is an embedding f of S^1 into S^3, although it is usually identified with its image $f(S_1)$. In other words, a knot is a subset of $\mathbf{R}^3$ which is homeomorphic to a circle.

The simplest knot has no crossings. It is called the *unknot*, and is shown in Fig. 12.2(a). The *trefoil knot*, shown in Fig. 12.2(b), has no representation with fewer than three crossings.

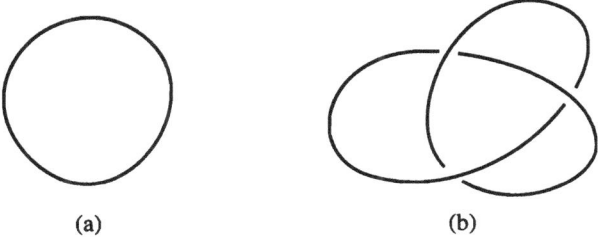

(a)   (b)

FIG. 12.2

A *link* with k components is a subset of $\mathbf{R}^3$ that is homeomorphic to the disjoint union of k circles. Figure 12.3 illustrates a link with three components.

Fig. 12.3

Two knots K and L are (*ambient*) *isotopic* if there exists a homotopy $h_t : S^3 \to S^3$ ($0 \le t \le 1$) such that $h_0 = 1$, each h_t is a homeomorphism and $h_1(K) = L$.

Topological embeddings of S^1 into S^3 may have strange features. In particular, there exist knots that have no piecewise linear representation. A knot K is *tame* if it is isotopic to a simple closed polygon in S^3, and *wild* if it is not tame. An example of a wild knot is shown in Fig. 12.4.

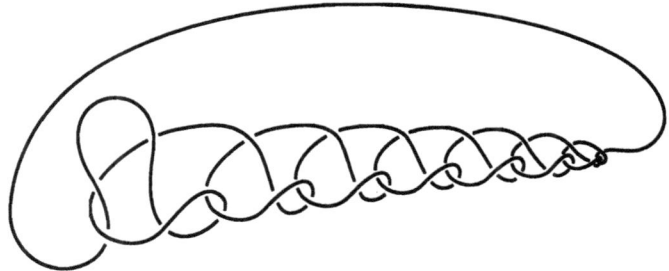

Fig. 12.4

Henceforth, when we refer to a knot, we shall assume it is tame and consequently that it has a piecewise linear representation. With this representation in mind, we can think of an *isotopy* as being any topological operation that deforms a polygon made up of vertices joined by elastic band edges; anything is allowed except cutting edges, subject to the proviso that distinct edges do not share points, except possibly end-points, and vertices remain distinct.

A knot is usually described by a plane projection containing only finitely many multiple points, such that all multiple points are double points, and these are crossing points. Such a projection determines the knot, provided that each crossing point has specified which is the over/under crossing. This information is conveyed by leaving a gap in the string of the undercrossing, and this then constitutes the *knot diagram*, as illustrated in Figs 12.1 and 12.2.

Two knot diagrams are *equivalent* if they are connected by a finite sequence of moves known as *Reidemeister moves*; these are the moves shown in Fig. 12.5, and their inverses. In each case, away from the local area of the move, the diagrams are unchanged.

Knots

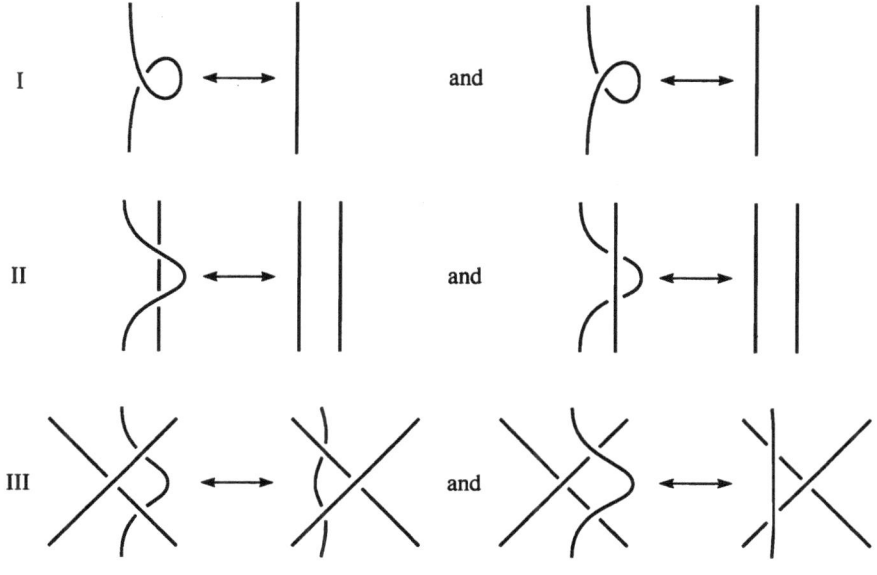

FIG. 12.5

The fundamental theorem of Reidemeister [16] is as follows.

Theorem 12.1 (Reidemeister's Theorem) *Two knots are ambient isotopic if and only if any diagram of one is equivalent to any diagram of the other under Reidemeister moves.*

The proof is based on the idea of knots being combinatorially equivalent and is described in Burde and Zieschang [3].

The two fundamental algorithmic problems about knots are:

the *unknotting problem*: to decide whether a knot really is knotted;

the *recognition problem*: given two knots, to decide whether they are isotopic.

Algorithms exist for both of these problems (see Hemion [9]); however, their status in the complexity hierarchy seems unclear.

12.3   Tait colourings

Given any link diagram D, it is easy to prove that we can colour the faces black and white in such a way that no two faces with a common edge have the same colour. By convention, we colour the boundary faces black, giving the *checkerboard* or *Tait colouring* of the diagram D.

From this we can get a canonical signed graph $S(D)$; its vertices are the black faces of the Tait colouring, and two vertices are joined by a signed edge if they share a crossing. The sign of the edge is positive or negative, according to the (conventional) rule shown in Fig. 12.6.

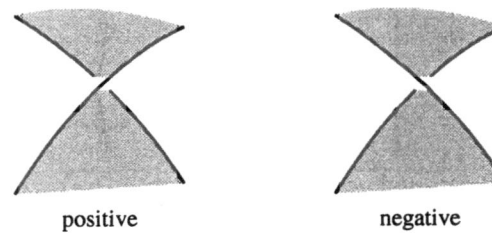

positive   negative

FIG. 12.6

Clearly $S(D)$ is a plane graph, and it is easy to prove that every signed plane graph can arise in this way. More precisely, we have the following result.

Proposition 12.2   *There is a one-to-one correspondence between link diagrams and signed plane graphs.*

Proof   Given any signed plane graph S, we must find a link diagram D such that $S(D) = S$. Given any plane graph G, the *medial graph* $m(G)$ has vertices at the mid-point of each edge of G, and two vertices are joined by an edge in $m(G)$ if they are consecutive edges of a face (including the infinite face). Thus $m(G)$ is a 4-regular graph. The vertices of $m(G)$ are the crossings of the link diagram D, and the sign of the underlying edge of S determines which string goes over/under in D. □

For example, the Tait colouring and associated signed graph of a diagram are illustrated in Fig. 12.7.

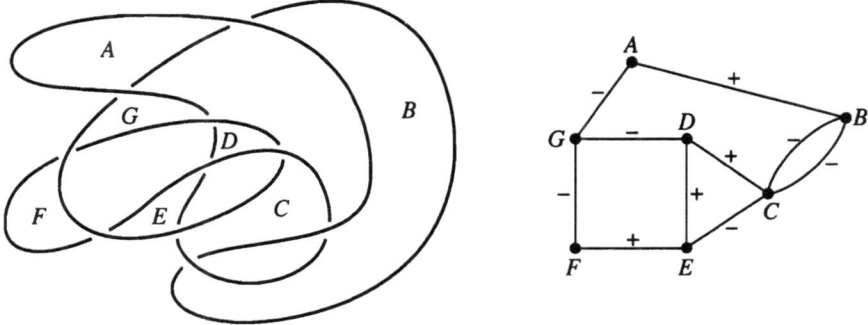

FIG. 12.7

In the same way as infinitely many link diagrams can represent equivalent links, there are infinitely many signed plane graphs that represent each link.

It is routine to check that the Reidemeister moves I, II and III can be translated into local moves on signed graphs, and thus we obtain an equivalent formulation of Reidemeister's Theorem.

Theorem 12.3 *Two plane signed graphs S_1 and S_2 represent the same link if and only if S_1 can be transformed into S_2 by some finite sequence of the moves I', II' and III' shown in Fig. 12.8.*

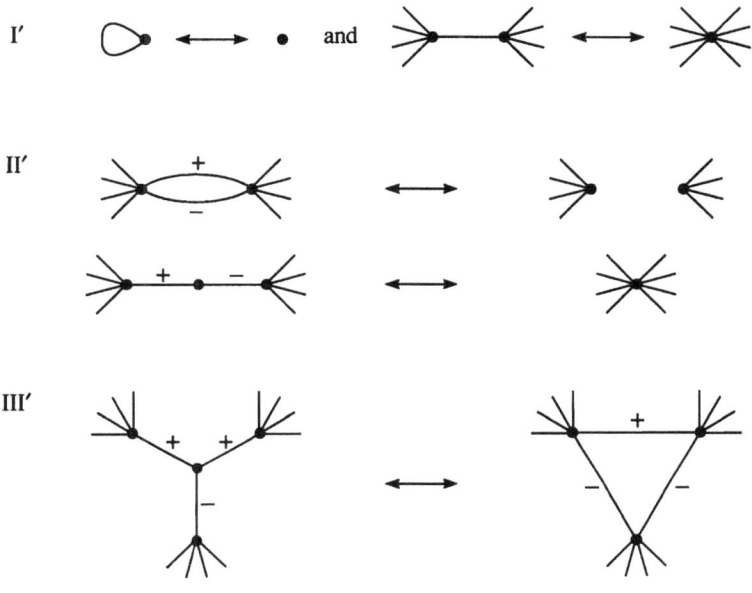

FIG. 12.8

For more on the exact relationship between combinations of Reidemeister moves and their graph equivalents, see Yajima and Kinoshita [26].

12.4 Classifying knots

Two link diagrams with different numbers of components obviously represent non-isotopic links. This is the simplest example of a link invariant and is used as a first subdivision of links.

A less trivial invariant is the *crossing number*, defined as the minimum number of crossings in any diagram representing the link. Determining the crossing number seems to be difficult; certainly no polynomial-time algorithm is known.

All knots with crossing number less than or equal to 7 can be represented by link diagrams in which the crossings are alternately over/under/over .... Links with this property are called *alternating* and are an important (but relatively small) subclass of the class of all links. It is easy to see that the following results are true.

Proposition 12.4 *A link L is alternating if and only if it has a signed graph representation in which all edges have the same sign.*

The *mirror image* of a link L is the link $\overline{L}$ obtained from any link diagram D of L by interchanging the over/under nature of the strings at each crossing. A link is *amphicheiral* or *achiral* if it is isotopic to its mirror image. No fast and infallible test of amphicheirality is known.

Proposition 12.5 *If G is a plane signed graph and G^* is its plane dual with the sign of each edge multiplied by -1, then the links $L(G)$ and $L(G^*)$ are isotopic.*

The proof is an easy consequence of the star triangle transformation, shown in Fig. 12.8(III'); alternatively, it can be seen by considering the two representations of the associated links on a sphere.

A fundamental operation on two knots K_1 and K_2 is to join them together as shown in Fig. 12.9. The resulting knot is called a *product* of K_1 and K_2 and is denoted by $K_1 \# K_2$.

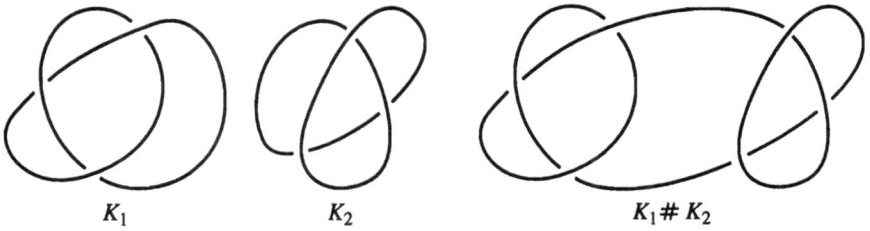

FIG. 12.9

A knot is *prime* if it is not the unknot and cannot be expressed as the product of two non-trivial knots. In classifying knots it is usual to list only those that are prime; unfortunately, it is not easy to decide whether a knot is prime.

Many other partial invariants of these types exist and have graphical formulations that are easy to obtain.

12.5 Braids and the Seifert graph

A *braid* on m strings is constructed as follows: take m distinct points $P_1, \ldots, P_m$ and m distinct points $Q_1, \ldots, Q_m$ on two horizontal lines, and link them by m simple arcs (strings) in $\mathbf{R}^3$, each starting at P_i and ending at $Q_{\pi(i)}$, where π is a permutation of $(1, 2, \ldots, m)$. The strings are required to 'run downwards' as in Fig. 12.10. The collection of strings constitutes an *m-braid*. The map $i \mapsto \pi(i)$ is the *permutation* of the braid. The braid is *closed* by joining the points P_i, Q_i as illustrated in Fig. 12.10.

Clearly each closed braid defines a link; conversely, we have the following theorem of Alexander [2].

Theorem 12.6 *Every link can be represented as a closed braid.*

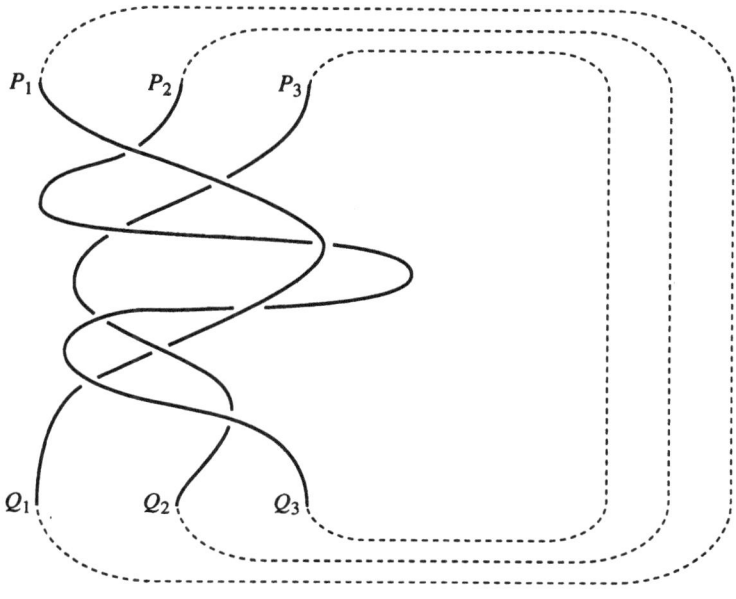

FIG. 12.10

The minimum number of strings in any braid representation of L is known as the *braid index* of L, and is denoted by $\beta(L)$. It characterizes the unknot, in the sense that K is the unknot if and only if $\beta(K) = 1$. Thus any polynomial-time algorithm that determines the braid index would be of great interest.

A classical theorem about links is that any oriented link L is the boundary of a compact connected orientable surface called the *Seifert surface*, constructed from an oriented diagram D by 'splitting' each crossing of D as shown in Fig. 12.11, and then glueing together the resulting set of disjoint discs using twisted bands to preserve orientability.

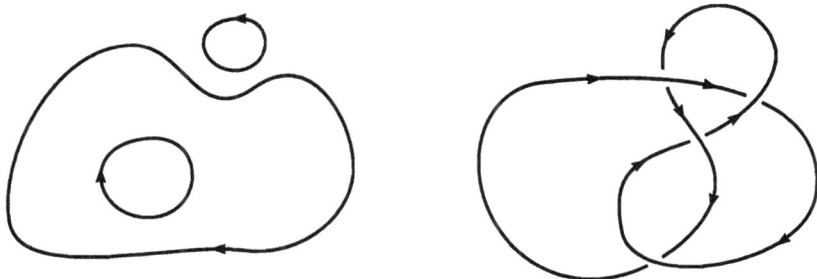

FIG. 12.11

The *genus* $g(K)$ of a knot is the minimum genus of an oriented surface Σ in S^3 which has K as its boundary. Thus, by taking any Seifert surface of a knot K and then determining its genus, we can obtain an upper bound on $g(K)$.

Moreover, this can be done quickly and easily. However, determining $g(K)$ seems to be hard. It is certainly as hard as the unknotting problem, since $g(K) = 0$ if and only if K is equivalent to the unknot.

The *Seifert graph* $\Gamma(D)$ of an oriented link diagram D is a signed graph whose vertices are the Seifert circles (or discs) constructed in the above splitting process, with signed edges joining two circles whenever they share a crossing. The sign of the crossing is determined by the convention shown in Fig. 12.12.

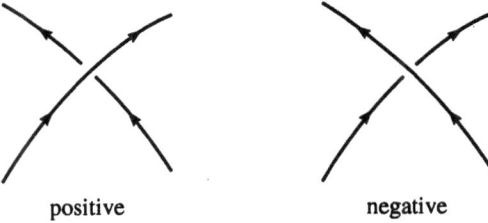

FIG. 12.12

It is easy to see that any Seifert graph is planar and bipartite. Another important property, observed by Murasugi and Przytycki [15], is the following.

Theorem 12.7 *If the Seifert graph is non-separable, then it uniquely determines the underlying link.*

Conversely, given any Seifert graph of an oriented link, it bounds the braid index. More precisely, by using a key theorem of Yamada [27], we have the following result.

Theorem 12.8 *For any link L, the braid index is given by*

$$\beta(L) = \min s(D),$$

where the minimum is taken over all link diagrams D of L.

12.6 The Jones and Kauffman bracket polynomials

The Alexander polynomial is a classical long-standing invariant of oriented knots. It is somewhat surprising (particularly with hindsight) that it was not until 1984 that the Jones polynomial, formally very little different from the Alexander polynomial, was discovered and sparked off the discovery of a host of similar polynomial invariants such as the (Kauffman) bracket, Homfly and Dubrovnik polynomials. There exist several excellent easily accessible surveys of what has been done in this field over the last few years (see, for example, Lickorish [11] or Kauffman [10]), so here we concentrate on those aspects of knot polynomial theory that are of particular interest in combinatorics.

Like the Alexander polynomial, the Jones polynomial $V_k(t)$ is an invariant of *oriented* knots. Because of this, we concentrate on the Kauffman bracket polynomial. It is very close to the Jones polynomial and to other familiar combinatorial polynomials.

The *polynomial* $[D]$ of a link diagram D of the link L is obtained by applying the equations

$$[\times] = A[\,)\,(\,] + B[\,\asymp\,], \quad [DU] = d[D] \quad \text{and} \quad [U] = 1$$

locally at crossings. It is not difficult to show that $[D]$ is well defined on link diagrams, in the sense that the result is independent of the order in which the crossings are treated, and is a polynomial in the three variables A, B and d, which are assumed to commute.

As it stands, this polynomial is not an invariant of isotopy. However, if we restrict it by insisting that

$$B = A^{-1} \quad \text{and} \quad d = -(A^2 + A^{-2}),$$

then the new one-variable polynomial $\langle D \rangle$ is invariant under Reidemeister moves II and III, and is called the (*Kauffman*) *bracket polynomial* of D. It is *not* an invariant under Reidemeister move I. However, provided that the link is oriented and then suitably normalized, the resulting invariant is the Jones polynomial of the link.

More precisely, we define the *writhe* $w(L)$ of an *oriented* link L to be the sum of the signs at crossings, calculated according to the convention shown in Fig. 12.12, and then define a (Laurent) polynomial f on oriented link diagrams by

$$f_D(A) = (-A^3)^{-w(D)} \langle D \rangle,$$

where $w(D)$ is the writhe of D, and where $\langle D \rangle$ for an oriented diagram is obtained by forgetting about the orientation. We can then define the *Jones polynomial* $V_L(t)$ of an oriented link L by

$$V_L(t) = (-t)^{-3w(L)/4} \langle L \rangle_{A = t^{-1/4}}.$$

Thus, for most purposes, the Kauffman bracket is as good an invariant as the Jones polynomial. For example, we have the following result.

Theorem 12.9 *If L_1 and L_2 are isotopic, then $\langle L_1 \rangle = A^\alpha \langle L_2 \rangle$ for some integer α.*

In particular, we have the following consequence.

Corollary 12.10 *If L is isotopic to the unknot, then $\langle L \rangle$ is a power of A.*

Thus we have a reasonably effective method of telling some knots apart—namely, calculate their respective bracket polynomials, and if their ratio is not a power of A then the knots must be different. For example, consider the unknot U and the trefoil T in Fig. 12.2. Their bracket polynomials are given by $\langle U \rangle = 1$, by definition, and

$$\langle T \rangle = A^7 - A^3 - A^{-5}.$$

Hence the two knots are not isotopic. Unfortunately, this is far from being an infallible test. There are many examples of knots which are not isotopic but have the same Jones polynomials.

Another way of looking at the bracket polynomial, and one which is certainly easier in hand calculations, is by using signed graphs. Given a Tait colouring of a link diagram D, we form the associated signed graph and note the actions on the graph corresponding to the operation of the bracket rules.

The rule

$$\langle \rangle = A \langle \rangle + A^{-1} \langle \rangle$$

on the link diagram corresponds to

$$\langle G \rangle = A \langle G/e \rangle + A^{-1} \langle G - e \rangle$$

on the Tait graph G of D; here, $G - e$ and G/e denote the usual deletion and contraction of the edge e from G, where e is the edge corresponding to the crossing.

Thus the bracket rules are equivalent to

$$\langle G \rangle = A \langle G - e \rangle + A^{-1} \langle G/e \rangle \quad \text{if } e \text{ is positive,}$$
$$\langle G \rangle = A \langle G/e \rangle + A^{-1} \langle G - e \rangle \quad \text{if } e \text{ is negative,}$$

where e is the edge joining the black faces in each case.

Combining these two rules with the rules

$$\langle C^+ \rangle = -A^{-3}, \quad \langle C^- \rangle = -A^3, \quad \langle L^+ \rangle = -A^3, \quad \langle L^- \rangle = -A^{-3},$$

where C^+ and C^- represent the positive and negative signed isthmuses, and L^+ and L^- represent the positive and negative signed loops, we see that the bracket polynomial can be regarded as a specialization of a 'Tutte polynomial' on signed graphs.

The *Tutte polynomial* of an (unsigned) graph G is the two-variable polynomial $T(G; x, y)$ defined by

$$T(G; x, y) = \sum_{S \subseteq E(G)} (x-1)^{r(E) - r(S)} (y-1)^{|S| - r(S)},$$

where the *rank* $r(A)$ of a set of edges is the (matroid) rank of A—namely, the cardinality of a maximal circuit-free subset of A.

In the special case that L is an alternating link, it can be represented as a link diagram in which each edge has positive sign. In this case, Thistlethwaite [22] noticed that the following is true.

Theorem 12.11 *If L is an alternating link, then its bracket polynomial is given by the relation*

$$\langle L \rangle = A^{2|V(G)|-|E(G)|-2} T(G; -A^{-4}, -A^4),$$

where G is the unsigned graph of any positive link diagram representing L, and $T(G; x, y)$ is its Tutte polynomial.

The above result is of use only when L is an alternating link presented in the form of an alternating diagram. More general is the following *states model* representation of $\langle L \rangle$ in terms of concepts more familiar in combinatorics. The following result is from Schwärzler and Welsh [18].

Theorem 12.12 *Let D be any link diagram representing the link L, and suppose that the associated signed graph $G = G(D)$ is connected. For any subset S of $E(G)$, let S^+ denote the positive signed part and let S^- denote the negative signed part. Then, with r denoting the rank function in G, we have*

$$\langle D \rangle = A^{|E^-|-|E^+|-2r(E)} \sum_{X \subseteq E} A^{4(r(X)-|X^-|)} (-A^4 - 1)^{r(E)+|X|-2r(X)}.$$

As we mentioned earlier, there are many examples of non-isotopic knots with the same bracket (and Jones) polynomial. However, it is an outstanding open problem to decide if there exists a knot that has the same Jones polynomial as the unknot but is not isotopic to it. Recent work of Dasbach and Hougardy [4] implies that any such example must have at least 17 crossings.

12.7 Bivariate polynomials

The discovery of the Jones polynomial led to the discovery of many other such invariants. In this section we briefly describe one of these; for more details, see Kauffman [10] and Lickorish [11]. Our description is in terms of their defining 'skein relations'. These relations are the knot counterparts of the deletion/ contraction formulas of matroids and graphs, and in each case represent local changes at a crossing of the knot or link that can be carried out in an arbitrary order to give the same polynomial. Moreover, each of these changes is designed so as to preserve equivalence under the Reidemeister moves, so that links that are isotopic have the same polynomial.

Kauffman's two-variable polynomial $\Lambda(D; a, z)$ in $Z[a^{\pm 1}, z^{\pm 1}]$ is a Laurent polynomial of regular isotopy of a diagram D of an unoriented knot or link. In other words, it is invariant under Reidemeister moves II and III. It is defined by the following formulas.

For a simple closed curve O, $\Lambda(O) = 1$.
$a^{-1}\Lambda(C_+) = a\Lambda(C_-) = \Lambda(C_0)$, where the C_i are as shown:

$$\underset{C_+}{\text{⌒}} \quad \underset{C_-}{\text{⌒}} \quad \underset{C_0}{|}$$

$\Lambda(D_+) + \Lambda(D_-) = z(\Lambda(D_0) + \Lambda(D_\infty))$, where the D_i are as shown:

$$\underset{D_+}{\times} \quad \underset{D_-}{\times} \quad \underset{D_0}{)(} \quad \underset{D_\infty}{\asymp}$$

For example, the trefoil in Fig. 12.2 has

$$\Lambda(D; a, z) = 2a^{-1} - a + (1 + a^2)z + (a^{-1} + a)z^2.$$

The two-variable Kauffman polynomial $F(L; a, z)$ of oriented isotopy is then defined by

$$F(L; a, z) = a^{-w}\Lambda(D; a, z),$$

where w is the writhe of the oriented link L formed by the orientation of D (see Fig. 12.12).

The relationship between Λ, or (almost equivalently) F, and other polynomials we have considered is summarized in the following:

$$\langle D \rangle = \Lambda(D; -A^3, A + A^{-1}) \quad \text{and} \quad V_L(t) = F(L; -t^{3/4}, t^{-1/4} + t^{1/4}).$$

However, although F and Λ are quite strong link invariants, it is known that again there are many examples of non-isotopic links that have the same Kauffman polynomial.

12.8 The Tait conjectures

We now use the representation of the bracket obtained in Section 12.6 to settle some problems that had remained open since the end of the last century and were known as the *Tait conjectures*. For an account of their history, we refer to de la Harpe, Kervaire and Weber [7] and Thistlethwaite [21]. A diagram is *reduced* if it has no crossing of the type shown in Fig. 12.13.

FIG. 12.13

Tait's first conjecture is as follows.

Theorem 12.13 *If a link L has an alternating diagram D which is reduced and has n crossings, then there is no diagram representing L which has fewer than n crossings.*

In other words, given an alternating reduced diagram D of L, the crossing number of L is the number of crossings in D.

Proof The key idea of the proof is to notice that since $(A^3)^{-w(D)}\langle D \rangle$ is an invariant of isotopy, so is span $\langle D \rangle$, where the *span* of a Laurent polynomial is the difference between its maximum and minimum degrees. Now since D is an alternating link diagram, we may assume that the black-face graph G defined by its Tait colouring has only positive edges; otherwise, take the dual. But using the representation

$$\langle D \rangle = A^\alpha T(G; -A^{-4}, A^4)$$

and the representation of T as a rank polynomial, it is easy to verify that span $\langle D \rangle = 4|E|$, provided that G has no loop or isthmus. In other words, if D is reduced, then span $\langle D \rangle = 4|E(G)|$ and $|E(G)|$ equals the number of crossings of D. But span $\langle D \rangle$ is an invariant of isotopy. Hence this equality must hold for *any* reduced alternating representation of the link L. □

A stronger version of Theorem 12.13 is the following.

Theorem 12.14 *If L is an alternating prime link, then any non-alternating diagram representing L has crossing number strictly greater than the span of V_L.*

To see that the property of being prime is needed above, consider the following construction. Take any pair of alternating knots K_1 and K_2, and obtain two alternating representations of K_2, one with a positive Tait graph G_2 and the other with a negative Tait graph G_2^*, the plane dual of G_2. Now form $K_1 \# K_2$ in two distinct ways, corresponding to this direct sum. The one from G_1 and G_2 is an alternating representation, but the one from G_1 and G_2^* is not.

Tait and his co-workers in the 19th century seemed to discount the existence of amphicheiral links of odd crossing number. It is still not known whether any such links exist.

What does follow from the above is the following statement.

Theorem 12.15 *Every alternating amphicheiral link has even crossing number.*

Proof If L is isotopic to its mirror image, then span $\langle L \rangle$ is an even integer, since $\langle L; A \rangle = \langle \overline{L}; A^{-1} \rangle$. But for alternating links, we have just shown that span $\langle L \rangle$ is the crossing number. □

In contrast to the two previous theorems, whose proofs (with hindsight) are relatively easy, Tait's flyping conjecture is of a different order of magnitude. A *flype* is a move on a tangle with two input and two output strings, obtained by rotating the tangle by 180° (see Figure 12.14).

FIG. 12.14

Towards the end of the 19th century, Tait conjectured the following remarkable result, known for many years as *Tait's flyping conjecture*.

Theorem 12.16 *Any two reduced alternating projections of the same link can be obtained from each other by a sequence of flypes.*

In 1990 Schrijver proved Tait's conjecture for link diagrams that have no cutsets of two edges and the only ones with four edges are the sets of edges at a vertex. This was a major advance, and so it was not too suprising when the full conjecture was proved in 1991 by Menasco and Thistlethwaite [14]. Neither proof is easy; both involve intricate geometric reasoning, and we do not give even a sketch here. We do, however, note the following useful consequence.

Corollary 12.17 *If a link L has an alternating link diagram D for which the associated positive graph $G(D)$ is 3-connected, then D is the only diagram with positive graph representing L.*

If L and K are two alternating links related by a flype, then it is easy to see that their positive graphs $G(L)$ and $G(K)$ are 2-isomorphic. This immediately implies

$$T(G(L); x, y) = T(G(K); x, y);$$

in other words, the Tutte polynomial is invariant under flypes, and therefore is an invariant of alternating knots.

12.9 Two applications

We close by describing two other recent applications of graph theory in combinatorial knot theory.

Thistlethwaite [23] has found a striking combinatorial relation linking the two-variable Kauffman polynomial Λ of a link with the chromatic and flow polynomials of certain minors of the checkerboard graph $G = G(D)$, where D is any diagram of the link. This leads to a powerful test of non-triviality. We sketch the details.

If D has n crossings, we write

$$\Lambda(D; a, z) = \sum u_{rs} a^r z^s,$$

and then define the two *outer polynomials* of D by

$$\phi_D^+(t) = u_{0,n} + u_{1,n-1}t + u_{2,n-2}t^2 + \cdots + u_{n,0}t^n,$$

$$\phi_D^-(t) = u_{0,n} + u_{-1,n-1}t + u_{-2,n-2}t^2 + \cdots + u_{-n,0}t^n.$$

If we now let E^+ and E^- denote the subgraphs of E containing just the positive and negative edges of E, respectively, then we obtain the following result.

Theorem 12.18 *The outer polynomials of any link diagram D satisfy*

$$\phi_D^+(t) = k^+ F(G|E^+; \lambda) P(G.E^-; \lambda) \quad \text{and} \quad \phi_D^-(t) = k^- F(G|E^-; \lambda) P(G.E^+; \lambda),$$

where F and P are the flow and chromatic polynomials, $\lambda = 1 - t$, $k^{\pm}$ are easily determined constants, and $G|A$ and $G.A$ denote the deletion and contraction of the edges of G not in A.

This result leads directly to the following easily checked sufficient condition for non-triviality. A graph has non-zero chromatic polynomial unless it has a loop; similarly, it has non-zero flow polynomial unless it has an isthmus.

We deduce the following results.

Corollary 12.19 *A link diagram represents a non-trivial link if its graph G is such that*

either $G|E^-$ has no isthmus and $G.E^+$ has no loop,

or $G|E^+$ has no isthmus, and $G.E^-$ has no loop.

This is a remarkably effective test, in as much as all non-trivial knot diagrams on eleven or fewer crossings satisfy it; however, there does exist a non-trivial knot for which no 12-crossing diagram satisfies either of these criteria. For more on this, see Lickorish and Thistlethwaite [13] and Schwärzler and Welsh [18].

Another application was motivated by the problem of enzyme classification discussed in the first section, which leads naturally to the question of how many prime knots and links have crossing number n.

Over time, knot theorists have listed all prime knots with up to thirteen crossings. However, this is an extremely difficult task (see Thistlethwaite [21]).

Let $k(n)$ and $l(n)$ denote the numbers of prime knots and links with crossing number n. Ernst and Sumners showed in [5] that $k(n)$ grows at least exponentially. Their bound gives

$$\liminf_{n \to \infty} k(n)^{1/n} \geq 2.$$

The first upper bound was given by Welsh [24], who used arguments and results from graph theory to obtain the bounds

$$4 \leq \liminf_{n \to \infty} l(n)^{1/n} \leq \limsup_{n \to \infty} l(n)^{1/n} \leq \tfrac{27}{2}.$$

The problem of showing that the limit exists remains open. It would follow immediately from the above upper bound if, as I conjecture is true,

$$l(m+n) \geq l(m)l(n) \quad \text{for } m,n \geq 3.$$

However, proving this does not seem easy.

For more on this and other topics treated above, see Welsh [25], and for a beautiful introduction to the wider aspects of knot theory, see Adams [1].

References

1. C. C. Adams, *The Knot Book*, W. H. Freeman, 1994.
2. J. W. Alexander, Topological invariants of knots and links, *Trans. Amer. Math. Soc.* **30** (1928), 275–306.
3. G. Burde and H. Zieschang, *Knots*, de Gruyter, 1985; *MR* **87b**: 57004.
4. O. T. Dasbach and S. Hougardy, Does the Jones polynomial detect unknottedness?, *Experiment. Math.*, to appear.
5. C. Ernst and D. W. Sumners, The growth of the number of prime knots, *Math. Proc. Cambridge Philos. Soc.* **102** (1987), 303–315; *MR* **88m**: 57006.
6. C. F. Gauss, Zur mathematischen Theorie der electrodynamischen Wirkungen, *Werke Königl. Gesell. Wiss. Göttingen* **5** (1833), 605.
7. P. de la Harpe, M. Kervaire and C. Weber, On the Jones polynomial, *Enseign. Math. (2)* **32** (1986), 271–335; *MR* **88f**: 57004.
8. G. Hemion, On the classification of homeomorphisms of 2-manifolds and the classification of 3-manifolds, *Acta Math.* **42** (1979), 123–155; *MR* **80f**: 57003.
9. G. Hemion, *The Classification of Knots and 3-Dimensional Spaces*, Oxford University Press, 1992; *MR* **94g**: 57015.
10. L. H. Kauffman, An invariant of regular isotopy, *Trans. Amer. Math. Soc.* **318** (1990), 417–471.
11. W. B. R. Lickorish, Polynomials for links, *Bull. London Math. Soc.* **20** (1988), 558–588; *MR* **90d**: 57004.
12. W. B. R. Lickorish, The panorama of polynomials for knots, links and skeins, *Contemp. Math.* **78** (1988), 399–414; *MR* **90d**: 57003.
13. W. B. R. Lickorish and M. B. Thistlethwaite, Some links with nontrivial polynomials and their crossing-numbers, *Comment. Math. Helv.* **63** (1988), 527–539; *MR* **90a**: 57010.
14. W. W. Menasco and M. B. Thistlethwaite, The Tait flyping conjecture, *Bull. Amer. Math. Soc.* **25** (1991), 403–412; *MR* **92b**: 57017.
15. K. Murasugi and J. H. Przytycki, The index of a graph with applications to knot theory, Preprint, 1991.
16. K. Reidemeister, Homotopieringe und Linsemraume, *Abh. Math. Sem. Univ. Hamburg* **2** (1935), 102–109.
17. A. Schrijver, Tait's flyping conjecture for well connected links, Preprint, 1990.

18. W. Schwärzler and D. J. A. Welsh, Knots, matroids and the Ising model, *Math. Proc. Cambridge Philos. Soc.* **113** (1993), 107–139; *MR* **94c**: 57019.
19. D. W. Sumners, The role of knot theory in DNA research, *Geometry and Topology* (ed. C. McCrory and T. Schifrin), Marcel Dekker, 1987, pp. 297–318; *MR* **88c**: 57012.
20. D. W. Sumners, The knot theory of molecules, *J. Math. Chem.* **1** (1987), 1–14; *MR* **88j**: 92014.
21. M. B. Thistlethwaite, Knot tabulations and related topics, *Aspects of Topology* (ed. I. M. James and E. H. Kronheimer), London Math. Soc. Lecture Note Ser. **93**, Cambridge University Press, 1995, pp. 1–76; *MR* **86j**: 57004.
22. M. B. Thistlethwaite, A spanning tree expansion of the Jones polynomial, *Topology* **26** (1987), 297–309; *MR* **88h**: 57007.
23. M. B. Thistlethwaite, On the Kauffman polynomial of an adequate link, *Invent. Math.* **93** (1988), 285–296; *MR* **89g**: 57009.
24. D. J. A. Welsh, On the number of knots and links, *Colloq. Math. Soc. János Bolyai* **60** (1991), 713–718; *MR* **94f**: 57010.
25. D. J. A. Welsh, *Complexity: Knots, Colourings and Counting*, London Math. Soc. Lecture Note Ser. **186**, Cambridge University Press, 1993; *MR* **94m**: 57027.
26. T. Yajima and S. Kinoshita, On the graphs of knots, *Osaka Math. J.* **9** (1957), 155–163; *MR* **20**#4845.
27. S. Yamada, The minimal number of Seifert circles equals the braid index of a link, *Invent. Math.* **89** (1987), 347–356; *MR* **88f**: 57015.

13
Probability

COLIN MCDIARMID

What are 'typical' properties of a graph? How can random methods yield existence results for graphs? What will randomized graph algorithms do? What have isoperimetric inequalities for graphs to do with probability? This chapter gives an introductory taste of the fertile area of graphs and probability.

13.1 Introduction

Since Paul Erdős brought together the subjects of graph theory and probability some forty years ago, the relationship has flowered marvellously. Here we attempt to give something of the flavour of this still burgeoning area—no attempt is made at a complete survey!

We discuss first some typical properties of random graphs and of algorithms on random graphs; for example, what can we say about the chromatic number and about methods to find good colourings? Then we see how to use random graphs in order to obtain deterministic results, and continue along this line with the help of the *Lovász Local Lemma*. The last two sections concern deterministic graphs. First we consider randomized algorithms on graphs—that is, procedures that operate on a deterministic graph and involve flipping coins. Finally, we discuss the relationship between results on concentration of measure and isoperimetric inequalities for graphs, focussing on the n-cube Q_n.

13.2 Random graphs: usual behaviour

There are several natural models of random graphs: we focus on two of these. The random graph $G_{n,p}$ has vertices $v_1, \ldots, v_n$, and the $\binom{n}{2}$ possible edges occur independently, each with probability p. When $p = \frac{1}{2}$, this corresponds to sampling uniformly from all labelled graphs on n vertices. The random graph $G_{n,m}$ also has vertices $v_1, \ldots, v_n$: each such graph with exactly m edges is equally likely. Its behaviour is very similar to that of $G_{n,p}$, with $p = m/\binom{n}{2}$.

When we investigate random graphs, sometimes it is because we want to know whether a property usually holds, or how well an algorithm usually works, and sometimes our aim is to prove a deterministic result. In this section we discuss the usual behaviour.

We focus on four topics: *colourings*, *Hamiltonian cycles*, *minimum spanning trees* and *perfect graphs*. For references not cited here, and for many further results, see [2]. When considering colourings we shall use the model $G_{n,p}$ with the edge-probability p constant (not depending on n).

Vertex-colouring The natural place to start is with the independence number $\alpha(G_{n,p})$. We might guess that it should usually be close to r, where r is such that the expected number of independent r-sets—that is, $\binom{n}{r}(1-p)^{\binom{r}{2}}$—is about 1. Such an $r = r(n)$ is asymptotically equal to $2\log n / \log(1/(1-p))$, which equals $2\log_2 n$ when $p = \frac{1}{2}$. The following result shows that α is remarkably predictable.

Theorem 13.1 *The independence number satisfies $\alpha(G_{n,p}) = r - 1$ or r a.s.*

Here 'a.s.' (almost surely) means that the statement holds with probability that tends to 1 as $n \to \infty$. The upper bound in the theorem is easy to obtain: it follows directly from the 'first moment method'—that, if always $X \geq 0$, then the probability $\Pr(X \geq 1)$ is at most the expected value EX. The lower bound follows after some computations from the 'second moment method'—that, if $EX \geq 0$, then $\Pr(X > 0) \geq (EX)^2/E(X^2)$, or a similar inequality. Here the random variable X counts the number of independent k-sets in $G_{n,p}$, where we take k to be $r + 1$ or $r - 1$.

The chromatic number $\chi(G)$ is at least the number of vertices divided by the independence number $\alpha(G)$. Hence, by Theorem 13.1, $\chi(G_{n,p}) \geq n/r$ a.s. It was conjectured in 1975 that this bound is about the correct value, and in 1988 this was proved by Bollobás [4]. In fact, we have the following rather precise result (see [21]).

Theorem 13.2 $\chi(G_{n,p}) = n/(r + O(1))$ *a.s.*

The key step in the proof is to establish 'concentration of measure'. We find that a.s. the random graph $G_{n,p}$ has the property that *every* induced subgraph with at least $n/(\log n)^2$ vertices contains an independent set of nearly r vertices, and so we can repeatedly remove such a set and colour it. The last few vertices can each receive a new colour.

The simple greedy (or sequential) colouring algorithm usually uses about twice as many colours as necessary. It is not known whether any polynomial-time algorithm usually achieves a ratio strictly less than 2.

The situation is quite different for another model of random graphs. Consider a fixed k, and suppose that the random graph G_n is obtained by sampling uniformly at random from the graphs on vertices $v_1, \ldots, v_n$ which have no subgraph K_{k+1}. Then G_n is a.s. k-colourable, and there is a polynomial-expected-time algorithm that colours G_n optimally—see Prömel and Steger [30] and the references therein.

Edge colouring The *chromatic index* $\chi'(G)$ of a graph G is the smallest number of colours needed to colour the edges of G so that adjacent edges are coloured differently. *Vizing's Theorem* says that the chromatic index of a graph with maximum degree Δ is equal to Δ (Class 1) or $\Delta + 1$ (Class 2). It is of great interest (and NP-hard) to determine which case occurs for a given graph or class of graphs.

How common are Class 2 graphs? Almost all graphs $G_{n,p}$ have a unique vertex of maximum degree, and so are in Class 1. Thus $\Pr(\chi' = \Delta + 1) = o(1)$. In fact, this probability decreases to 0 very quickly (see [11]).

Theorem 13.3 *As $n \to \infty$,*
$$n^{-(\frac{1}{2}+o(1))n} \leq \Pr(\chi' = \Delta + 1) \leq n^{-(\frac{1}{8}+o(1))n}.$$

The lower bound here arises from regular graphs of odd order. The upper bound arises from a polynomial-time algorithm that attempts to edge-colour the graph with Δ colours, and only rarely fails on $G_{n,p}$. It works by repeatedly removing near-perfect matchings. Very recently, Perkovic and Reed [28] have proposed an algorithm that will edge-colour $G_{n,p}$ optimally in polynomial expected time.

Total colouring The *total chromatic number* $\chi''(G)$ of a graph G is the smallest number of colours needed to colour the vertices and edges of G so that adjacent vertices, adjacent edges, and incident vertices and edges are coloured differently. It is clear that $\chi'' \geq \Delta + 1$. The well-known *Total Colouring Conjecture* asserts that $\chi'' = \Delta + 1$ or $\Delta + 2$.

Theorem 13.4 *As $n \to \infty$,*
$$\Pr(\chi'' > \Delta + 1) \leq n^{-(\frac{1}{8}+o(1))n},$$
and, for some constant c with $0 < c < 1$,
$$\Pr(\chi'' > \Delta + 2) = o(c^{n^2}).$$

In the second inequality above, the bound on the probability is very small, and so we obtain some support for the Total Colouring Conjecture, but we would prefer the bound to be zero!

Connectedness and Hamiltonian cycles How many edges do we need to make a random graph connected or Hamiltonian? The threshold for connectivity in $G_{n,m}$ was given by Erdős and Rényi [10] in 1960 in a seminal paper on random graphs. The proof involves estimating expected numbers of small components.

Theorem 13.5 *If $m = \frac{1}{2}n(\log n + c_n)$ and $c_n \to c$ as $n \to \infty$, then*
$$\Pr(G_{n,m} \text{ is connected}) \to e^{-e^{-c}} \quad \text{as } n \to \infty.$$

It follows that if $c_n \to -\infty$ then the probability $\to 0$, and if $c_n \to \infty$ then the probability $\to 1$.

Determining the threshold for the existence of a Hamiltonian cycle was a major open problem for many years. The key idea turned out to be *rotation-extension*.

Suppose that we have a path $v_0 v_1 \ldots v_k$ of length k. If v_0 or v_k has a neighbour not in the path, then we can extend the path by simply adding such a neighbour. Failing this, suppose that v_k has a neighbour v_i, where $0 \leq i \leq k-2$. If $i = 0$ and there is an edge $v_j w$ joining the cycle $v_0 v_1 \ldots v_k v_0$ to the rest of the graph, then the path $w v_j v_{j+1} \ldots v_k v_0 \ldots v_{j-1}$ has length $k+1$. If $i \neq 0$, then we perform a *rotation* and construct the path $v_0 v_1 \ldots v_i v_k v_{k-1} \ldots v_{i+1}$; this path still has length k, but has a different end-point v_{i+1}, from which we can again look for an extension. Using rotations and extensions, Komlós and Szemerédi eventually solved the threshold problem in 1983—see, for example, [2].

Theorem 13.6 *If $m = \frac{1}{2} n (\log n + \log \log n + c_n)$ and $c_n \to c$ as $n \to \infty$, then*

$$\Pr(G_{n,m} \text{ is Hamiltonian}) \to e^{-e^{-c}} \quad \text{as } n \to \infty.$$

There is, in fact, a polynomial-time algorithm, based on the rotation-extension idea, that finds a Hamiltonian cycle with probability as in the theorem above (see Bollobás, Fenner and Frieze [5]). It has recently been shown that, for any fixed $r \geq 3$, a random r-regular graph on n vertices is a.s. Hamiltonian (see Robinson and Wormald [31, 32]). Their proofs are quite different, and involve a delicate use of the second moment method.

The probabilities in the last two theorems are, in fact, the limiting probabilities that the minimum degree δ satisfies $\delta \geq 1$ and $\delta \geq 2$, respectively. Indeed, there is a beautifully precise result, as follows.

Consider the *random graph process* that starts with n isolated vertices and adds the $\binom{n}{2}$ possible edges one at a time, uniformly at random. The *hitting time* $\tau_n(\delta \geq 1)$ is the random number of edges that have to be added to make each degree at least 1, and similarly for other graph properties, such as being connected or Hamiltonian. The next result shows that, in the random graph process, the edge whose addition kills the last isolated vertex usually makes the graph connected, and similarly the edge whose addition brings the last vertex up to degree 2 usually makes the graph Hamiltonian.

Theorem 13.7

$$\tau_n(\text{connected}) = \tau_n(\delta \geq 1) \quad a.s.$$

$$\tau_n(\text{Hamiltonian}) = \tau_n(\delta \geq 2) \quad a.s.$$

Minimum spanning trees Let G be any connected graph. Suppose that, for each edge of G, we have a random length (or weight) uniformly distributed on $[0, 1]$, and these random variables are independent. Let $msp(G)$ denote the expected value of the minimum total length of a spanning tree.

The asymptotic behaviour of this quantity for the complete graph K_n was determined by Frieze in 1985, using early results of Erdős and Rényi on the emergence of a 'giant component' in sparse random graphs.

Theorem 13.8 *As $n \to \infty$, $msp(K_n) \to \zeta(3)$, where $\zeta(3) = \sum_{k=1}^{\infty} k^{-3} \sim 1.2$.*

For corresponding results for $msp(K_{n,n})$, see [12].

Almost all Berge graphs are perfect Recall that a graph is *perfect* if the chromatic number of each induced subgraph equals the maximum clique size. A *Berge graph* is a graph with no induced subgraph isomorphic to an odd cycle or the complement of an odd cycle. All perfect graphs are Berge graphs: the *Strong Perfect Graph Conjecture* is that *all Berge graphs are perfect*.

Let P_n be the number of perfect graphs, and let B_n be the number of Berge graphs, on vertices $v_1, \ldots, v_n$. Then $P_n \leq B_n$, and the Strong Perfect Graph Conjecture asserts that $P_n = B_n$. Prömel and Steger [29] have recently shown that $P_n/B_n \to 1$ as $n \to \infty$.

13.3 Random graphs: deterministic results

We now consider some examples of the use of random graphs to obtain deterministic results—that is, results that do not involve randomness.

Ramsey numbers The *Ramsey number* $R(k,l)$ is the least number n such that in any red–blue colouring of the edges of the complete graph K_n, there is either a red K_k or a blue K_l. An easy induction shows that these numbers are finite, and that $R(k,k) \leq (4 + o(1))^k$.

What about lower bounds? Consider a random 2-colouring of the edges of K_n. The probability that there is a monochromatic K_k is at most the expected number of such subgraphs, which is $2\binom{n}{k} \times 2^{-\binom{k}{2}}$. But this quantity is less than 1 if $n \leq 2^{k/2}$, for $k \geq 3$. It follows that in this case there must exist a 2-colouring with no monochromatic K_k, and so $R(k,k) > 2^{k/2}$. Indeed, the argument yields $R(k,k) \geq ck2^{k/2}$ for a positive constant c; see also Section 13.4 below. This beautifully simple proof is from Erdős in 1947 (see, for example, [1]). The lower bound has never been improved, apart from the constant factor c.

Graphs with large chromatic number and large girth The proof of the following theorem is another classic early example of the probabilistic method, due to Erdős in 1959 (see, for example, [1]). Deterministic constructions had to await Lovász in 1968.

Theorem 13.9 *For each k and g, there is a graph H with chromatic number $\chi(H)$ at least k and girth at least g.*

To prove this theorem, it suffices to show that there is a graph G on n vertices such that

(a) each set of $s = \lceil n/k \rceil$ vertices spans at least $\sqrt{n}$ edges;
(b) fewer than $\sqrt{n}$ cycles have length less than g.

For, in this case, we may obtain H by deleting an edge from each cycle in G of length less than g. Clearly, then, H has girth at least g, and $\alpha(H) < s$, so

$$\chi(H) \geq n/\alpha(H) \geq k.$$

Is there such a graph G? Consider $G_{n,p}$, where $p = 2k^2/n$ and n is large. The probability that property (a) fails is at most the expected number of s-sets spanning fewer than $\sqrt{n}$ edges, and this quantity is $o(1)$. For property (b), observe that if E_n denotes the expected number of cycles of length less than g, then

$$E_n \leq \sum_{i=3}^{g-1} n^i p^i \leq (2k^2)^g.$$

Thus the probability that (b) fails is at most $E_n/\sqrt{n} = o(1)$. Hence, for n sufficiently large, there is a graph G as required.

The Hajós conjecture In 1943, Hadwiger conjectured that *if $\chi(G) = k$, then G has a subgraph contractible to K_k*. A natural strengthening of this was Hajós' conjecture of 1961, that *if $\chi(G) = k$, then G has a subgraph homeomorphic to K_k*. In 1979, Catlin produced counter-examples to the Hajós conjecture, but the conjecture was blown out of the water shortly afterwards by Erdős and Fajtlowicz (see [2]). Let $p = \frac{1}{2}$ and $k = \lceil n^{2/3} \rceil$. We have already seen that $\chi(G_{n,p}) \geq n/(2\log_2 n)$ a.s. The probability that $G_{n,p}$ has a subgraph homeomorphic to K_k is at most the probability that there is a k-set of vertices 'missing' at most $n - k$ edges: but this is at most the expected number of such k-sets, which is $o(1)$. Thus a.s. $G_{n,p}$ has no subgraph homeomorphic to K_k. Hence, for all large n, there are graphs with chromatic number at least $n/(2\log_2 n)$ and with no subgraph homeomorphic to K_k for any $k \geq n^{2/3}$ (which is much smaller).

Expanders Expanders are important in theoretical computer science—for example, in sorting networks and for economizing in space or randomness. Roughly speaking, an *expander* is a graph in which each set S of vertices that is not too big has many neighbours outside S; there are many variations on the precise conditions. Generally, it is easy to show that a suitable random graph is an expander with high probability, but hard to give an explicit construction (see, for example, [1, 2]).

13.4 The Lovász Local Lemma

This section continues in the direction of the previous one. The *Lovász Local Lemma* [9] is remarkably useful for proving existence results in graph theory and elsewhere. Let $A_1, \ldots, A_n$ be events in a probability space, and let G be a graph with vertex set $V = \{1, \ldots, n\}$. We call G a *dependency graph* for these events if, for each $i \in V$, the event A_i is independent of the family of all the other events A_j, for j not adjacent to i.

Theorem 13.10 (Lovász Local Lemma: symmetric version)
Let $A_1, \ldots, A_n$ be events with a dependency graph G. Suppose that each event A_i has probability at most p, and that each vertex degree in G is at most d. If $ep(d+1) \leq 1$, then $\Pr(\bigcap_{i=1}^{n} \overline{A_i}) > 0$.

We discuss two applications of this result below: see [1] for extensions, references not given here, and further applications.

Hypergraph-colouring The classic application [9] of the above result is to 2-colouring of hypergraphs.

Theorem 13.11 *Let (E_i) be a family of subsets of a set S, where each subset E_i has size at least k and meets at most d other subsets E_j. If $e(d+1) \leq 2^{k-1}$, then there is a 2-colouring of S such that none of the subsets E_i is monochromatic.*

This result gives a slight improvement in the lower bound for the Ramsey number $R(k,k)$ discussed earlier: we take S to be the edge set of K_n, and the subsets E_i of S to be the edge sets of the $\binom{n}{k}$ possible subgraphs K_k. We obtain only a factor of 2 improvement, but it gives the best bound known.

To prove Theorem 13.11, we colour the points of S blue or red independently and with equal probability. For each subset E_i, let A_i be the event that E_i is monochromatic. Then $\Pr(A_i) \leq 2^{-(k-1)}$. We may form a natural dependency graph G by joining two distinct vertices i and j whenever the corresponding subsets E_i and E_j intersect. Then each vertex in G has degree at most d. But $e(d+1)2^{-(k-1)} \leq 1$, and so by the Local Lemma there must be a colouring as desired.

The condition in Theorem 13.11 can be improved slightly, to $e(d+2) \leq 2^k$ (see [23]). Recently, Beck caused excitement when he gave an efficient general method for *finding* a colouring as in Theorem 13.11, although under more restrictive conditions on k and d.

Linear arboricity A *linear forest* is a forest in which each component is a path. The *linear arboricity* $la(G)$ of a graph G is the minimum number of linear forests needed to cover its edges. The *Linear Arboricity Conjecture* asserts that if G has maximum degree d, then $la(G) \leq \lceil (d+1)/2 \rceil$. It is clear that $la(G) \geq d/2$, and for regular graphs it is easy to check that $la(G) \geq \lceil (d+1)/2 \rceil$. The best asymptotic upper bound known is due to Alon (see [1]).

Theorem 13.12 *There is a constant c such that for every graph G with maximum degree d,*

$$la(G) \leq d/2 + cd^{2/3}(\log d)^{1/3}.$$

The proof uses the Local Lemma in different places. The following result supplies a key step. Note that any graph with n vertices and maximum degree d has an independent set of size at least $n/(d+1)$. At the cost of decreasing the size of the independent set by a constant factor, we can insist that it has a certain structure.

Theorem 13.13 *Let H be a graph with maximum degree d, and let $V_1, V_2, \ldots$ be pairwise disjoint subsets of the vertex set V, each of size $s \geq 2ed$. Then there is an independent set W that contains a vertex from each set V_i.*

To prove this theorem, suppose that from each V_i independently we pick exactly one vertex uniformly at random, and let W be the random set of vertices picked. For each edge f of G, let A_f be the event that W contains both end-vertices of f. Then each event A_f has probability at most s^{-2}, and there is a natural dependency graph G in which each degree is less than $2sd$; in particular, there is a vertex of G for each edge of H, and the vertices of G corresponding to distinct edges f and f' of H are adjacent if some V_i contains an end-vertex of both these edges. But $es^{-2}2sd = 2ed/s \le 1$. So, by the Local Lemma, there is some set W such that none of the events A_f holds—that is, W is an independent set.

By using Alon's methods and the Local Lemma again, we can obtain a result on random graphs that lends some support to the Linear Arboricity Conjecture (see [25, 27]).

Theorem 13.14

(a) For fixed degree r, a random r-regular graph has linear arboricity equal to $\lceil (r+1)/2 \rceil$ a.s.

(b) For fixed edge probability p, the random graph $G_{n,p}$ has linear arboricity equal to $\lceil \Delta/2 \rceil$ a.s.

13.5 Deterministic graphs: random methods

Now we suppose that we are given some (deterministic) graph, and consider some random things we can do with it!

Finding a minimum cut   Let G be a connected graph of order n (≥ 3). There are well-known efficient deterministic methods for finding a minimum cut based on network flow ideas, but here is a far simpler randomized method (see [15, 17]):

while at least three vertices remain
choose an edge uniformly at random and contract it
(keeping any multiple edges formed, but not loops);
output the set of edges remaining.

We claim that the probability that we obtain a minimum cut is at least $2/n^2$, and so the probability that a minimum cut is not found in tn^2 independent attempts is at most $(1 - 2/n^2)^{tn^2} \le e^{-2t}$.

To see why this claim is true, consider a fixed minimum cut C of size k. The algorithm returns C if none of the edges in C is chosen. Now G has at least $nk/2$ edges, since each vertex degree is at least k, and so the probability that the first edge chosen is *not* in C is at least $1 - (2/n)$. Conditional on this happening, the probability that the second edge chosen is not in C is at least $1 - 2/(n-1)$, and so on. Hence the probability that C is returned is at least

$$\prod_{i=1}^{n-2}\left(1 - \frac{2}{n-i+1}\right) = \frac{2}{n(n-1)}.$$

Finding a maximum cut   Suppose that we have a graph $G = (V, E)$ with non-negative weights $w_{ij} = w_{ji}$ on its edges. The *maximum (weighted) cut problem* is to find a set S of vertices that maximizes the weight of the edges between S

and $V \setminus S$. This problem is NP-hard. The following elegant random rounding idea has recently been shown to give good approximations (see [13]).

The maximum cut problem may be modelled by the quadratic integer program

$$\text{find } \max \tfrac{1}{2} \sum_{i<j} w_{ij}(1 - y_i y_j), \quad \text{subject to } y_i \in \{-1, 1\} \text{ for all } i \in V,$$

where we take G to be K_n, without loss of generality. Here $y_i = 1$ corresponds to the vertex i being in S, say. We relax this program by replacing the variables $y_i \in \{-1, 1\}$ by variables $\mathbf{v}_i \in S^n$, where S^n is the unit Euclidean ball in $\mathbf{R}^n$. This gives the semi-definite program

$$\text{find } \max \tfrac{1}{2} \sum_{i<j} w_{ij}(1 - \mathbf{v}_i \cdot \mathbf{v}_j), \quad \text{subject to } \mathbf{v}_i \in S^n \text{ for all } i \in V,$$

which can be solved in polynomial time.

The randomized method takes a (near) optimal solution $\mathbf{v}_1, \ldots, \mathbf{v}_n$ to the relaxed program, picks a random vector $\mathbf{r} \in S^n$, and outputs $S = \{i : \mathbf{v}_i \cdot \mathbf{r} \geq 0\}$. It turns out that the expected weight of the corresponding cut is at least 0.878 times the optimal value. Similar ideas lead to an algorithm that properly colours a 3-colourable graph using $O(n^{1/4} \log n)$ colours (see [16]). Both of these algorithms are naturally randomized, but may be de-randomized to run in polynomial time and beat the best previously known such approximation algorithms.

Random walks on graphs A *random walk* on a graph G starts at some vertex v_0, picks a neighbour v_1 uniformly at random, from v_1 picks a neighbour v_2 uniformly at random, and so on. There are intimate connections between random walks and electrical networks. Given vertices u and v in G, let the *commute time* C_{uv} be the expected number of steps starting from u to reach v plus the expected number of steps starting from v to reach u, and let the *effective resistance* R_{uv} be the overall electrical resistance between u and v when each edge has resistance 1.

Seemingly unrelated quantities like commute times and effective resistances satisfy the same harmonic relations, and we find that if G is connected and has n vertices and m edges, then $C_{uv} = 2m R_{uv}$. It follows that $C_{uv} < n^3$ (see [6]).

Now suppose that we wish to test whether u and v are in the same component of G, when n is very large. Starting at the vertex u, perform a random walk for $2n^3$ steps. If we ever reach v we say 'yes' (correctly), and if not we say 'no'. By Markov's inequality, the probability that we say 'no' incorrectly is less than $\tfrac{1}{2}$. This leads to a randomized algorithm that works in logarithmic space, since we need remember only where we are and how many steps we have taken.

Random walks on large expander graphs G are particularly important. The random walk is 'rapidly mixing', in that it converges exponentially fast to the stationary distribution. If G is also regular, then the stationary distribution is uniform. But then we can sample almost uniformly from the vertices of G, and from this we can approximately count the number of vertices. This approach allows us to estimate, for example, the number of perfect matchings in a dense graph (see [14]) and the volume of a convex body (see [8]).

The triangle problem It is NP-hard to determine whether there is a 2-colouring of the vertices of a graph G such that no triangle is monochromatic. But suppose that G has a proper vertex 3-colouring. Then clearly there must exist a 2-colouring of the vertices as above—just amalgamate two of the colours in a proper 3-colouring; the *triangle problem* for G is to find such a 2-colouring.

Consider the following randomized 'bit-flipping' method (see [22]):

start with an arbitrary 2-colouring.
while we find a monochromatic triangle
    pick a random vertex in the triangle, and flip its colour.

What could be simpler? It is also effective: if G has n vertices, then the expected number of iterations until we find a 2-colouring with no monochromatic triangles is at most $\frac{3}{8}n^2$. To see this, fix a proper vertex-colouring of G with the colours -1, 0 and 1, and use the colours -1 and 1 for the 2-colouring; then half of the dot product of the two colourings traces a symmetric random walk bounded within an interval of length n. There are good deterministic ways to solve the triangle problem using linear algebra, but there is no known 'graphical' method (see [24]).

Total colouring Let us return briefly to total colourings, and prove an upper bound for $\chi''(G)$ by a simple randomized method (see [20, 26]).

Theorem 13.15 *Let G be a graph with n vertices and maximum vertex degree Δ, and let k be an integer with $k! \geq n$. Then $\chi''(G) \leq \Delta + k + 2$.*

The algorithm starts with a vertex-colouring with $\Delta+1$ colours and an edge-colouring with $\Delta + 1$ different colours. We pick a random bijection π between the two sets of colours, identify the corresponding colours, and let the 'rejection graph' R contain those edges that would give a colour clash in G. Now

$$\chi''(G) \leq \Delta + 1 + \chi'(R).$$

But simple manipulations show that, for each vertex, the probability that its degree in R is at least $k+1$ is less than $1/k!$. Hence

$$\Pr(\Delta(R) \geq k+1) < n/k! \leq 1,$$

and so for some bijection π we have

$$\chi'(R) \leq \Delta(R) + 1 \leq k + 1.$$

The bound can in fact be tightened (see [7]) to

$$\chi''(G) \leq \chi'(G) + k.$$

To block or to direct? Two town planners, Blocker and Director, are bored and play a game on a graph G. Blocker flips a coin independently for each edge e, and blocks (deletes) it with probability $\frac{1}{2}$. Let $P_B(hc)$ and $P_B(s \to t)$ denote the corresponding probabilities that there remains an unblocked Hamiltonian cycle or path from s to t. Director similarly flips a coin for each edge e, and orients it so that each orientation is equally likely. Let $P_D(hc)$ and $P_D(s \to t)$ denote the corresponding probabilities that there is a directed Hamiltonian cycle or path from s to t.

How do these probabilities compare? It turns out (see [18]) that

$$P_B(hc) \leq P_D(hc) \quad \text{and} \quad P_B(s \to t) = P_D(s \to t).$$

We can prove these results by an induction involving the edges, using the following idea. Consider a particular edge $e = uv$ of G, and fix any pattern of blocking or orienting on all the other edges. If there is a Hamiltonian cycle not using e (and respecting any orientations), then what happens with e is irrelevant, so we assume that this does not happen. Similarly, we assume that there is a Hamiltonian path from u to v that does not use e, or one from v to u, or both. But then the probability that Blocker permits a Hamiltonian cycle is $\frac{1}{2}$, and the probability that Director does is either $\frac{1}{2}$ or 1.

13.6 Concentration of measure and isoperimetric inequalities

In 1979 Maurey used a 'concentration of measure' inequality to prove an isoperimetric inequality for 'permutation graphs' (see below). That started a ball rolling which has had exciting effects in the study of normed spaces, random graphs, the mathematics of operational research and theoretical computer science (see [19], which contains references not given here). This ball is still on the move (see [33]).

We start with a special case of the 'independent bounded differences inequality'.

Theorem 13.16 *Let $X_1, \ldots, X_n$ be independent random variables and, for each k, let X_k take values in a set A_k. Suppose that the measurable function $f : \prod_k A_k \to \mathbf{R}$ satisfies $|f(\mathbf{x}) - f(\mathbf{x}')| \leq 1$, whenever the vectors $\mathbf{x}$ and $\mathbf{x}'$ differ only in one coordinate. Let Y be the random variable $f(X_1, \ldots, X_n)$. Then, for any $t > 0$, both $\Pr(Y - EY \geq t)$ and $\Pr(Y - EY \leq -t)$ are at most $\exp(-2t^2/n)$.*

This result can be proved by elementary manipulations with the moment generating function $M_Y(s) = E(e^{sY})$ of Y. It extends the standard Chernoff bound for binomial random variables: to see this, take each set $A_k = \{0, 1\}$ and let $f(x_1, \ldots, x_n) = \sum_k x_k$. In the random graph $G_{n,p}$, let the random variable X_k be the set of edges of the form $v_i v_k$, with $i < k$. Then we obtain the following result.

Theorem 13.17 *Suppose that the graph function f satisfies $|f(G) - f(G')| \leq 1$ whenever G' can be obtained from G by switching edges incident with a single vertex. Then the random variable $Y = f(G_{n,p})$ satisfies*

$$\Pr(|Y - EY| \geq t) \leq 2 \exp(-2t^2/n), \quad \text{for any } t > 0.$$

If we take f as the chromatic number χ, we see that $\chi(G_{n,p})$ is concentrated around its mean.

Consider a finite metric space (V, d). We shall be interested in the case when V is the vertex set of a graph and d measures distance in the graph. For $A \subseteq V$ and $t \geq 0$, the *t-neighbourhood* A_t of A is the set $\{v \in V : d(v, A) \leq t\}$, where $d(v, A) = \min\{d(v, x) : x \in A\}$. A 'discrete isoperimetric inequality' gives a lower bound on $|A_t|$ that depends on $|A|$ and t.

We now focus on the n-cube Q_n (see Chapter 1), and the *permutation graph* associated with the symmetric group S_n, in which the vertices are the $n!$ permutations of $\{1, \ldots, n\}$, and two vertices g and h are adjacent if $g^{-1}h$ is a transposition.

Theorem 13.18 *Let A be any set of half of the 2^n vertices of the n-cube Q_n, and let $t_0 = \sqrt{n/2}$. Then, for any $t \geq t_0$,*

$$|A_t|/2^n \geq 1 - \exp(-2(t - t_0)^2/n).$$

We can deduce this isoperimetric inequality from Theorem 13.16 by taking each X_k uniformly distributed on $A_k = \{0, 1\}$ and letting $f(x) = d(x, A)$ (see [19]). From a similar bounded differences inequality, we can obtain what is essentially the result of Maurey with which we started this section: we simply replace the phrase 'the 2^n vertices of the n-cube Q_n' in Theorem 13.18 by 'the $n!$ vertices of the permutation graph of S_n'.

In the cube Q_n, a *Hamming ball* of radius r centred at a vertex v consists of all vertices w such that the distance $d(v, w) < r$, together with some vertices at distance r. A seminal theorem of Harper in 1966 (see [3]) concerns two sets A and C of vertices in Q_n. The theorem asserts that there exist Hamming balls B_0 centred at the all-0 vector and B_1 centred at the all-1 vector, such that we have $|B_0| = |A|$, $|B_1| = |C|$ and $d(B_0, B_1) \geq d(A, C)$. This yields the precise solution to the isoperimetric problem for the n-cube. From Harper's theorem and a Chernoff bound, we can obtain the following cleaner version of Theorem 13.18.

Theorem 13.19 *Let A be any set of 2^{n-1} vertices of the n-cube Q_n. Then, for any $t \geq 0$,*

$$|A_t|/2^n \geq 1 - \exp(-2t^2/n).$$

The isoperimetric inequality in Theorem 13.18 was obtained from Theorem 13.16 on concentration of measure. It is interesting that we can also reverse this process. Recall that a number L is a *median* of a random variable Y if both $\Pr(Y \leq L) \geq \frac{1}{2}$ and $\Pr(Y \geq L) \geq \frac{1}{2}$.

Theorem 13.20 *Let X be uniformly distributed over $V = \{0, 1\}^n$, let the function f on V satisfy $|f(\mathbf{x}) - f(\mathbf{x}')| \leq 1$ whenever $\mathbf{x}$ and $\mathbf{x}'$ differ only in one coordinate, and let L be a median of $Y = f(X)$. Then, for any $t \geq 0$, both $\Pr(Y - L > t)$ and $\Pr(Y - L < -t)$ are at most $\exp(-2t^2/n)$.*

This result is similar to Theorem 13.16 with 'mean' replaced by 'median'. To see how it follows from Theorem 13.19, let $A = \{x \in V : f(x) \leq L\}$. Then

$$\Pr(Y - L \leq t) \geq \Pr(X \in A_t) \geq 1 - \exp(-2t^2/n).$$

References

1. N. Alon and J. H. Spencer, *The Probabilistic Method*, Wiley-Interscience, 1992; *MR* **93h**: 60002.
2. B. Bollobás, *Random Graphs*, Academic Press, 1985; *MR* **87f**: 05152.
3. B. Bollobás, *Combinatorics*, Cambridge University Press, 1986; *MR* **88g**: 05001.
4. B. Bollobás, The chromatic number of random graphs, *Combinatorica* **8** (1988), 49–55; *MR* **89i**: 05244.
5. B. Bollobás, T. I. Fenner and A. M. Frieze, An algorithm for finding Hamilton paths and cycles in random graphs, *Combinatorica* **7** (1987), 327–341; *MR* **89h**: 05049.
6. A. K. Chandra, P. Raghavan, W. L. Ruzzo, R. Smolensky and P. Tiwari, The electrical resistance of a graph captures its commute and cover times, *Proc. 21st Annual ACM Symp. Theory of Computing*, ACM Press, 1989, pp. 574–586.
7. A. G. Chetwynd and R. Häggkvist, Some upper bounds on the total and list chromatic numbers of multigraphs, *J. Graph Theory* **16** (1992), 503–516; *MR* **93i**: 05060.
8. M. E. Dyer, A. M. Frieze and R. Kannan, A random polynomial-time algorithm for approximating the volume of convex bodies, *J. Assoc. Comput. Mach.* **38** (1991), 1–17; *MR* **91m**: 68162.
9. P. Erdős and L. Lovász, Problems and results on 3-chromatic hypergraphs and some related questions, *Infinite and Finite Sets* (ed. A. Hajnal et al.), North-Holland, 1975, pp. 609–627; *MR* **52#2938**.
10. P. Erdős and A. Rényi, On the evolution of random graphs, *Publ. Math. Inst. Hungar. Acad. Sci.* **5** (1960), 17–61; *MR* **23#A2338**.
11. A. Frieze, B. Jackson, C. McDiarmid and B. A. Reed, Edge-colouring random graphs, *J. Combin. Theory (B)* **45** (1988), 135–149; *MR* **89m**: 05101.
12. A. Frieze and C. McDiarmid, On random minimum length spanning trees, *Combinatorica* **9** (1989), 363–374; *MR* **91j**: 05092.
13. M. X. Goemans and D. R. Williamson, 0.878 approximation algorithms for max cut and max 2SAT, *Proc. 26th Annual ACM Symp. Theory of Computing*, ACM Press, 1994, pp. 422–431.
14. M. R. Jerrum and A. Sinclair, Approximating the permanent, *SIAM J. Comput.* **18** (1989), 1149–1178; *MR* **91a**: 05075.
15. D. R. Karger, Global min-cuts in RNC, and other ramifications of a simple min-cut algorithm, *Proc. 4th ACM-SIAM Symp. Discrete Algorithms*, ACM Press, 1993; *MR* **93m**: 68134.

16. D. R. Karger, R. Motwani and M. Sudan, Approximate graph coloring by semidefinite programming, *Proc. 35th Symp. Foundations of Computer Science*, 1994, pp. 2–13.
17. D. R. Karger and C. Stein, An $O(n^2)$ algorithm for minimum cuts, *Proc. 26th Annual ACM Symp. Theory of Computing*, ACM Press, 1994, pp. 757–765.
18. C. McDiarmid, General percolation and random graphs, *Adv. in Appl. Probab.* **13** (1981), 40–60.
19. C. McDiarmid, On the method of bounded differences, *Surveys in Combinatorics, 1989* (ed. J. Siemons), London Math. Soc. Lecture Note Ser. **141**, Cambridge University Press, 1989, pp. 148–188; *MR* **91e**: 05077.
20. C. McDiarmid, Colourings of random graphs, *Graph Colouring* (ed. R. Nelson and R. J. Wilson), Pitman Res. Notes Math. Ser. **218**, Longman, 1990, pp. 79–86; *MR* **93k**: 05072.
21. C. McDiarmid, On the chromatic number of random graphs, *Random Structures Algorithms* **1** (1990), 435–442; *MR* **92m**: 05172.
22. C. McDiarmid, A random recolouring method for graphs and hypergraphs, *Combin. Probab. Comput.* **2** (1993), 363–365; *MR* **94k**: 05190.
23. C. McDiarmid, Hypergraph colouring and the Lovász Local Lemma, *Discrete Math.*, to appear.
24. C. McDiarmid, A random bit-flipping method for seeking agreement, *Random Structures Algorithms*, to appear.
25. C. McDiarmid and B. Reed, Linear arboricity of random regular graphs, *Random Structures Algorithms* **1** (1990), 443–445; *MR* **92m**: 05173.
26. C. McDiarmid and B. Reed, On total colourings of graphs, *J. Combin. Theory (B)* **57** (1993), 122–130; *MR* **93j**: 05058.
27. C. McDiarmid and B. Reed, Almost every graph can be covered by $\lceil \Delta/2 \rceil$ linear forests, *Combin. Probab. Comput.* **4** (1995), 257–268; *MR* **96f**: 05147.
28. L. Perkovic and B. A. Reed, Edge coloring in polynomial expected time, to appear.
29. H. J. Prömel and A. Steger, Almost all Berge graphs are perfect, *Combin. Probab. Comput.* **1** (1992), 53–79; *MR* **93e**: 05089.
30. H. J. Prömel and A. Steger, Coloring clique-free graphs in linear expected time, *Random Structures Algorithms* **3** (1992), 375–402; *MR* **93m**: 68083.
31. R. W. Robinson and N. C. Wormald, Almost all cubic graphs are Hamiltonian, *Random Structures Algorithms* **3** (1992), 117–125; *MR* **93d**: 05105.
32. R. W. Robinson and N. C. Wormald, Almost all regular graphs are Hamiltonian, *Random Structures Algorithms* **5** (1994), 363–374; *MR* **95g**: 05092.
33. M. Talagrand, Concentration of measure and isoperimetric inequalities in product spaces, *Inst. Hautes Études Sci. Publ. Math.* **81** (1995), 73–205.

14
Statistics

PETER WILD

In this chapter we give several examples of the close connection between statistical design and graph spectra. Focusing on block designs, we describe measures of efficiency of an experimental design in terms of certain functions of eigenvalues of the Fisher information matrix. The most efficient designs correspond to graphs with some extreme spectral behaviour.

14.1 Introduction

One of the main aims of mathematics is to model problems that arise in the real world—that is, to represent a real-life situation as a mathematical system. By this means, many different instances of the same problem can be tackled using a single theory. This theory may be descriptive, providing a clearer understanding of the problem. It may also be a tool that can be used to prove previously unsuspected results that may be applied to the real-life situation to make predictions about its behaviour.

Much of our knowledge of the real world comes from systematic observation of its behaviour and the collection of data from experiments. Statistics is the science that deals with the analysis and interpretation of data—in particular, the nature and significance of error in our facts and figures. Our models of the real world and the deviation (error) of our observations from these must be well understood, in order that our conclusions about the world's behaviour and decisions based on our observations may be well founded. Mathematical models, therefore, form an important part of statistics.

There is a natural relationship between graph theory and statistics. Statistics is often interested in making comparisons between objects, and graphs are a useful and powerful way of representing relations between objects. We mention just two of the many statistical methods that demonstrate this. Hypothesis testing allows the comparison of possible alternatives for a probability distribution underlying the observations and, in experimental design, comparison between parameters or factors may be made by estimation of treatment contrasts.

A direct way in which graphs may play a role in statistics is to provide a descriptive apparatus (model) that identifies those relationships of importance and aids the collection of relevant data for a meaningful statistical analysis. However, there is an even deeper connection between graph theory and certain aspects of statistics—in particular, experimental design. Here the theory that has been developed through the study of graphs contributes to the analysis itself.

This is nowhere more apparent than in the connection between graph theory and experimental design, through the use of matrix algebra techniques. It is this connection that we are most concerned with in this chapter. A design matrix may be associated with an experimental design, and when this matrix is related to an adjacency matrix of a graph, information about the experimental design may be obtained from properties of the graph. In particular, it is the eigenvalues of these matrices that interest us.

It was the work of Fisher [5] that instigated much research in experimental design and its relation to spectral properties. The eigenvalues of the Fisher information matrix of an experimental design are related to variances of certain estimates of contrasts. The efficient and optimal design of statistical experiments involves minimizing these variances, and so is concerned with those matrices having extreme spectral properties.

There have been many applications of matrix algebra to graph theory, and algebraic graph theory has proved to be a fruitful area of research (see Chapter 6). Results in this area may be related to problems in statistics. The connections that we consider concern spectral properties of graphs and their relation to various optimality criteria for statistical design. Properties of graphs that are of interest include connectedness, numbers of spanning trees, numbers of closed walks, and characterizations of line graphs.

14.2 Experimental design

Often an experiment is carried out to determine the effect of some variable or other on a system. This is sometimes achieved by fixing all other variables and observing the response when different levels of one particular variable, or *treatment*, are applied. As it is difficult to control every aspect of the environment under which an experiment is performed, when experiments are repeated there is an essential probabilistic variation in the outcome from that which is expected. It is the aim of good experimental design to provide valid measures of this error. Of course, experimental error could also arise because the experiment was performed poorly or was badly designed. But here we consider only error due to random, rather than biased, fluctuations.

The basic unit of an experimental design is the *plot*. This is the experimental unit to which some level of the variable under consideration is applied. Often many different treatments are under consideration, and the *design* is an assignment of treatments, or *varieties*, to the plots. From each plot is obtained a *yield*, or result of the experiment. This yield depends on which treatment was assigned to it, and is subject to random error. Here we concentrate on *block designs* in which the plots are grouped into *blocks*, so that plots within a block are homogeneous, and subject to the same random error.

An example from agricultural trials is as follows. The aim is to compare yields from several different varieties of wheat. These are the treatments and they are grown in several fields. As one field may be more fertile than another, the fields (blocks) are divided into strips (plots) and each strip is sown with a different variety. Each variety of wheat (treatment) is sown in several plots in various blocks. The aim of the design is to distribute the varieties among the fields so that a fair comparison may be made between them.

In order to describe an experimental design, we record which treatments are assigned to which plots and which plots are in which blocks, by means of matrices. The *plot × treatment* matrix $\mathbf{X} = (x_{ij})$ is the $n \times v$ matrix such that $x_{ij} = 1$ if plot i is assigned treatment j, and $x_{ij} = 0$ otherwise. Since any plot can be assigned only one treatment, each row of $\mathbf{X}$ has one 1 and $v - 1$ zeros. The *plot × block* matrix $\mathbf{Z} = (z_{ij})$ is the $n \times b$ matrix such that $z_{ij} = 1$ if plot i is in block j, and $z_{ij} = 0$ otherwise. Each row of $\mathbf{Z}$ has one 1 and $b - 1$ zeros. The *incidence* matrix of the design is the $v \times b$ matrix $\mathbf{N} = \mathbf{X}^T \mathbf{Z} = (n_{ij})$, where n_{ij} is the number of plots of block j that are assigned variety i. If $n_{ij} = 0$ or 1 for all i, j, then the design is called *binary*.

The following example will be considered throughout the chapter. (Where we leave the example and return to the main text, you will see the symbol ◊.)

Example 14.1 Consider the design $\mathcal{D}$ with 3 treatments $\{A, B, C\}$ which are applied to the 6 plots of 3 blocks as follows:

$$\text{block } 1 = \{A, B\}; \quad \text{block } 2 = \{A, C\}; \quad \text{block } 3 = \{B, C\}.$$

Then the plot × treatment matrix $\mathbf{X}$, the plot × block matrix $\mathbf{Z}$, and the incidence matrix $\mathbf{N} = \mathbf{X}^T \mathbf{Z}$, are as follows:

$$\mathbf{X} = \begin{pmatrix} 1 & 0 & 0 \\ 0 & 1 & 0 \\ 1 & 0 & 0 \\ 0 & 0 & 1 \\ 0 & 1 & 0 \\ 0 & 0 & 1 \end{pmatrix}, \quad \mathbf{Z} = \begin{pmatrix} 1 & 0 & 0 \\ 1 & 0 & 0 \\ 0 & 1 & 0 \\ 0 & 1 & 0 \\ 0 & 0 & 1 \\ 0 & 0 & 1 \end{pmatrix} \quad \text{and} \quad \mathbf{N} = \begin{pmatrix} 1 & 1 & 0 \\ 1 & 0 & 1 \\ 0 & 1 & 1 \end{pmatrix}. \quad \Diamond$$

When an experiment is performed, we obtain a yield from each plot. Thus there is an $n \times 1$ real vector of yields $\mathbf{y} = (y_i) \in \mathbf{R}^n$, where y_i is the result of the experiment in plot i. If the experiment were repeated, we would expect fluctuations in the results dependent on the ambient conditions. Thus we have a random variable $\mathbf{Y}$, of which $\mathbf{y}$ is one possible outcome.

In our analysis, we assume that the yield depends linearly on the treatments; this is often a good approximation, possibly after some transformation of the data. Thus there is a $v \times 1$ vector τ of treatment parameters such that the expected value of the yield in plot i is $E(\mathbf{Y}_i) = \tau_j$ if plot i is assigned treatment j. In matrix form, we have

$$E(\mathbf{Y}) = \mathbf{X}\tau.$$

Thus $E(\mathbf{Y})$ belongs to a v-dimensional subspace spanned by the columns of $\mathbf{X}$; column j of $\mathbf{X}$ has 1s in positions corresponding to plots assigned treatment j, and 0s elsewhere. Thus our assumed model is that $\mathbf{Y}$ is a random variable with expectation $\mathbf{X}\tau$, for some unknown treatment parameters τ. We wish to estimate the treatment parameters, using the data $\mathbf{y}$ collected from our experiment.

We assume that plots within a block are subject to the same random error, and that different blocks may be subject to different conditions, so that the variance of $\mathbf{Y}$ may depend on the blocks. The model may be written $y_i = \tau_j + \beta_k + \varepsilon_i$, where β is a $b \times 1$ vector of block effects, ε is an $n \times 1$ vector of *errors*, and plot i is in block k and has been assigned treatment j. Thus

$$\mathbf{Y} = \mathbf{X}\tau + \mathbf{Z}\beta + \varepsilon.$$

We assume that β and ε are independent random variables, with $E(\beta) = 0$ and $E(\varepsilon) = 0$, and variances

$$\text{var}(\beta) = E(\beta\beta^T) = \sigma_b^2 \mathbf{I}, \quad \text{var}(\varepsilon) = \sigma^2 \mathbf{I}.$$

If we assume the simpler model that all plots are alike, independent of which block they are in, then $\beta = 0$ and $\mathbf{Y} = \mathbf{X}\tau + \varepsilon$. In this case, the *dispersion matrix* of $\mathbf{Y}$ is

$$\mathbf{V} = \mathbf{D}(\mathbf{Y}) = E((\mathbf{Y} - E(\mathbf{Y}))(\mathbf{Y} - E(\mathbf{Y}))^T) = E(\varepsilon\varepsilon^T) = \sigma^2 \mathbf{I}.$$

If we have the result $\mathbf{y}$ of one experiment, then the least squares estimate $\hat{\tau}$ of the treatment parameters is the one that minimizes the sum of squares of the deviation of the yield from the expected value $\mathbf{X}\hat{\tau}$. This provides the best straight line fit to the data. Thus the least squares estimate is such that

$$\sum \varepsilon_i^2 = \|\mathbf{y} - \mathbf{X}\hat{\tau}\|^2 = (\mathbf{y} - \mathbf{X}\hat{\tau})^T(\mathbf{y} - \mathbf{X}\hat{\tau})$$

is a minimum. Setting the partial derivatives of $(\mathbf{y} - \mathbf{X}\tau)^T(\mathbf{y} - \mathbf{X}\tau)$ with respect to $\tau_1, \ldots, \tau_v$ to zero, we obtain the *normal equations*

$$\mathbf{X}^T\mathbf{y} - \mathbf{X}^T\mathbf{X}\hat{\tau} = \mathbf{0}.$$

The least squares estimate is $\hat{\tau} = (\mathbf{X}^T\mathbf{X})^{-1}\mathbf{X}^T\mathbf{y}$, and $\mathbf{X}\hat{\tau} = \mathbf{X}(\mathbf{X}^T\mathbf{X})^{-1}\mathbf{X}^T\mathbf{y}$ is the orthogonal projection of $\mathbf{y}$ onto the subspace spanned by the columns of $\mathbf{X}$.

Example 14.1 (continued) Suppose that an experiment is conducted using design $\mathcal{D}$, and that the yields are $\mathbf{y} = (2, 13, 4, 21, 15, 23)^T$. Now $\mathbf{X}^T\mathbf{X} = 2\mathbf{I}$ and $\mathbf{X}^T\mathbf{y} = (6, 28, 44)^T$, so $\hat{\tau} = (3, 14, 22)^T$. In this case, the estimate of a treatment parameter is simply the average of the yields for the plots assigned that treatment. ◊

With the model $\mathbf{Y} = \mathbf{X}\tau + \mathbf{Z}\beta + \varepsilon$ and $E(\mathbf{Y}) = \mathbf{X}\tau$, the dispersion matrix of $\mathbf{Y}$ is

$$\mathbf{V} = \mathbf{D}(\mathbf{Y}) = E((\mathbf{Z}\beta + \varepsilon)(\mathbf{Z}\beta + \varepsilon)^T) = \sigma_b^2 \mathbf{Z}\mathbf{Z}^T + \sigma^2 \mathbf{I}.$$

If we know σ_b and σ, or their ratio, then we can write $\mathbf{V} = \sigma_1^2 \mathbf{P}\mathbf{P}^T$, for some matrix $\mathbf{P}$, since $\mathbf{V}$ is positive definite. Let $\mathbf{W} = \mathbf{P}^{-1}\mathbf{Y}$. Then

$$E(\mathbf{W}) = \mathbf{P}^{-1} E(\mathbf{Y}) \quad \text{and} \quad D(\mathbf{W}) = \mathbf{P}^{-1} \mathbf{V} (\mathbf{P}^{-1})^T = \sigma_1^2 \mathbf{I}.$$

Now, finding the least squares estimate using $\mathbf{w} = \mathbf{P}^{-1}\mathbf{y}$, we obtain the generalized least squares estimate $\hat{\tau}$ that minimizes $(\mathbf{w} - E(\mathbf{w}))^T (\mathbf{w} - E(\mathbf{w}))$; that is,

$$(\mathbf{y} - \mathbf{X}\hat{\tau})^T (\mathbf{P}^{-1})^T \mathbf{P}^{-1} (\mathbf{y} - \mathbf{X}\hat{\tau}) = (\mathbf{y} - \mathbf{X}\hat{\tau}) \mathbf{V}^{-1} (\mathbf{y} - \mathbf{X}\hat{\tau})$$

is a minimum, and the normal equations are

$$\mathbf{X}^T \mathbf{V}^{-1} \mathbf{X} \hat{\tau} = \mathbf{X}^T \mathbf{V}^{-1} \mathbf{y}.$$

In this case, $\mathbf{X}\hat{\tau}$ is the projection of $\mathbf{y}$ onto the subspace spanned by the columns of $\mathbf{X}$, orthogonal with respect to the inner product $\langle \mathbf{x}, \mathbf{y} \rangle = \mathbf{x}^T \mathbf{V}^{-1} \mathbf{y}$.

Example 14.1 (continued) Suppose that $\sigma = \sigma_b$. Then the dispersion matrix is

$$\mathbf{V} = \sigma(\mathbf{Z}\mathbf{Z}^T + \mathbf{I}) = \begin{pmatrix} 2 & 1 & 0 & 0 & 0 & 0 \\ 1 & 2 & 0 & 0 & 0 & 0 \\ 0 & 0 & 2 & 1 & 0 & 0 \\ 0 & 0 & 1 & 2 & 0 & 0 \\ 0 & 0 & 0 & 0 & 2 & 1 \\ 0 & 0 & 0 & 0 & 1 & 2 \end{pmatrix}, \quad \text{and} \quad \mathbf{V}^{-1} = \tfrac{1}{27} \begin{pmatrix} 2 & -1 & 0 & 0 & 0 & 0 \\ -1 & 2 & 0 & 0 & 0 & 0 \\ 0 & 0 & 2 & -1 & 0 & 0 \\ 0 & 0 & -1 & 2 & 0 & 0 \\ 0 & 0 & 0 & 0 & 2 & -1 \\ 0 & 0 & 0 & 0 & -1 & 2 \end{pmatrix}.$$

The coefficient matrices of the normal equations are

$$\mathbf{X}^T \mathbf{V}^{-1} \mathbf{X} = \tfrac{1}{27} \begin{pmatrix} 4 & -1 & -1 \\ -1 & 4 & -1 \\ -1 & -1 & 4 \end{pmatrix} \quad \text{and} \quad \mathbf{X}^T \mathbf{V}^{-1} = \tfrac{1}{27} \begin{pmatrix} 2 & -1 & 2 & -1 & 0 & 0 \\ -1 & 2 & 0 & 0 & 2 & -1 \\ 0 & 0 & -1 & 2 & -1 & 2 \end{pmatrix}.$$

For the yields $\mathbf{y} = (2, 13, 4, 21, 15, 23)^T$, we have the solution

$$\hat{\tau} = (3.4, 14, 21.6)^T. \quad \diamond$$

If we do not know σ_b and σ, then we cannot get a direct single estimate. However, σ results from the variation within blocks, and σ_b results from the variation between blocks, and so we get a 'within-block' and a 'between-block' estimate. The block averaging operator $\mathbf{B} = \mathbf{Z}(\mathbf{Z}^T \mathbf{Z})^{-1} \mathbf{Z}^T$ is the projection onto the column space of $\mathbf{Z}$. The vector $\mathbf{B}\mathbf{y}$ is the vector whose ith entry is the average of the yields over the plots in the block containing plot i. The operator $\mathbf{I} - \mathbf{B}$ is the projection onto the orthogonal subspace—this operator eliminates block effects.

Example 14.1 (continued) The block averaging operator is

$$\mathbf{B} = \tfrac{1}{2}\begin{pmatrix} 1 & 1 & 0 & 0 & 0 & 0 \\ 1 & 1 & 0 & 0 & 0 & 0 \\ 0 & 0 & 1 & 1 & 0 & 0 \\ 0 & 0 & 1 & 1 & 0 & 0 \\ 0 & 0 & 0 & 0 & 1 & 1 \\ 0 & 0 & 0 & 0 & 1 & 1 \end{pmatrix}. \quad \diamond$$

Now $(\mathbf{I} - \mathbf{B})\mathbf{Y}$ has dispersion matrix

$$(\mathbf{I} - \mathbf{B})\mathbf{D}(\mathbf{Y})(\mathbf{I} - \mathbf{B})^T = \sigma^2(\mathbf{I} - \mathbf{B}).$$

Hence, in the subspace spanned by the columns of $(\mathbf{I} - \mathbf{B})\mathbf{X}$, generalized least squares is the usual least squares, since in this subspace the projector $\mathbf{I} - \mathbf{B}$ acts as the identity. Thus the estimate $\hat{\tau}$ *in this subspace* minimizes the expression

$$\|(\mathbf{I} - \mathbf{B})\mathbf{y} - (\mathbf{I} - \mathbf{B})\mathbf{X}\tau\|^2.$$

The *reduced normal equations* are

$$\mathbf{X}^T(\mathbf{I} - \mathbf{B})\mathbf{X}\hat{\tau} = \mathbf{X}^T(\mathbf{I} - \mathbf{B})\mathbf{y},$$

and $\mathbf{X}\hat{\tau}$ is the orthogonal projection of $\mathbf{y}$ onto the column space of $(\mathbf{I} - \mathbf{B})\mathbf{X}$. The matrix

$$\mathbf{C} = \mathbf{X}^T(\mathbf{I} - \mathbf{B})\mathbf{X} = \mathbf{X}^T\mathbf{X} - \mathbf{N}(\mathbf{Z}^T\mathbf{Z})^{-1}\mathbf{N}^T$$

is called the *Fisher information matrix*.

Example 14.1 (continued) The coefficient matrices of the normal equations are

$$\mathbf{X}^T(\mathbf{I} - \mathbf{B}) = \tfrac{1}{2}\begin{pmatrix} 1 & -1 & 1 & -1 & 0 & 0 \\ -1 & 1 & 0 & 0 & 1 & -1 \\ 0 & 0 & -1 & 1 & -1 & 1 \end{pmatrix} \quad \text{and} \quad \mathbf{C} = \tfrac{1}{2}\begin{pmatrix} 2 & -1 & -1 \\ -1 & 2 & -1 \\ -1 & -1 & 2 \end{pmatrix}. \quad \diamond$$

The $v \times 1$ unit vector $\mathbf{j} = (1, 1, \ldots, 1)^T$ is an eigenvector of $\mathbf{C}$ with eigenvalue 0. Thus $\mathbf{C}$ is a singular matrix, and the normal equations can be solved only for a linear function $\mathbf{a}^T\tau$, where $\mathbf{a}^T$ is in the row span of $\mathbf{C}$. If $\mathbf{a} \in \mathbf{R}^v$ is orthogonal to $\mathbf{j}$, then the linear combination of the treatment parameters $\mathbf{a}^T\tau$ is called a *contrast*. The estimation of such a linear function provides a comparison between the treatments. A contrast $\mathbf{a}^T\tau$ is *estimable* in the column space of $(\mathbf{I} - \mathbf{B})\mathbf{X}$ if there is a linear combination of the projected yields $\mathbf{u}^T(\mathbf{I} - \mathbf{B})\mathbf{Y}$ such that $\mathbf{a}^T\tau = E(\mathbf{u}^T(\mathbf{I} - \mathbf{B})\mathbf{Y})$. In this case, we have $\mathbf{a}^T\hat{\tau} = \mathbf{u}^T(\mathbf{I} - \mathbf{B})\mathbf{y}$.

Example 14.1 (continued) We have

$$\hat{\tau}_3 - \hat{\tau}_2 = \tfrac{2}{3}(0,-1,1)\mathbf{X}^T(\mathbf{I}-\mathbf{B})\mathbf{X}\hat{\tau}$$
$$= \tfrac{2}{3}(0,-1,1)\mathbf{X}^T(\mathbf{I}-\mathbf{B})\mathbf{y}$$
$$= \tfrac{1}{3}(1,-1,-1,1,-2,2)\mathbf{y}.$$

With $\mathbf{y} = (2,13,4,21,15,23)^T$, we have

$$\hat{\tau}_3 - \hat{\tau}_2 = \tfrac{22}{3}; \quad \text{similarly,} \quad \hat{\tau}_3 - \hat{\tau}_1 = \tfrac{53}{3}. \quad \diamond$$

A contrast of the form $\tau_i - \tau_j$ is called a *simple contrast*. The design is *connected* if and only if all simple contrasts are estimable, and this holds if and only if $\mathbf{C}$ has rank $v-1$. In this case, $\mathbf{C}$ has $v-1$ non-zero eigenvalues $\lambda_1, \ldots, \lambda_{v-1}$ with corresponding mutually orthogonal unit eigenvectors $\mathbf{u}_1, \ldots, \mathbf{u}_{v-1}$. Let $\mathbf{P}$ be the orthogonal matrix with columns $(\mathbf{u}_1, \ldots, \mathbf{u}_{v-1}, \mathbf{j})$. Then

$$\mathbf{P}^T \mathbf{C} \mathbf{P} = \text{diag}(\lambda_1, \ldots, \lambda_{v-1}, 0)$$

and

$$\mathbf{C} = \mathbf{P}\, \text{diag}(\lambda_1, \ldots, \lambda_{v-1}, 0)\, \mathbf{P}^T = \sum \lambda_i \mathbf{u}_i \mathbf{u}_i^T.$$

Put $\mathbf{C}^- = \sum (1/\lambda_i) \mathbf{u}_i \mathbf{u}_i^T$. Then the solutions for $\hat{\tau}$ are $\mathbf{C}^- \mathbf{X}^T(\mathbf{I}-\mathbf{B})\mathbf{y} + \alpha \mathbf{j}$, for arbitrary α. Hence, for a contrast $\mathbf{a}^T \tau$, we have $\mathbf{a}^T \hat{\tau} = \mathbf{a}^T \mathbf{C}^- \mathbf{X}^T (\mathbf{I}-\mathbf{B})\mathbf{y}$.

The variance of the contrast $\mathbf{a}^T \tau$ is $\sigma^2 \mathbf{a}^T \mathbf{C}^- \mathbf{a}$. If $\mathbf{a}^T \tau = \tau_i - \tau_j$ is a simple contrast, then

$$\text{var}(\mathbf{a}^T \tau) = \sigma^2 (c_{ii}^- + c_{jj}^- - 2 c_{ij}^-),$$

and the average variance of the pairwise treatment contrasts is

$$\frac{1}{v(v-1)} \sum_{i \ne j} \text{var}(\tau_i - \tau_j) = \sigma^2 \frac{2}{v-1} \text{tr}(\mathbf{C}^-) = \sigma^2 \frac{2}{v-1} \sum \lambda_i^{-1}.$$

Example 14.1 (continued) The two non-zero eigenvalues of $\mathbf{C}$ are $\tfrac{3}{2}$ and $\tfrac{3}{2}$, and $\mathbf{C}^- = \tfrac{4}{9} \mathbf{C}$. Thus the average variance of the pairwise treatment contrasts is $\tfrac{4}{9} \sigma^2$. $\diamond$

A good estimate is one that has small variance. In an experiment we should like to make as many good estimates as possible, and avoid the possibility that some estimate in which we are interested has large variance. Thus two reasonable objectives of the design of a statistical experiment are to minimize the average variance of the simple contrasts and to minimize the maximum variance of all contrasts. For a given matrix $\mathbf{Z}$, a design ($\mathbf{X}$ matrix) that achieves the first objective is called *A-optimal*, and is characterized by minimizing $\sum \lambda_i^{-1}$. A design that achieves the second objective is called *E-optimal* and is characterized by maximizing the minimum of the eigenvalues $\lambda_1, \ldots, \lambda_{v-1}$. A third optimality criterion is that of *D-optimality*, which relates to the volume of a confidence ellipsoid of the estimate, under the assumption that the yields follow a multivariate Gaussian distribution. A design is *D-optimal* if $\prod \lambda_i$ is maximized; see Kiefer [13] for a discussion of optimality criteria.

14.3 A-optimality and closed walks

We consider designs in which there are v treatments and b blocks, each block contains k plots, and each treatment is assigned to r plots. Designs with these parameters are called (v, b, k, r)-*designs*. Thus $\mathbf{X}^T\mathbf{X} = r\mathbf{I}$, $\mathbf{Z}^T\mathbf{Z} = k\mathbf{I}$, and the number of plots is $vr = bk$. A *concurrence* between two treatments i and j is a pair of plots within a block to which the treatments i and j have been assigned. The *variety concurrence graph* of a design is the graph whose vertex set is the set of treatments, where an edge joins a pair of treatments for each concurrence between them.

Example 14.1 (continued) The design $\mathcal{D}$ is a $(3, 3, 2, 2)$-design. There is exactly one concurrence between each pair of treatments, and the variety concurrence graph is a triangle. ◊

The variety concurrence graph of a design is a connected graph if and only if, for each pair of vertices i and j, there is a sequence of plots $\alpha_1, \ldots, \alpha_{2m}$ such that:

treatment i is assigned to α_1;
treatment j is assigned to α_m;
for $l = 1, \ldots, m$, the plots α_{2l-1} and α_{2l} are in the same block;
for $l = 1, \ldots, m-1$, the plots α_{2l} and α_{2l+1} are assigned the same treatment.

But this is exactly the condition that ensures that the contrast $\tau_i - \tau_j$ is estimable by the estimate $\mathbf{u}^T(\mathbf{I} - \mathbf{B})\mathbf{y}$, where $u_{\alpha_{2l-1}} = 1$ and $u_{\alpha_{2l}} = -1$, for $l = 1, \ldots, m$. Hence we have the following result.

Theorem 14.2 *A design is connected if and only if its variety concurrence graph is connected.*

A design is *A-optimal* if it minimizes $\sum \lambda_i^{-1}$, over all designs with the same parameters. Now the sum of the non-zero eigenvalues for designs with these parameters is

$$\sum \lambda_i = \text{tr}(\mathbf{C}) = \text{tr}(r\mathbf{I} - (1/k)\mathbf{N}\mathbf{N}^T) = vr - (1/k)\sum n_{ij}^2.$$

Since $\mathbf{N}$ has given row-sum r and column-sum k, the average of the non-zero eigenvalues is a maximum when $n_{ij} = 0$ or 1, for all i and j. Thus binary designs minimize the average of the eigenvalues. Considering those designs with a given value μ of the average of the non-zero eigenvalues, we see that $\sum \lambda_i^{-1}$ is a maximum when $\lambda_i = \mu$, for $i = 1, \ldots, v - 1$. So a design that minimizes $\sum \lambda_i^{-1}$ is a binary design with all eigenvalues equal, if such a design exists.

When there is only one non-zero eigenvalue, we have

$$\mathbf{C} = \mu(\mathbf{I} - (1/v)\mathbf{J}),$$

where $\mathbf{J}$ is the all-1 matrix. Then

$$\mathbf{NN}^T = k(r\mathbf{I} - \mathbf{C}) = rk\left(1 - \frac{\mu}{r}\right)\mathbf{I} + \frac{\mu k}{v}\mathbf{J} = (r - \lambda)\mathbf{I} - \lambda \mathbf{J},$$

where

$$\lambda = \frac{r(k-1)}{v-1} = \frac{k\mu}{v}.$$

In this case, λ is an integer satisfying $\lambda(v - 1) = r(k - 1)$, and all pairs of treatments are assigned to plots in the same block exactly λ times. Hence the design is a 2-(v, k, λ) design or *balanced incomplete block design* (*BIBD*). We deduce that for those parameters for which there exists a BIBD, a design is A-optimal if and only if it is a BIBD.

Theorem 14.3 *A BIBD is an A-optimal design.*

Since BIBDs do not exist for some sets of parameters, it is of interest to characterize the next best case. Now BIBDs are A-optimal, because the eigenvalues $\lambda_1, \ldots, \lambda_{v-1}$ are all equal. These λ_i give a measure of variance—the average variance of the treatment contrasts $\tau_i - \tau_j$. This variance is reduced if there are many comparisons between the treatments i and j—that is, they are assigned many times to plots in the same block. A BIBD has all $\binom{v}{2}$ comparisons equally often.

Note that if a treatment is assigned to two plots within the same block, then the comparison between these two plots is wasted. Therefore we seek binary designs, those with a $(0, 1)$-incidence matrix $\mathbf{N}$. Such a design maximizes $\mu = \sum \lambda_i/(v - 1)$. If $k \leq v$, then there always exists a binary design for any parameter set b, v, k, r satisfying $vr = bk$. In a BIBD, each of the $v(v-1)/2$ pairs of treatments are compared equally often. Thus there exists an integer λ such that each pair of treatments is assigned to plots together in λ blocks. Since $k(k-1)/2$ comparisons between pairs of treatments are made in each block, we have $v(v-1)\lambda/2 = bk(k-1)/2$, and hence

$$(v - 1)\lambda = r(k - 1).$$

This is a non-trivial restriction on the parameters of a BIBD.

If, for a given set of parameters, there does not exist a BIBD, then the next most equitable assignment of treatments to plots is obtained when each pair of treatments is assigned to plots in either λ or $\lambda + 1$ blocks, for some integer λ. A design for which this holds is called a *regular graph design*.

The incidence matrix $\mathbf{N}$ of a regular graph design satisfies the equation

$$\mathbf{NN}^T = (r - \lambda)\mathbf{I} + \lambda \mathbf{J} + \mathbf{A},$$

where $\mathbf{A} = (a_{ij})$ is a $(0,1)$-matrix such that $a_{ij} = 1$ if and only if treatments i and j are assigned $\lambda + 1$ times to plots in the same block. $\mathbf{A}$ is the adjacency matrix of a regular graph of degree $r(k-1) - \lambda(v-1)$.

Now the sum of the squares of the eigenvalues of the Fisher information matrix is given by

$$\sum \lambda_i^2 = \operatorname{tr}\left(\left(r\mathbf{I} - \frac{1}{k}\mathbf{N}\mathbf{N}^T\right)^2\right) = 2(v-1)\mu - v + \frac{1}{r^2 k^2}\sum \lambda_{ij}^2,$$

where the entries of $\mathbf{N}\mathbf{N}^T$ are λ_{ij}. Among binary designs this is a minimum for regular graph designs.

A design is (M,S)-*optimal* if it maximizes $\sum \lambda_i$ among designs of the same parameters, and it minimizes $\sum \lambda_i^2$ among designs satisfying this condition. Thus regular graph designs are (M,S)-optimal. It is a long-standing conjecture (see [9, 12]) that an (M,S)-optimal design that maximizes $\sum \lambda_i^{-1}$ among (M,S)-optimal designs is A-optimal—that is, it maximizes this sum among all designs with the same parameters. The concept of (M,S)-optimality and its relation to A-optimality may be extended by considering the variety concurrence graph.

In the variety concurrence graph of a design, a treatment assigned to l plots in a block gives rise to $l(l-1)$ loops at the corresponding vertex of the variety concurrence graph. Maximizing the sum of the eigenvalues of the Fisher information matrix of the design, thereby making the design binary, corresponds to minimizing the number of loops in the variety concurrence graph. Minimizing $\sum \lambda_i^2$, making each pair of treatments appear together in a block as nearly equally often as possible, corresponds to minimizing the number of closed walks of length 2 in the variety concurrence graph. Thus a regular graph design minimizes the number of loops and the number of closed walks of length 2 in the variety concurrence graph.

In a regular graph design, some pairs of treatments are compared $\lambda+1$ times, while others are compared only λ times. The variance of the treatment contrast $\tau_i - \tau_j$ for the former is usually smaller than for the latter. If treatments i and j are compared only λ times, we would like more *second-hand* comparisons between them—that is, treatments l such that i and l are compared $\lambda + 1$ times and treatments j and l are compared $\lambda+1$ times. Thus, in the graph with adjacency matrix $\mathbf{A}$, we would like a pair of non-adjacent vertices i and j to be connected by as many paths of length 2 as possible. Maximizing the number of paths of length 2 between vertices that are not joined is the same as minimizing the number of cycles of length 3 in the graph.

The matrix $\hat{\mathbf{A}} = \mathbf{N}\mathbf{N}^T - r\mathbf{I} = (\hat{a}_{ij})$ records the number of edges $\hat{a}_{ij}$ joining any two vertices i and j of the variety concurrence graph of a design with incidence matrix $\mathbf{N}$. The number of closed walks of length h in the variety concurrence graph is $\Gamma_h = \operatorname{tr}(\hat{\mathbf{A}}^h) = \sum \nu_i^h$, where the eigenvalues of $\hat{\mathbf{A}}$ are $\nu_0, \ldots, \nu_{v-1}$. Now $\hat{\mathbf{A}} = r(k-1)\mathbf{I} - k\mathbf{C}$, where $\mathbf{C}$ is the Fisher information matrix of the design. Hence we may relate the eigenvalues of the variety concurrence graph to those of **C**. We have

$$\nu_0 = r(k-1) \quad \text{and} \quad \nu_i = r(k-1) - k\lambda_i, \quad \text{for } i = 1, \ldots, v-1.$$

It follows from [14] that

$$\sum \lambda_i^{-1} = \frac{k}{r(k-1)} \sum \left(\frac{\Gamma_h}{r^h(k-1)^h} - 1 \right).$$

Closed walks of length h are weighted $1/r^h(k-1)^h$ in this sum, so those of shorter length are more important. Indeed, Paterson [14] has made the following conjecture.

Conjecture *A connected binary (v, b, k, r)-design is A-optimal if and only if it minimizes sequentially Γ_h, for $h = 2, 3, \ldots$.*

Let D be an A-optimal design, and let Γ_h be as above. The conjecture says that, for any other design with the same parameters as D, $\Gamma_t < \Gamma'_t$ where t is the first index h for which Γ'_h, the number of closed walks in its variety concurrence graph, is not equal to Γ_h. Conversely, if two designs X and X' have numbers of closed walks Γ_h and Γ'_h, respectively, and if $\Gamma_h = \Gamma'_h$ for $h < t$, and $\Gamma_t < \Gamma'_t$, then X' is not A-optimal.

Paterson [14] has an example of two binary designs, the one with larger Γ_2 having the smaller value of $\sum \lambda_i^{-1}$. So, in this comparison of numbers of closed walks, one of the designs must be A-optimal. If the conjecture is true, then this approach to A-optimal designs is a subproblem of determining, among all graphs with n vertices and m edges, those graphs with the fewest closed walks.

14.4 Bounds

If a design is not a BIBD, it is not easy to determine whether it is A-optimal. In practice, we settle for a design that is reasonably efficient, providing a small average variance of pairwise treatment contrasts. In order to know how efficient a design is, we require bounds on the average variance.

A *complete block design* is one in which each treatment has been assigned to exactly one plot in each block. Thus the parameters satisfy $v = k$ and $b = r$. The incidence matrix and Fisher information matrix are $\mathbf{N} = \mathbf{J}$ and $\mathbf{C} = r\mathbf{I} - \mathbf{J}$. The eigenvalues of $\mathbf{C}$ satisfy $\lambda_i = r$, for $i = 1, \ldots, v-1$, and the average variance of pairwise treatment contrasts is $2\sigma^2/r$.

The *efficiency factor* of a design is the ratio of the average variance of pairwise treatment contrasts of the complete design with the same number of plots, to that of the given design. The *canonical efficiency factors* $e_i = \lambda_i/r$ are the non-zero eigenvalues of the matrix $(1/r)\mathbf{C} = \mathbf{I} - (1/rk)\mathbf{NN}^T$. The canonical efficiency factors satisfy $0 < e_i \leq 1$, and the *efficiency factor* (EF) of a design is the harmonic mean of the canonical efficiency factors—that is, $EF = (v-1)/\sum e_i^{-1}$. A design with small average variance of pairwise treatment contrasts has an efficiency factor close to 1.

Hence comparisons between designs may be made via their efficiency factors, and it is of interest to determine bounds on the efficiency factor of a design with given parameters. Such bounds may be given in terms of the numbers of closed walks of length h, in the variety concurrence graph.

Put $a_{-1} = 0$ and $a_n = \sum e_i^n/(v-1)$, for $n \geq 0$. For $1 \leq m \leq n+2$, let $\mathbf{A}(m,n)$ be the $m \times m$ matrix

$$\mathbf{A}(m,n) = \begin{pmatrix} a_{n+1-m} & a_{n+2-m} & \cdots & a_n \\ a_{n+2-m} & a_{n+3-m} & \cdots & a_{n+1} \\ \vdots & \vdots & & \vdots \\ a_n & a_{n+1} & \cdots & a_{n+m-1} \end{pmatrix},$$

and let $D(m,n) = \det \mathbf{A}(m,n)$. If there are at least $m+1$ distinct canonical efficiency factors, then

$$\frac{1}{v-1}\sum e_i^{-1} \geq -\frac{D(m+2,m)}{D(m+1,m+1)},$$

with equality if and only if there are exactly $m+1$ distinct canonical efficiency factors [6]. Thus

$$EF \leq c_m = -\frac{D(m+1,m+1)}{D(m+2,m)}$$

and

$$c_m = c_{m-1} - \frac{D(m+1,m+1)^2}{D(m+2,m)D(m+1,m-1)},$$

and so $c_0, \ldots, c_m$ is a decreasing sequence of upper bounds for the efficiency factor, with $EF = c_m$ when there are exactly $m+1$ distinct values among $e_1, \ldots, e_{v-1}$. Put

$$\gamma_h = \frac{1}{v-1}\left(\frac{\Gamma_h}{r^h(k-1)^h} - 1\right) = \sum_{n=0}^{h}\left(-\frac{k}{k-1}\right)^n a_n.$$

Then

$$c_m = \frac{k-1}{k} \times \frac{\begin{vmatrix} 1 & 1 & \cdots & 1 \\ \gamma_0 & \gamma_1 & \cdots & \gamma_{m+1} \\ \vdots & \vdots & & \vdots \\ \gamma_m & \gamma_{m+1} & \cdots & \gamma_{2m+1} \end{vmatrix}}{\begin{vmatrix} 0 & \gamma_0 & \gamma_0+\gamma_1 & \cdots & \gamma_0+\cdots+\gamma_m \\ \gamma_0 & \gamma_1 & \gamma_2 & \cdots & \gamma_{m+1} \\ \vdots & \vdots & \vdots & & \vdots \\ \gamma_m & \gamma_{m+1} & \gamma_{m+2} & \cdots & \gamma_{2m+1} \end{vmatrix}}.$$

Thus c_m depends on the numbers of closed walks $\Gamma_0, \ldots, \Gamma_{2m+1}$. Minimizing $\gamma_0, \ldots, \gamma_{2m+1}$ makes c_m large. Now c_m is an upper bound, so under the assumption that we can find a design with efficiency factor close to this bound, we would attempt to construct a design with minimal numbers of closed walks. The first bound is

$$c_0 = a_1 = \frac{k-1}{k}\left(\frac{v - (\Gamma_1/r(k-1))}{v-1}\right).$$

Γ_1 is the number of loops in the variety concurrence graph. Whenever there are l plots in a block all assigned the same treatment, we get $l(l-1)$ loops. Thus $\Gamma_1 = 0$ for binary designs, and $\Gamma_1 \geq 2$ for non-binary designs.

The efficiency factor of a design achieves the bound c_0 if there is exactly one canonical efficiency factor, which must then equal c_0. Such a design is called *efficiency balanced*. In this case, $\mathbf{NN}^T = \alpha \mathbf{I} + \beta \mathbf{J}$, for some integers α and β satisfying $\alpha + v\beta = rk$. In the case of a binary design, $\alpha + \beta = r$, $EF = c_0 = (k-1)v/k(v-1)$ and the design is a BIBD; for a non-binary design, $\alpha + \beta \geq r+2$ and $\Gamma_1 = \text{tr}(\mathbf{NN}^T - r\mathbf{I}) \geq 2v$, so that

$$EF \leq \frac{(k-1)v}{k(v-1)}\left(1 - \frac{2}{r(k-1)}\right).$$

A design with more than one distinct canonical efficiency factor has $EF \leq c_1 < c_0$. The best possible efficiency factor for binary design in this case is of the order of

$$\frac{(k-1)v}{k(v-1)}\left(1 - \frac{1}{r^2 k^2}\right).$$

So it is likely, even when there exists no BIBD, that there exists a binary design with efficiency factor higher than that for any non-binary efficiency balanced design. Whether this is so in all cases remains an open question. If it is so, there still remains the question of whether there exists a non-binary design, not attaining the bound c_0, but with an efficiency factor higher than that for any binary design. The question is whether the fact that the variety concurrence graph of a non-binary design has many loops can be compensated for by having sufficiently few closed walks of length 2 and 3 in comparison with the variety concurrence graphs of binary designs.

Now, upper bounds for the efficiency factor are used to evaluate how good a proposed design is. One can then make a decision on whether to use the design if it is close to optimal, or to try to find a better design if it is not. We may write the upper bound c_1 as follows:

$$c_1 = c_0 + \frac{k}{k-1}\left(\frac{(\gamma_2 - \gamma_1)^2}{-\gamma_3 + \gamma_2(2\gamma_1 + 1) - \gamma_1^3 - \gamma_1^2}\right).$$

For a regular graph design, γ_1 and γ_2 are determined by the parameters, and the problem of establishing a good bound on the efficiency factor of a regular graph design becomes that of determining a lower bound on the number of triangles in the corresponding graph.

When the design is a BIBD, the lower bound 0 is attained. If the degree Δ of the graph satisfies $2\Delta - v > 0$, then each edge must belong to at least $2\Delta - v$ triangles. The minimum number of triangles arises when equality holds for every edge, and in this case the graph is strongly regular. Paterson and Wild [15] have given lower bounds on the number of triangles in the variety concurrence graph, and hence on the efficiency factor of the design, for some special classes of designs. It is of interest to characterize graphs according to the number of triangles they contain.

14.5 Spanning trees and D-optimality

D-optimal designs correspond to designs that minimize the confidence ellipsoid when estimating treatment parameters under the assumption that the yields have a multivariate Gaussian distribution. Here we see that there is a connection between identifying D-optimal designs and determining which graphs with n vertices and m edges have the maximum number of spanning trees.

Let G be a graph, possibly with multiple edges. We label the edges of G with indeterminates, so that $x_{ij}^{(k)}$, $k = 1, \ldots, l_{ij}$, label the edges joining i and j if there are l_{ij} edges joining them. Let $\hat{\mathbf{K}}$ be the matrix with off-diagonal ij-entry equal to $-\sum x_{ij}^{(k)}$ and with the ith diagonal entry equal to minus the sum of the other entries in row i. Now each spanning tree of G is associated with a monomial of degree $n - 1$, the product of the indeterminates corresponding to the edges of the spanning tree. The enumeration of the spanning trees of G is closely related to the matrix $\hat{\mathbf{K}}$ (see [4]).

Theorem 14.4 *Let G be a graph of order n, and let $\hat{\mathbf{K}}_1$ be an $(n-1) \times (n-1)$ submatrix of $\hat{\mathbf{K}}$. Then the monomials appearing in the multinomial $\det \hat{\mathbf{K}}_1$, after expansion and cancellation, are all square-free and give an exact listing of the spanning trees of G.*

By setting all the indeterminates $x_{ij}^{(k)}$ equal to 1, we may associate with any graph its *Kirchhoff matrix* $\mathbf{K} = (c_{ij})$, where $-c_{ij}$ = the number of edges joining vertices i and j, for $i \neq j$, and $c_{ii} = -\sum_{j \neq i} c_{ij}$. If the graph is simple, then each off-diagonal entry of $\mathbf{K}$ is the negative of the corresponding off-diagonal entry of the adjacency matrix $\mathbf{A}$, and each diagonal entry of $\mathbf{K}$ is the degree of the corresponding vertex.

Example 14.1 (continued) The variety concurrence graph of the design is a triangle. The vertex set is $V = \{A, B, C\}$, and each pair of vertices is joined by an edge. Each vertex has degree 2 and is adjacent to each of the other vertices. The adjacency matrix $\mathbf{A}$ and the Kirchhoff matrix $\mathbf{K}$ are

$$\mathbf{A} = \begin{pmatrix} 0 & 1 & 1 \\ 1 & 0 & 1 \\ 1 & 1 & 0 \end{pmatrix} \quad \text{and} \quad \mathbf{K} = \begin{pmatrix} 2 & -1 & -1 \\ -1 & 2 & -1 \\ -1 & -1 & 2 \end{pmatrix}. \quad \Diamond$$

The Kirchhoff matrix $\mathbf{K}$ satisfies the following theorem, known as the *Matrix-Tree Theorem* (see Chapter 1).

Theorem 14.5 (Matrix-Tree Theorem) *Let $\mathbf{K}$ be the Kirchhoff matrix of a graph G of order n. Then $\mathbf{K}$ has rank $n - 1$ if and only if G is connected. Further, if $\mathbf{K}$ has rank $n - 1$, then all the cofactors of $\mathbf{K}$ are equal, and their common value is the number of spanning trees of G.*

It turns out that, in the case of block designs with k plots in each block, the Fisher information matrix of the design is related to the Kirchhoff matrix of the variety concurrence graph. If $\mathbf{N}$ is the incidence matrix of the block design, then the Kirchhoff matrix of the variety concurrence graph is $\mathbf{K} = k\mathbf{R} - \mathbf{NN}^T$, where $\mathbf{R}$ is a diagonal matrix with diagonal entries equal to the number of plots that have been assigned the corresponding treatment. This is simply k times the Fisher information matrix.

Recall from Section 14.2 that a block design is D-optimal if it maximizes $\prod \lambda_i$, where $\lambda_1, \ldots, \lambda_{v-1}$ are the non-zero eigenvalues of the Fisher information matrix. But this product is equal to the sum of v cofactors of this matrix. Hence determining the D-optimality of a block design is equivalent to determining the optimality of the variety concurrence graph with respect to the maximum number of spanning trees it contains.

It is therefore of interest, from a statistical point of view, to determine, among those graphs with a given number of vertices and edges, which ones have a maximum number of spanning trees.

14.6 Line graphs and E-optimality

We consider E-optimality of regular graph designs. For such a design, the Fisher information matrix $\mathbf{C}$ has the form

$$\frac{r(k-1) + \lambda + 1}{k}\mathbf{I} - \frac{\lambda + 1}{k}\mathbf{J} + \mathbf{A}$$

for some binary matrix $\mathbf{A}$. The matrix $\mathbf{A}$ is symmetric and has zeros on the diagonal. It is, therefore, the adjacency matrix of some graph on v vertices. Now $\mathbf{Aj} = r(k-1)\mathbf{j}$, where $\mathbf{j}$ is the all-1 vector, and so the graph is regular of degree $r(k-1)$.

There is, clearly, a direct relationship between the eigenvalues of $\mathbf{A}$ and of $\mathbf{C}$, and the problem of determining E-optimal designs (in the above class) is related to the problem of determining which simple Δ-regular graphs of order n have the greatest minimum eigenvalue. If a graph is a solution to this problem and arises from a design as described above, then the design is E-optimal.

Of interest, therefore, are the regular graphs with large minimum eigenvalue. A simple graph with at least one edge has minimum eigenvalue $\lambda_n \leq -1$. Indeed, we may assume that the adjacency matrix of such a graph has principal 2-minor

$$\begin{pmatrix} 0 & 1 \\ 1 & 0 \end{pmatrix},$$

whose eigenvalue -1 is necessarily an upper bound for λ_n.

The *triangular graph* $T(k)$ ($k \geq 3$) is the graph whose vertex set consists of all 2-subsets of $\{1, \ldots, k\}$, and where two vertices are adjacent if and only if they have an element in common. For example, $T(4)$ has six vertices, $\{1,2\}$, $\{1,3\}$, $\{1,4\}$, $\{2,3\}$, $\{2,4\}$ and $\{3,4\}$, and only the pairs of vertices $\{1,2\}$ and $\{3,4\}$, $\{1,3\}$ and $\{2,4\}$, and $\{1,4\}$ and $\{2,3\}$ are not adjacent; thus $T(4)$ is the octahedron graph—that is, K_6 with a 1-factor removed.

The triangular graphs have minimum eigenvalue -2. In fact, these graphs are characterized by their spectra (see [3, 7, 16]).

Theorem 14.6 *The triangular graph $T(k)$ is a regular connected graph on $\binom{k}{2}$ vertices with distinct eigenvalues -2, $2k - 4$ and $k - 4$.*

Conversely, if $k \neq 8$ and G is a regular graph on $\binom{k}{2}$ vertices with distinct eigenvalues -2, $2k - 4$ and $k - 4$, then G is isomorphic to $T(k)$.

The triangular graphs are an example of a family of graphs with minimum eigenvalue at least -2. These are the line graphs as defined in Chapter 1. The triangular graph $T(n)$ is the line graph of the complete graph K_n.

Suppose that the simple graph G has n vertices and m edges, and let $\mathbf{M} = (m_{ij})$ be its incidence matrix—that is, $m_{ij} = 1$ if vertex i is incident with edge j, and $m_{ij} = 0$ otherwise. Then an adjacency matrix for $L(G)$ is $\mathbf{M}^T\mathbf{M} - 2\mathbf{I}$. Since $\mathbf{M}^T\mathbf{M}$ is positive semi-definite, the following result is immediate.

Theorem 14.7 *Let G be a simple graph, and let H be the line graph of G. Then the minimum eigenvalue of H is at least -2.*

The *m-dimensional octahedron* $K_{m(2)}$ is the complete multipartite graph with m partite sets of size 2. It has $n = 2m$ vertices, and is the complement of a 1-factor. $K_{m(2)}$ has an adjacency matrix of the form

$$\begin{pmatrix} \mathbf{J} - \mathbf{I} & \mathbf{J} - \mathbf{I} \\ \mathbf{J} - \mathbf{I} & \mathbf{J} - \mathbf{I} \end{pmatrix}.$$

It is easy to check that the distinct eigenvalues of $K_{m(2)}$ are -2, 0 and $2m - 2$. It is also easy to see that, for $m > 3$, no $K_{m(2)}$ is a line graph. Conversely, there are only finitely many connected regular graphs with minimum eigenvalue $\lambda_n \geq -2$, besides the m-dimensional octahedra, that are not line graphs. This follows from a result of Hoffman [8], who characterized sufficiently large graphs with minimum eigenvalue $\lambda_n \geq -2$ as generalized line graphs. Cameron et al. [2] have given an alternative approach to this problem, which relates the graphs to root systems, and identifies the exceptions with subsets of lines of the root system E_8.

In summary, we have the following result on E-optimal regular graph designs.

Theorem 14.8 *Let D be an E-optimal regular graph design. Then, with finitely many exceptions, the graph determined by D is a perfect matching, a line graph, or an m-dimensional octahedron.*

14.7 Row–column designs

A *row–column design* is a statistical design in which the plots are partitioned into blocks of two types, *rows* and *columns*. For example, in an agricultural field trial, weedkillers of various types may be applied in one direction (along rows of the field) and insecticides may be applied in the orthogonal direction (along columns of the field). In a row–column design, the usual model is $y_i = \tau_j + \beta_k + \gamma_l + \varepsilon_i$, where β_k is the row effect, γ_l is the column effect, and treatment j is assigned to plot i which is in row k and column l.

As before, the design is connected if all simple treatment contrasts are estimable. This occurs if the coefficient matrix of the reduced normal equations

$$\mathbf{C} = r\mathbf{I} - \frac{1}{c}\mathbf{N}_k\mathbf{N}_k^T - \frac{1}{k}\mathbf{N}_c\mathbf{N}_c^T + \frac{r}{v}\mathbf{J}$$

has rank $v-1$, where $\mathbf{N}_k$ is the $v \times k$ treatment/row incidence matrix, and $\mathbf{N}_c$ is the $v \times c$ treatment/column incidence matrix. Also, the efficiency of the design may be expressed in terms of the eigenvalues of this matrix.

Properties of row–column designs can also be established using graphs. Let G be the labelled digraph whose vertex set is the set of treatments, and where arc i joins treatment x to treatment y if x and y are assigned to plots in the same row and in columns i and $i+1$, respectively. We consider paths in this graph, ignoring the direction of the arcs.

Let $e_1, \ldots, e_{c-1}$ be the standard basis of the real vector space $\mathbf{R}^{c-1}$. For each closed walk γ in G, we define the vector $\mathbf{u}(\gamma) \in \mathbf{R}^{c-1}$ by

$$\mathbf{u}(\gamma) = \sum (a_j - b_j)\mathbf{e}_j,$$

where a_j is the number of times that arc j is traversed in γ in the direction of the arc, and b_j is the number of times that arc j is traversed in the opposite direction. A collection of closed walks of G is *independent* if the corresponding vectors are linearly independent in $\mathbf{R}^{c-1}$. Butz [1] has established the following result.

Theorem 14.9 *A row–column design is connected if and only if the corresponding graph G is connected and contains $c-1$ independent closed walks.*

John et al. [10] have related this result to graph-theoretical properties of certain classes of row–column designs and their component designs, taking rows as blocks and ignoring columns, or vice versa. John and Street [11] have considered efficiency factors of row–column designs, and have established upper bounds using techniques similar to those of Section 14.4.

14.8 Conclusion

The efficiency of a statistical design may be evaluated using the eigenvalues of the Fisher information matrix. There are various measures that are of interest, and designs that are optimal with respect to these measures correspond to graphs with extreme spectral behaviour. The problem of identifying optimal designs remains open in the general case. Some of the related problems in graph theory are the following:

determine which graphs with n vertices and m edges have the fewest closed walks;

characterize graphs according to the number of triangles they contain;

determine which graphs with n vertices and m edges have the most spanning trees;

determine which simple Δ-regular graphs of order n have the greatest minimum eigenvalue.

These are but a few of the many problems in graph theory that may find application in statistics. Graphs and statistics is a fruitful area of research, and the interested reader is referred to Constantine [4] for further details on the connection between the design of experiments and discrete mathematics.

References

1. L. Butz, *Connectivity in Multi-factor Designs*, Heldermann-Verlag, 1982.
2. P. J. Cameron, J. M. Goethals, J. J. Seidel and E. E. Shult, Line graphs, root systems, and elliptic geometry, *J. Algebra* **43** (1976), 305–327; *MR* **56**#182.
3. S. Chang, The uniqueness and nonuniqueness of the triangular association scheme, *Sci. Record* **3** (1959), 604–613.
4. G. M. Constantine, *Combinatorial Theory and Statistical Design*, Wiley, 1987; *MR* **88k**: 05001.
5. R. A. Fisher, *The Design of Experiments*, Oliver and Boyd, 1949.
6. S. Fitzpatrick and R. G. Jarrett, Upper bounds for the harmonic mean, with an application to experimental design, *Austral. J. Statist.* **28** (1986), 220–229; *MR* **88d**: 62126.
7. A. J. Hoffman, On the uniqueness of the triangular association scheme, *Ann. Math. Statist.* **31** (1960), 492–497; *MR* **32**#7949.
8. A. J. Hoffman, $-1-\sqrt{2}$?, *Combinatorial Structures and Their Applications* (ed. R. Guy *et al.*), Gordon and Breach, 1970, pp. 173–176.
9. J. A. John and T. J. Mitchell, Optimal incomplete block designs, *J. Roy. Statist. Soc. (B)* **39** (1977), 39–43; *MR* **58**#18933.

10. J. A. John, L. J. Paterson and P. Wild, Connectedness of generalized-cyclic row-and-column designs, *Utilitas Math.* **30** (1986), 109–122; *MR* **88e**: 05025.
11. J. A. John and D. J. Street, Bounds for the efficiency factor of row-column designs, *Biometrika* **79** (1992), 658–661; *MR* **94a**: 62118.
12. J. A. John and E. R. Williams, Conjectures for optimal block designs, *J. Roy. Statist. Soc. (B)* **44** (1982), 221–225; *MR* **83m**: 62133.
13. J. Kiefer, Optimum experimental design, *J. Roy. Statist. Soc. (B)* **21** (1951), 272–304.
14. L. J. Paterson, Circuits and efficiency in incomplete block designs, *Biometrika* **70** (1983), 215–225; *MR* **85k**: 62170.
15. L. J. Paterson and P. Wild, Triangles and efficiency factors, *Biometrika* **73** (1986), 289–299; *MR* **87m**: 62241.
16. S. S. Shrikhande, On a characterization of the triangular association scheme, *Ann. Math. Statist.* **30** (1959), 39–47; *MR* **21**#1673.

15
Computing

ROBIN WHITTY

Connections between graphs and computers are explored in this chapter—in particular, the kinds of digraph that arise in the analysis of computer programs. We examine program structuredness, complexity and testing, using ideas from formal language theory and compiler design: deterministic finite automata, attributed grammars and parse trees.

15.1 Introduction

Two important kinds of graph arise out of the theory of formal languages. *Finite state machines* model computation as a process of recognizing certain strings of symbols; in particular, the Turing machine has become familiar to graph theorists through its role in computational complexity (see Downey [9]). But the most basic machine, the *deterministic finite automaton*, should be even more familiar: it is a labelled digraph. Formal language theory is applied to programming languages such as Pascal or C, in order to translate them efficiently into machine code. This translation process is called *compilation* and is based on language parsing (Aho et al. [1] provides a comprehensive introduction); the associated *parse trees* are arguably the most pervasive of all the many kinds of trees in computing.

Our aim is to review the basic facts about finite automata and parsing, and to show how they conspire to produce some interesting graph theory in the study of program control flow. This provides a logical, but necessarily limited, view of graphs in computing. Parsing can be applied to graphs as well as to programs: *graph grammars* offer another connection (see Courcelle [8]). The more computers are networked together, the more computing has in common with communications, a very rich area of problems and techniques in graph theory (see Bermond et al. [5] for a good survey). Most influential of all has been computational complexity in terms of raising problems and uncovering deep results (see Downey [9] for a most readable introduction), even to the extent that it has *identified* a large part of graph theory and computing. The connections in this chapter are more tenuous; I hope that some of the unanswered questions may motivate someone to strengthen the links.

15.2 Languages

Let Σ be a finite non-empty set of symbols. Strings of symbols over Σ are called *words*, and the *length* $|w|$ of a word w is the number of symbols in w. The empty word of length 0 is denoted by ε. A *language* is any set of words over Σ. If L_1 and L_2 are languages over Σ, then we can construct new languages using the following operations:

the *product* $L_1 L_2$ of L_1 and L_2 is the set $\{w_1 w_2 \mid w_i \in L_i,\ i=1,2\}$, where $w_1 w_2$ denotes the concatenation of w_1 and w_2;

the *union* $L_1 \cup L_2$ of L_1 and L_2 is the usual set union;

the *closure* L^* of L is the set $\bigcup_{i=0}^{\infty} L^i$, where $L^0 = \{\varepsilon\}$ and $L^{i+1} = LL^i$, for $i \geq 1$; we write $L^+ = L^* - \{\varepsilon\} (= LL^*)$;

the *quotient* $L_1 \backslash L_2$ of L_1 by L_2 is the set $\{w \in \Sigma^* \mid w_1 w \in L_2 \text{ for some } w_1 \in L_1\}$; that is, $L_1 \backslash L_2$ is obtained from the set of words in L_2 with prefixes in L_1 by removing those prefixes.

An important class of languages is the set of *regular languages*, defined to be the smallest class containing all finite languages and closed under product, union and closure. Languages over Σ in this class may be represented, not usually uniquely, as set expressions, called *regular expressions*; these are built using the product, union and closure operations and any finite number of finite subsets of Σ^*. For example, if Σ consists of the three symbols 0, 1 and the decimal point, then the regular expression

$$(\{0\} \cup \{1\}(\{1\} \cup \{0\})^*)(\{\varepsilon\} \cup \{.\}(\{1\} \cup \{0\})^*\{1\}) \tag{15.1}$$

represents a regular language L_B over Σ. This language may be written more succinctly as $(0 \cup 1(1 \cup 0)^*)(\varepsilon \cup .(1 \cup 0)^* 1)$. The language L_B contains precisely the real numbers represented in binary form—for example, 0, 10, 10.1 and 101.011.

15.3 Grammars

A grammar $\Gamma = (V, \Sigma, \rightarrow, s)$ over Σ is a device for generating the set of words comprising a language. It augments a set Σ of *terminal symbols* with a set V, disjoint from Σ, of variables called *non-terminal symbols*. One non-terminal symbol s is called the *start symbol*. A finite relation $\rightarrow$ is defined in $Q^+ \times Q^*$, where Q is the set $V \cup \Sigma$; its pairs are the *production rules* of the grammar.

Let $\Rightarrow$ be the infinite relation formed by making $\rightarrow$ left and right invariant—that is, if $w \rightarrow w'$, then $xwy \rightarrow xw'y$ for all $x, y \in Q^*$. The closure $\Rightarrow^*$ of $\Rightarrow$ is defined in the usual way, and the language $L(\Gamma)$ generated by the grammar Γ is the set $\{w \in \Sigma^* \mid s \Rightarrow^* w\}$. We say that s *derives* w in this set, and so $L(\Gamma)$ is the set of words in Σ^* derived by s. Obviously, $L(\Gamma) = \varnothing$ unless $s \rightarrow w$ for some $w \in Q^*$, so this is usually assumed.

Grammars were classified by Chomsky into four types according to the exact form allowed for production rules, as shown in Table 15.1. Thus we find that type 3 grammars generate precisely the regular languages—those that can be represented as regular expressions. Clearly, regular languages are a subclass of the class of context-free languages, which in turn are a subclass of the context-sensitive languages, and so on. This inclusive hierarchy is known as the *Chomsky hierarchy*.

Table 15.1 The Chomsky hierarchy

type	allowed productions	languages generated
0	no restriction	recursively enumerable
1	$w \to w'$, where $|w| \geq |w'|$	context-sensitive
2	$v \to w$, where $v \in V$ and $w \in Q^*$	context-free
3	$v \to a$ or $v \to av'$, where $v, v' \in V$ and $a \in \Sigma$	regular

For example, the grammar $\Gamma = (V, \Sigma, \to, s)$, where $V = \{s, a, b, c, d\}$, $\Sigma = \{0, 1, .\}$ and

$$\to = \{ (s, 0a), (s, 1b), (a, \varepsilon), (a, .c), (b, 0b), (b, 1b), (b, \varepsilon), (b, .c),$$
$$(c, 0c), (c, 1d), (d, 0c), (d, 1d), (d, \varepsilon) \},$$

generates the regular language L_B defined in expression (15.1). For instance,

$$s \to 1b \Rightarrow 10b \Rightarrow 101b \Rightarrow 101.c \Rightarrow 101.0c$$
$$\Rightarrow 101.01d \Rightarrow 101.011d \Rightarrow 101.011\varepsilon, \qquad (15.2)$$

and so $s \Rightarrow^* 101.011 \in L(\Gamma)$.

A *parse* of a word $w \in L(\Gamma)$ is a deviation of w from s. Each step in the derivation of w corresponds to a production rule. The point of parsing is that actions associated with the production rules may be applied as the successive steps in a derivation are identified. The pre-eminent example is when the parse of a computer program is used to generate machine code: each construct in the grammar of the programming language (IF statement, FOR loop, procedure call, or whatever) is given the corresponding sequence of machine code instructions.

Our interest focuses on the calculation of certain values associated with language words. Such values are called *attributes*, and attaching attribute definition rules to the production rules yields an *attributed grammar*. Table 15.2 provides an example in which an attribute is used to calculate the values of words generated by the binary number grammar defined earlier. Subscripts are used to distinguish between symbols occurring on both sides of a production rule.

An attribute is *inherited* when attribute values on the right-hand side of the production rule are defined using attribute values from the left-hand side. This happens in the fifth and sixth rules of the first column, where $v(b_1)$ is defined. Otherwise, the attribute is *synthesized*.

Table 15.2 *Binary real number grammar*

production rule	attribute definition	production rule	attribute definition
$s \to 0a$	$v(a) = 0$	$b \to .c$	output $v(b) + v(c)$
$s \to 1b$	$v(b) = 1$	$c \to 0c_1$	$v(c) = \frac{1}{2}v(c_1)$
$a \to \varepsilon$	$v(a) = 0$	$c \to 1d$	$v(c) = \frac{1}{2} + \frac{1}{2}v(d)$
$a \to .c$	output $v(a) + v(c)$	$d \to 0c$	$v(d) = \frac{1}{2}v(c)$
$b \to 0b_1$	$v(b_1) = 2v(b)$	$d \to 1d_1$	$v(d) = \frac{1}{2} + \frac{1}{2}v(d_1)$
$b \to 1b_1$	$v(b_1) = 1 + 2v(b)$	$d \to \varepsilon$	$v(d) = 0$
$b \to \varepsilon$	$v(b) = 0$		

The parsing of a word is naturally represented as an ordered rooted labelled tree (see Knuth [19]), in which the symbols of the word are end-vertices and internal vertices are non-terminal symbols. For any internal vertex v with children $c_1, \ldots, c_k$, the pair $(v, c_1 \cdots c_k)$ is in the relation $\to$. Figure 15.1 depicts the parsing of the word 101.011, as given in the implications (15.2). Edges in the tree are annotated to show the values of attributes being passed around the tree. Arrows indicate whether the attribute is being synthesized (from below) or inherited (from above).

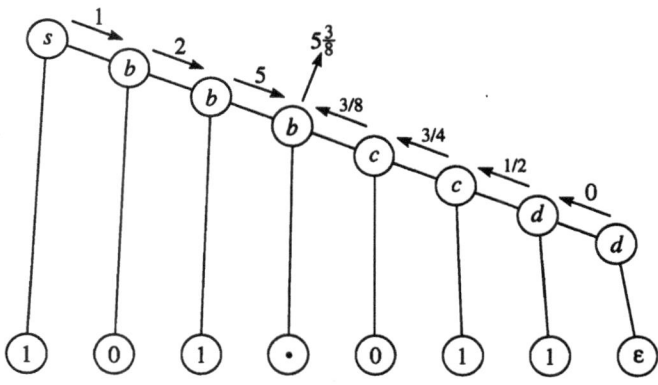

FIG. 15.1

The simplicity of the parse tree in Fig. 15.1 reflects the very restricted nature of type 3 grammars. In fact, a regular language is not normally dealt with using a parser based on an attributed grammar, but rather using a *lexical analyser* which can be built from a digraph model called a *deterministic finite automaton*. This is the second fundamental graph in computing; it is discussed in the next section.

To illustrate further the different domains of parsing and lexical analysis, Table 15.3 gives an attributed type 2 grammar for evaluating arithmetic expressions; the start symbol is e. The production rule '$e \to$ num' makes 'num' a terminal symbol—namely, $\Sigma = \{+, \times, (,), -, \text{num}\}$. We could provide an explicit definition of the format of 'num' as part of the grammar by appending to this table the thirteen rules of Table 15.2, but this would be clumsy. Instead, as suggested above, we assign the task of recognizing numbers to a lexical analyser which is also responsible for providing the value of the number (the function 'val' in '$v(e) = \text{val(num)}$').

Table 15.3 *Arithmetic expression grammar*

production rule	attribute definition
$e \to e_1 + e_2$	$v(e) = v(e_1) + v(e_2)$
$e \to e_1 \times e_2$	$v(e) = v(e_1) \times v(e_2)$
$e \to -e_1$	$v(e) = -v(e_1)$
$e \to (e_1)$	$v(e) = v(e_1)$
$e \to$ num	$v(e) = \text{val(num)}$

15.4 Finite automata

A *finite automaton* $M = (V, \Sigma, \delta, s, T)$ consists of a finite non-empty set V of *states*, a finite non-empty input alphabet Σ, a ternary relation δ in $\Sigma \times V \times V$, a unique *start state* $s \in V$, and a non-empty set $T \subseteq V$ of *termination states*.

The relation δ is a digraph with vertex set V and arc labels in Σ; this digraph may have loops and multiple arcs. If, for any vertex $v \in V$ and symbol $a \in \Sigma$, there is at most one arc with label a leaving vertex v, then δ is a partial mapping from $\Sigma \times V$ to V. In this case, M is said to be a *deterministic finite automaton*. An example of a deterministic finite automaton M_B with states $V = \{s, a, b, c, d\}$ and $T = \{a, b, d\}$ is given in Fig. 15.2.

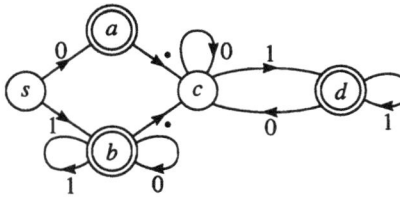

FIG. 15.2

For a deterministic finite automaton, the mapping δ extends in a natural way to a mapping $\delta^* : \Sigma^* \times V \to V$ given by

$$\delta^*(\varepsilon, v) = v \quad \text{and} \quad \delta^*(aw, v) = \delta(a, \delta^*(w, v)).$$

In the digraph interpretation of M, the mapping δ^* extends δ from arcs to arbitrary walks.

Finite automata are machines for recognizing languages. For each state $v \in V$, let $L(v)$ be the set of words $w \in \Sigma^*$ such that $\delta^*(w, v) \in T$. Then a deterministic finite automaton M *accepts* a word $w \in \Sigma^*$ if $w \in L(s)$. The set $L(M) (= L(s))$ of words accepted by M is the language *recognized* by M. For example, the deterministic finite automaton M_B in Fig. 15.2 recognizes exactly the language L_B generated by the grammar in Table 15.2. In fact, the grammar was constructed from Fig. 15.2 in a canonical way: its production rules correspond to an adjacency table for the digraph. It is relatively easy to implement the function δ^* of a deterministic finite automaton as a computer program which can recognize the corresponding language. This is how the lexical analyser would be constructed for recognizing the 'num' symbols in Table 15.3.

The deterministic finite automaton of Fig. 15.2 is unique in the following sense. Define an equivalence relation $\sim$ on the states V of a deterministic finite automaton M by $u \sim v$ if and only if $L(u) = L(v)$. M is in *reduced form* if its equivalence classes under $\sim$ are all singletons. Clearly, any deterministic finite automaton may be transformed into a reduced-form deterministic finite automaton which recognizes the same language by replacing its states by the equivalence classes under $\sim$. An algorithm for this is provided by Aho *et al.* [1].

The following theorem is due to Huffman.

Theorem 15.1 *If a language L is recognized by some deterministic finite automaton, then there is a unique reduced form deterministic finite automaton recognizing L.*

Exactly which languages can be recognized by some deterministic finite automaton is specified in a famous theorem of Kleene.

Theorem 15.2 (Kleene's Theorem) *A language is recognized by some deterministic finite automaton if and only if it is regular.*

Proofs of these two theorems can be found in Hopcroft and Ullman [17]. Kleene's Theorem tells us that deterministic finite automata belong at the bottom of the Chomsky hierarchy with type 3 grammars. The other types of languages given in Table 15.1 are recognized by the following enhancements of the basic finite automaton.

Pushdown finite automata: these are allowed to maintain a stack while reading a word; the state transition function δ is allowed to refer to the top of the stack, as well as the current state and input symbol, to decide the next state (digraph arcs are labelled by a pair).

Linear bounded automata: these also allow storage, but this time the input word itself is used instead of a separate stack; symbols may be replaced from a larger alphabet, and the word can be read backwards and forwards instead of just from beginning to end, so the correspondence with digraph walks is lost.

Turing machines: these extend linear bounded automata by allowing storage to increase the input word arbitrarily in length.

The precise definitions can be found in Hopcroft and Ullman [17], where the power of these machines is discussed in detail. We summarize this discussion in Table 15.4 (compare with Table 15.1).

Table 15.4 *Chomsky hierarchy including finite state machines*

machine type	deterministic	language recognized	grammar type
finite automata	yes	regular	type 3
	no	regular	type 3
pushdown finite automata	yes	deterministic	LR(k)
	no	context-free	type 2
linear bounded automata	yes	unknown: see [9, pp. 47–48]	
	no	context-sensitive	type 1
Turing machines	yes	recursively enumerable	type 0
	no	recursively enumerable	type 0

The study of finite state machines in general belongs more to algebra than to graph theory (see, for example, Eilenberg [11]). Hopcroft and Ullman [17] remains the most appealing introduction from a graph-theoretical perspective. We note in passing that questions relating to graphical enumeration of deterministic finite automata are dealt with comprehensively by Harary and Palmer [14, Chapter 6].

We conclude the first part of this chapter by describing an isolated example in which graph theory plays a role in the study of deterministic finite automata. This example turns out to be useful later in studying program control flow. If e is a regular expression, its *star height* sh(e) is the deepest nesting of closure operators in e. For example, the star height of the regular expression $(0 \cup 1(1 \cup 0)^*)(\varepsilon \cup .(1 \cup 0)^*1)$ in expression (15.1) is 1. The definition of regular expressions given earlier may be formalized by specifying a grammar that generates all regular expressions. Then a precise definition of star height may be given in terms of an associated attributed grammar; this is done in Table 15.5.

The *star height* of a regular language L is defined as sh(L) = $\min_e$ sh(e), where the minimum is taken over all expressions e representing L.

Table 15.5 *Star height grammar*

production rule	attribute definition
$e \to e_1 \cup e_2$	$\mathrm{sh}(e) = \max(\mathrm{sh}(e_1), \mathrm{sh}(e_2))$
$e \to e_1 e_2$	$\mathrm{sh}(e) = \max(\mathrm{sh}(e_1), \mathrm{sh}(e_2))$
$e \to (e_1)$	$\mathrm{sh}(e) = \mathrm{sh}(e_1)$
$e \to e_1^*$	$\mathrm{sh}(e) = \mathrm{sh}(e_1) + 1$
$e \to e_1^+$	$\mathrm{sh}(e) = \mathrm{sh}(e_1) + 1$
$e \to$ symbol	$\mathrm{sh}(e) = 0$

If D is a digraph with strongly connected components $D_1, \ldots, D_k$, its *feedback vertex number* $\phi(D)$ is defined by

$$\phi(D) = \max_i \min\{X \subseteq V(D_i) \mid D_i - X \text{ is an acyclic digraph}\}.$$

The *weak feedback vertex number* $\hat{\phi}(D)$ is the minimum number of applications of the following digraph operation needed to reduce D to an acyclic digraph: *delete one vertex from each strongly connected component of the digraph*. In the case of Fig. 15.2, $\phi(D)$ and $\hat{\phi}(D)$ both have value 2, but it is easy to see that $\phi(D) \geq \hat{\phi}(D)$ in general. Determining $\phi(D)$ is NP-hard, but to the best of my knowledge the status of $\hat{\phi}(D)$ is unknown. The feedback vertex number is sometimes defined as a global invariant without the above localization to the strongly connected components. The weak feedback vertex number has also been called the *cycle rank* of a digraph. We shall meet both numbers again subsequently, but for the moment it is the latter that is of interest. Its relationship with star height was first discovered by Eggan [10].

Theorem 15.3 *If L is a regular language, then*

$$\mathrm{sh}(L) = \min\{\hat{\phi}(M) \mid M \text{ is a finite automaton recognizing } L\}.$$

Consider the language L_B generated by the regular expression (15.1). The machine M_B in Fig. 15.2 has $\hat{\phi}(M_B) = 2$, so that $\mathrm{sh}(L_B) \leq 2$. As remarked above, $\mathrm{sh}(L_B)$ must be 1, but M_B is the unique deterministic finite automaton recognizing L_B. Therefore L_B must also be recognized by some non-deterministic finite automaton M_B', with $\hat{\phi}(M_B') = 1$. Such a machine is shown in Fig. 15.3.

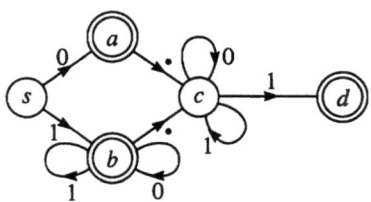

FIG. 15.3

Non-deterministic finite automata are much more complex than deterministic ones, even though they recognize the same class of languages (see Table 15.4); for instance, equivalence of deterministic finite automata can easily be decided using Theorem 15.1, but equivalence in the non-deterministic case is an undecidable problem. So it is impressive that the question of determining star height should have been completely solved (see Hashiguchi [15]). The solution is not graph-theoretical, however, and we shall content ourselves with describing a class of languages (discovered by Cohen [7]), for which star height may be determined graph-theoretically using only deterministic finite automata.

A regular language L over Σ has the *finite intersection property* if, for each pair of words $w, w' \in \Sigma^+$, either the quotients $w \backslash L$ and $w' \backslash L$ are identical, or $(w \backslash L) \cap (w' \backslash L)$ has finite cardinality. For example, consider the deterministic finite automaton formed from Fig. 15.2 by deleting all states except c and d, and making d the new start state and also the unique termination state. This machine recognizes the language $(1^* \cup 0^+ 1)^*$—that is, the set of all binary strings that are either empty or end in 1. This language clearly has the finite intersection property: deleting all occurrences of any given binary preface leaves the language unchanged. However, if c is chosen as the new start state, instead of d, then the machine recognizes the language $(1 \cup 0)^* 1$ consisting of all non-empty binary strings ending in 1 (compare with expression (15.1)). This language does not have the finite intersection property since $0 \backslash (1 \cup 0)^* 1 = (1 \cup 0)^* 1$, whereas $1 \backslash (1 \cup 0)^* 1 = (1^* \cup 0^+ 1)^*$.

The following theorem is due to Cohen [7].

Theorem 15.4 *Let L be a regular language with the finite intersection property, and let M be the unique minimum deterministic finite automaton recognizing L. Then $\text{sh}(L) = \hat{\phi}(M)$.*

For our previous example, we see that no expression of star height 1 can represent the set of all binary strings that are empty or end in 1. We can also see that the deterministic finite automaton M_B of Fig. 15.2 does not yield the star height of L_B, because M_B does not have the finite intersection property.

15.5 Flowgraphs

Control flow, the different execution paths that may be traced through a program's instructions, is modelled naturally by using a deterministic finite automaton. Program 'states' are either *action* states or *decision* states, the latter causing a branch in the flow of control. We make a corresponding partition Π of the alphabet Σ, and let $\Sigma(\Pi)$ denote the collection of partition sets of Σ. Singletons in $\Sigma(\Pi)$ represent actions (for example, input, output or assignment of values to variables); larger sets in $\Sigma(\Pi)$ represent the mutually exclusive outcomes of a decision (for example, '$x = 0$' and '$x \neq 0$').

A *labelled flowgraph* is a deterministic finite automaton $M = (V, \Sigma, \delta, s, t)$ with a single termination state $t \neq s$, such that there is a map m from V onto $\Sigma(\Pi) \cup \{\varnothing\}$ satisfying $m(t) = \varnothing$ and, for each $v \in V$,

$$\{a \in \Sigma \mid \delta(a, v) \text{ is defined}\} = m(v).$$

In addition, we require each state v to be reachable from the start state and to be able to reach the termination state: for each $v \in V$, there exist $w, w' \in \Sigma^*$ such that $\delta^*(w, s) = v$ and $\delta^*(w', v) = t$. Figure 15.4 shows a labelled flowgraph that models a Pascal program to check for counter-examples to the $3x+1$ conjecture—that every positive integer can be reduced to 1 by repeated halving, or multiplying by 3 and adding 1 whenever an odd number is reached (see Lagarias [21] for a fascinating survey).

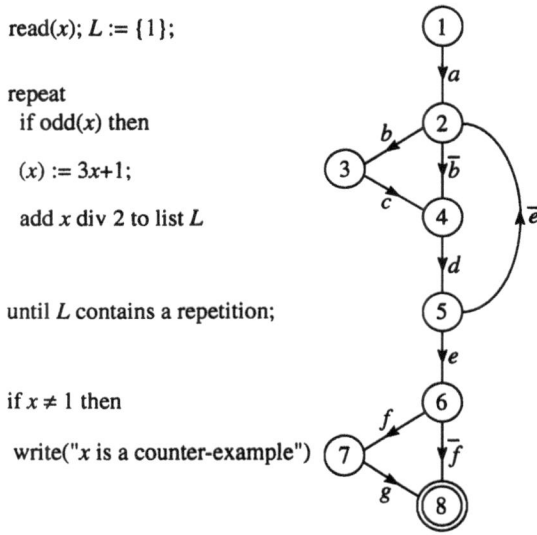

```
read(x); L := {1};

repeat
    if odd(x) then

        (x) := 3x+1;

    add x div 2 to list L

until L contains a repetition;

if x ≠ 1 then

    write("x is a counter-example")
```

FIG. 15.4

We digress at this point to mention that the general definition of a deterministic finite automaton has also been widely used as the basis for studying control flow; Hecht [16] provides admirable coverage. Most importantly, a deterministic finite automaton is *reducible* if it can be reduced to a single vertex by repeatedly applying two operations:

making the digraph simple by deleting loops and multiple arcs;

collapsing arcs that go to vertices of in-degree 1;

for instance, the deterministic finite automaton of Fig. 15.2 is reducible. Reducibility is a common property in computer programs, and reducible deterministic finite automata are very amenable to analysis. Many NP-complete problems have polynomial algorithms for reducible deterministic finite automata—notably, Shamir's algorithm for the feedback vertex number (see Rosen [25]).

The labelled flowgraph does not constitute a model of computation, since we have given no semantics to the actions and decisions. If we did this, we would end up with a strengthening of the basic deterministic finite automaton amounting to the Turing machine; the finiteness of Σ is an obstacle to be overcome. This is not of interest from a graph-theoretical point of view, however.

It is convenient to make explicit the underlying digraph of a labelled flowgraph as a model in its own right. An *unlabelled flowgraph* $F = (D, s, t)$ consists of a directed graph D with two distinguished vertices s and t, where t has outdegree 0 and each vertex of D lies on some s-t walk.

The graph-theoretical appeal of flowgraphs lies in their ability to capture two features of programs of inherent interest. First, the collection of different execution sequences through the program, modelled as a labelled flowgraph M, is precisely the language $L(M)$. Of course, it may happen that some of these sequences cannot be activated by any legal program input, but we are consistent in ignoring this since it is a semantic issue.

Secondly, programs tend to be written by combining prefabricated pieces of control flow—for instance, IF statements, or CASE statements, or WHILE loops. This has a natural counterpart as a type of digraph decomposition. Let $F = (D, s, t)$ and $F' = (D', s', t')$ be flowgraphs. Then F' is a *subflowgraph* of F if and only if

D' is a subgraph of D;

if $v \in V(D') \backslash \{s', t'\}$, then $v \neq s$ and $\text{indeg}_D(v) = \text{indeg}_{D'}(v)$;

if $v \in V(D') \backslash \{t'\}$, then $\text{outdeg}_D(v) = \text{outdeg}_{D'}(v)$.

The second condition guarantees that F' corresponds to a 'subprogram' of F, in the sense that s' is the only starting point of F' in F; note that s' must be reachable from s, since D is a flowgraph. Condition 3 ensures that the mapping onto $\Sigma(\Pi) \cup \{\varnothing\}$ for any labelled flowgraph with digraph D also works for D'.

A subflowgraph $F' = (D', s', t')$ of F is *trivial* if it is the unique unlabelled flowgraph with precisely one arc. If D' is collapsed to a single arc $s't'$ in F, then we obtain a new flowgraph $F'' = (D'', s, t)$. We express this *decomposition* by writing $F = F'' \uparrow F'$. For example, the flowgraph of Fig. 15.4 contains four trivial subflowgraphs (with start vertices 1, 3, 4 and 7) and six non-trivial proper subflowgraphs (one with start vertex 1, four with start vertex 2, and one with start vertex 6).

A flowgraph $F = (D, s, t)$ is *prime* if D is isomorphic to a path graph P_k ($k \geq 2$) or has no non-trivial, proper subflowgraphs. If F is not prime, it is *composite*. Prime flowgraphs may be thought of as the basic building blocks of programming and often appear as control structures in programming languages. The most common ones are shown in Fig. 15.5: P_k corresponds to building programs using the sequence operator ';'; D_1 is 'IF THEN'; D_2 is 'WHILE DO'; etc.

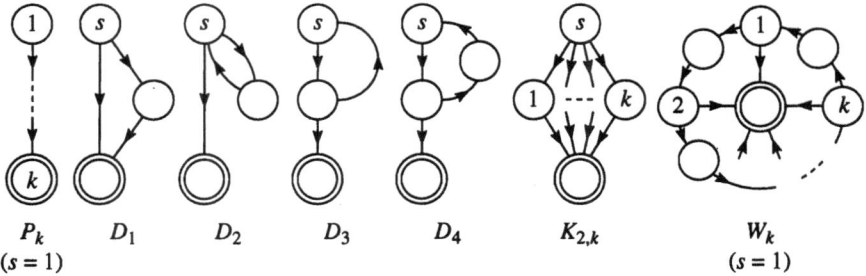

Fig. 15.5

A given composite flowgraph $F = (D, s, t)$ can be expressed (not necessarily uniquely) as a decomposition $F = P \uparrow F_1, \ldots, F_k$, where P is a prime flowgraph and the F_i are arc-disjoint subflowgraphs of F. For example, the flowgraph in Fig. 15.4 can be represented by the *decomposition expression*

$$(P_4, 1, 8) \uparrow ((D_3, 2, 6) \uparrow (P_2, 2, 5)) \uparrow (D_1, 2, 4)), (D_1, 6, 8)$$

or, if the vertices involved do not matter,

$$P_4 \uparrow (D_3 \uparrow P_2 \uparrow D_1), D_1.$$

Flowgraph decomposition as a technique for program analysis was first dealt with in detail by Linger, Mills and Witt [22]. The first consideration of decomposition in terms of decomposition expressions is probably due to Prather and Giulieri [24]. An analysis of the uniqueness of the decomposition in terms of the decomposition operation was made by Fenton *et al.* [12]. The best algorithm for flowgraph decomposition appears to be the linear algorithm of Tarjan and Valdes [26], based on decomposition into 3-connected components. However, their paper is not self-contained. Since flowgraph sizes, in practice, do not usually exceed two hundred vertices, it might seem worthwhile to look for a more self-contained (even if slower) approach.

The decomposition of a flowgraph, as represented by a decomposition expression, can be regarded as belonging to the language $L(\Gamma_F)$ generated by the type 2 grammar Γ_F shown in Table 15.6; the start symbol is F. The operator ',' is taken to have higher precedence than the operator $\uparrow$. The non-terminal E represents a collection of arc-disjoint subflowgraphs.

Table 15.6 *The flowgraph grammar*

$F \to P \uparrow E$
$E \to E, F$
$E \to F$
$F \to (F)$
$F \to P$
$P \to \text{prime}$

Let us define an equivalence relation on $L(\Gamma_F)$ by saying that two words w_1 and w_2 are *equivalent* if there is some flowgraph for which w_1 and w_2 are both legitimate decomposition expressions. For example, another decomposition expression representing Fig. 15.4 is

$$P_3 \uparrow P_3 \uparrow D_1, (D_3 \uparrow P_2 \uparrow D_1).$$

Let $[L(\Gamma_F)]$ denote the collection of equivalence classes under this relation. It is not hard to characterize $[L(\Gamma_F)]$: we ignore the order in which arc-disjoint subflowgraphs appear in a flowgraph (so that $D_1, (D_3 \uparrow P_2 \uparrow D_1)$ is the same as $(D_3 \uparrow P_2 \uparrow D_1), D_1)$, and we treat $P_k \uparrow P_{k'}, E$ as equivalent to $P_{k+k'-2} \uparrow E$. We may now define a mapping from the class of unlabelled flowgraphs to $[L(\Gamma_F)]$ that associates with each flowgraph a unique decomposition expression. This mapping is not onto, since not all words in $L(\Gamma_F)$ can be realized as valid flowgraphs. For instance, in $(F_1, a, b) \uparrow (F_2, c, d), (F_3, d, c)$, the vertices of F_2 and F_3 must induce a sink, contrary to the fact that flowgraph vertices must be able to reach the termination vertex. Characterizing the words in $L(\Gamma_F)$ that cannot be realized as flowgraphs is an open problem.

Incidentally, the grammar Γ_F is an example of a graph grammar, disguised as a string grammar by replacing graphs with symbols (a technique introduced by Bauderon and Courcelle [4]). In the next section, we use attributes of Γ_F to explore a number of graph-theoretical techniques of program analysis. Analysis of decomposition expressions using attributed grammars was introduced by Whitty [27].

15.6 Program analysis

Program structuredness Consider the attributed grammar in Table 15.7.

Table 15.7 *Definition of S-structuredness*

production rule	attribute definition
$F \to P \uparrow E$	$d_S(F) = \min(d_S(P), d_S(E))$
$E \to E_1, F$	$d_S(E) = \min(d_S(E_1), d_S(F))$
$E \to F$	$d_S(E) = d_S(F)$
$F \to (F_1)$	$d_S(F) = d_S(F_1)$
$F \to P$	$d_S(F) = d_S(P)$
$P \to$ prime	$d_S(P) = \begin{cases} 1, & \text{prime} \in S \\ 0, & \text{otherwise} \end{cases}$

This grammar calculates a 0-1 function $d_S(F)$ which takes the value 1 precisely when F decomposes entirely into primes from the set S. Flowgraphs taking the value 1 are called *S-structured*. Most simply, we may consider the set

$S_1 = \{P_k \ (k \geq 2), \ D_1, \ D_2\}$ (see Fig. 15.5). These are called *D-structures* after Dijkstra, who advocated the construction of programs from some such small set of basic control structures: the set corresponds to allowing only sequencing, selection and iteration of programming structures—interpreted differently by many as a puritanical ban on the GOTO statement.

Let us examine the properties of the nominal complexity measure $d_S(F)$. First, do there exist degrees of unstructuredness among flowgraphs F for which $d_S(F) = 0$? More specifically, for a given labelled flowgraph $M = (V, \Sigma, \delta, s, t)$ with $d_S(M) = 0$, does there exist another flowgraph $M' = (V', \Sigma, \delta', s', t')$ with $d_S(M') = 1$ and $L(M) = L(M')$? Two versions of this specific question were answered independently by Kosaraju [20] and Kasai [18]; we give the latter answer. A cycle C in a flowgraph $F = (D, s, t)$ is *t-separable* if $V(C)$ contains a vertex separating $V(C)$ from t. For example, vertex 5 in Fig. 15.4 separates the cycle $2 \to 4 \to 5 \to 2$ from the termination vertex 8; the cycle W_k in Fig. 15.5, however, is not t-separable.

Theorem 15.5 *Let $M = (V, \Sigma, \delta, s, t)$ be a labelled flowgraph. If each cycle is t-separable, then there exists a D-structured labelled flowgraph recognizing $L(M)$. Conversely, if M is in reduced form and contains some cycle that is not t-separable, then there is no D-structured labelled flowgraph recognizing $L(M)$.*

We digress to mention that the dual condition, that all cycles in a deterministic finite automaton are s-separable, is necessary and sufficient for a deterministic finite automaton to be reducible; this result is due to Hopcroft and Ullman—see Hecht [16, Section 4.2].

Kosaraju's result is slightly stronger than Kasai's, but its converse part requires not only that M is in reduced form, but that the mapping $V \to \Sigma(\Pi) \cup \{\varnothing\}$ in the definition of a labelled flowgraph is one-to-one and onto. In either case, if there *does* exist a D-structured flowgraph, this may be created by first putting M into reduced form and then repeatedly applying an operation called *vertex splitting*. This replaces a vertex x of in-degree k with k copies of x, each taking all of its outgoing arcs and precisely one of its incoming arcs. For the purpose of vertex splitting, the start vertex is treated as having a dummy incoming arc. Thus the prime (D_3, s, t) may be transformed by vertex splitting at s into the flowgraph $(P_3, s, t) \uparrow (D_2, a, t)$. Clearly, vertex splitting in a labelled flowgraph M preserves $L(M)$.

The converse part of Theorem 15.5 identifies a class of flowgraphs for which $d_S(F) = 0$ in a strong sense. How strong is determined by the next result, which is due essentially to Boehm and Jacopini [6], although our statement makes explicit a construction due to Ashcroft and Manna [2]. Recall that $\phi(D)$ denotes the maximum cardinality of a minimum feedback vertex set among the strong components of the digraph D.

Theorem 15.6 *For each program P with labelled flowgraph $M = (V, \Sigma, \delta, s, t)$, there is a program P' computing the same function as P, whose labelled flowgraph is D-structured and whose alphabet has cardinality $|\Sigma| + 4\phi(M)$.*

Proof Define a new partitioned alphabet $\Sigma_M(\Pi_M)$ by

$$\Sigma_M(\Pi_M) = \Sigma(\Pi) \cup \{\{A_i^0\}, \{A_i^1\}, \{B_i^0, B_i^1\} : i = 1, \ldots, \phi(M)\}.$$

Let the strongly connected components of M be $S_1, \ldots, S_p$, and suppose that these components are numbered in such a way that no arc joins S_i to S_j, for $i < j$; it is well known that this can be done. We consider each S_k in turn, starting with S_1, and produce a sequence of new labelled flowgraphs $M = M_0, M_1, \ldots, M_p$. From M_p, we shall construct the desired program P'.

Thus, let $k \geq 1$. If all cycles in S_k are t-separable, then set $M_{k+1} = M_k$ and proceed to S_{k+1}. If not, take a minimum feedback vertex set $\{v_1, \ldots, v_q\}$ for S_k, say, where $q \leq \phi(M)$. Add new vertices $u_1, \ldots, u_q, w_1, \ldots, w_q$ and $t_1, \ldots, t_q$, and replace each arc in M_k directed to t by an arc to t_1 with the same label. Next, delete each arc $v_i'v_i$ with label $\alpha \in \Sigma$ and add the new arcs $v_i'u_i$ with label α, $u_i t_1$ with label A_i^1, $t_i w_i$ with label B_i^1, $w_i v_i$ with label A_i^0, and $t_i t_{i+1}$ with label B_i^0, where $t_{q+1} = t$.

The resulting digraph M_{k+1} is again a labelled flowgraph. Any path in M_k that included an arc into v_i has been extended by inserting, in place of that arc, a subpath comprising the sequence of arc labels $\alpha A_i^1 B_1^0 \ldots B_{i-1}^0 B_i^1 A_i^0$, for some $\alpha \in \Sigma$. Similarly, any M_k-path that terminated in t has been extended by a subpath comprising the sequence of arc labels $B_1^0 \ldots B_q^0$. Each cycle in S_k has been replaced by a cycle passing $t_1, \ldots, t_i$ which is separated from t by t_i, for some $i \leq q$. Moreover, any cycle in M_k passing $S_{k'}$ and S_k, for $k' < k$, must also pass such a separating vertex, since no arcs pass from $S_{k'}$ to S_k. We conclude that M_p is a labelled flowgraph satisfying the condition of Theorem 15.5, and that we can therefore find a labelled flowgraph M' which is D-structured and recognizes $L(M_p)$.

Finally, we interpret each symbol A_i^j as the assignment action '$a_i = j$' and B_i^j as the outcome '$a_i = j$' of a test for the value of the variable a_i. Assuming that the a_i are automatically initialized to 0, we have realized M' as a program P' that computes the same function as P, as required. □

Nesting complexity For flowgraphs that are structured over the primes in Fig. 15.5, the grammar in Table 15.8 calculates the highest level at which a loop is 'nested'. The notion of program complexity exemplified in this grammar has been much studied (see Fenton [13]), the idea being that it may be possible to predict in some way the effort needed to debug programs, test them, make changes to them, or whatever. A specific example for this particular grammar might be its use as a basis for analysing performance, on the premise that loops within loops within loops are likely to take longer to run, on average or in the worst case, than programs in which loops are not nested.

Table 15.8 *Definition of loop nesting height $h(F)$*

production rule	attribute definition
$F \to P \uparrow E$	$h(F) = h(P) + h(E)$
$E \to E_1, F$	$h(E) = \max(h(E_1), h(F))$
$E \to F$	$h(E) = h(F)$
$F \to (F_1)$	$h(F) = h(F_1)$
$F \to P$	$h(F) = h(P)$
$P \to$ prime	$h(P) = \begin{cases} 1, & \text{prime} \in \{D_2, D_3, D_4, W_k \ (k \geq 2)\} \\ 0, & \text{prime} \in \{D_1, P_k, K_{2,k} \ (k \geq 2)\} \end{cases}$

As with structuredness, we can ask how robust is the measure $h(F)$: given a labelled flowgraph M, can we find a flowgraph M' with $L(M') = L(M)$ and $h(M') < h(M)$? Motoki [23] has provided a useful lemma for answering questions of this kind.

Lemma 15.7 *Let M be a labelled flowgraph in reduced form for which $L(M)$ has the finite intersection property, and let M' be a labelled flowgraph such that $L(M') = L(M)$. Then $h(M') \geq \hat{\phi}(M)$.*

Proof We have

$$\begin{aligned} h(M') &\geq \hat{\phi}(M') && \text{(by induction, using Table 15.8)} \\ &\geq \text{sh}(L(M')) && \text{(immediate from Theorem 15.3)} \\ &= \text{sh}(L(M)) && \text{(since } L(M') = L(M)) \\ &= \hat{\phi}(M) && \text{(by Theorem 15.4).} \end{aligned}$$
□

Lemma 15.7 shows that, for arbitrary depths, there are languages L for which any labelled flowgraph recognizing L and structured over the primes of Fig. 15.5 must have loops nested to at least that depth. In fact, Motoki used the same approach to show that this is true even if loops are allowed to have exit arcs bypassing arbitrarily many levels of nesting (available in the programming language Ada, for instance).

The contexts in which the invariants ϕ and $\hat{\phi}$ appear in Theorem 15.6 and Lemma 15.7 seem intriguingly close. If we could replace ϕ with $\hat{\phi}$ in the statement of Theorem 15.6, this would suggest a more efficient way of restructuring unstructured programs. This would be especially interesting if a good algorithm exists for calculating or approximating $\hat{\phi}$.

Program testability We conclude by looking at another measurement of program complexity, this time relating to how difficult a program is to test.

A standard procedure for testing a program is to run it on test data in such a way as to maximize the number of different routes through the code that have been exercised. Obviously, it is undesirable that someone who buys a program should be the first to exercise some route that harbours a fatal error! This is often the case, however, and it is easy to see why: given a flowgraph $F = (D, s, t)$, there are generally infinitely many s-t walks in D. Not all of these walks correspond to possible sequences of execution, but the number that do is likely to be infinite or at least too large to test exhaustively.

A basic measure of how hard a program is to test is to count the number of s-t trails; this is, of course, finite. For a flowgraph F, let this be denoted by $\tau(F)$. Then we ask, can $\tau(F)$ be computed by using some attribute of the grammar in Table 15.6? It is certainly not obvious that this can be done; for instance, $\tau(D_3) = 1$ and $\tau(D_1) = 2$, but $\tau(D_3 \uparrow D_1) = 4$, because the D_1 within the D_3 allows the previously unused back arc in D_3 to become part of a trail (see Fig. 15.6).

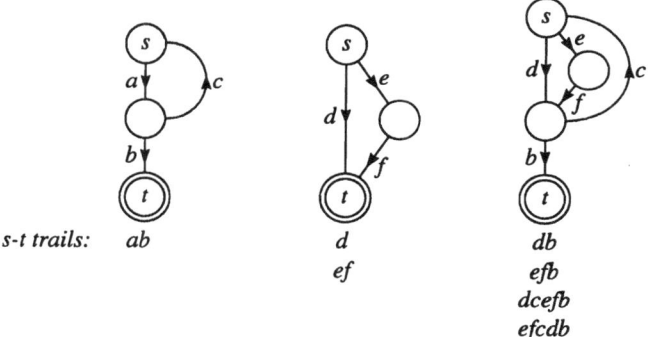

FIG. 15.6

Bainbridge [3] has discovered two sufficient conditions which make it possible to calculate $\tau(F)$ using an attributed grammar. In addition to $\tau(F)$, however, there has to be a second attribute; this is $\hat{\tau}(F)$, the number of ordered pairs of arc-disjoint s-t trails in F. As a useful example, the sufficient conditions are satisfied by flowgraphs that are structured over the primes of Fig. 15.5; an attributed grammar can therefore be constructed, and is shown in Table 15.9.

Bainbridge also shows that s-t walks that pass no vertex more than k times can be calculated in a similar manner. His grammar, however, clearly marks a considerable relaxation of the parsing rules imposed by the flowgraph grammar proposed in Table 15.6. For a start, the lists E of arc-disjoint subflowgraphs have been artificially represented, using the '$F_1, \ldots, F_k$' construction. Secondly, rows 5 and 7 require the F_i to be taken in the correct order around the cycle in the prime (D_4 and W_k, respectively). This is a requirement that cannot be captured by the grammar.

Perhaps the most interesting open question is to determine exactly which graph invariants can be calculated using attributed grammars, and under what restrictions. This is a general question in graph grammars, but it seems certain that most progress can be made for very specific types of grammar, such as the flowgraph grammar. One particular question is how much extra power can be achieved through the use of inherited, as well as synthesized, attributes (as illustrated in Fig. 15.1). All of the flowgraph grammars presented in this chapter have used synthesized attributes exclusively.

Table 15.9 *Definition of the number of s-t trails $\tau(F)$*

production rule	attribute definition	
$F \to P_k \uparrow F_1, \ldots, F_k$	$\tau(F) = \prod_{i=1}^{k} \tau(F_i)$	$\hat{\tau}(F) = \prod_{i=1}^{k} \hat{\tau}(F_i)$
$F \to D_1 \uparrow F_1$	$\tau(F) = 1 + \tau(F_1)$	$\hat{\tau}(F) = 2\hat{\tau}(F_1)$
$F \to D_2 \uparrow F_1$	$\tau(F) = 1 + \tau(F_1)$	$\hat{\tau}(F) = 0$
$F \to D_3 \uparrow F_1$	$\tau(F) = \tau(F_1) + \hat{\tau}(F_1)$	$\hat{\tau}(F) = 0$
$F \to D_4 \uparrow F_1, F_2$	$\tau(F) = \tau(F_1) + \hat{\tau}(F_1)\tau(F_2)$	$\hat{\tau}(F) = 0$
$F \to K_{2,k} \uparrow F_1, \ldots, F_k$	$\tau(F) = \sum_{i=1}^{k} \tau(F_i)$	$\hat{\tau}(F) = 2\sum_{i=1}^{k-1}\sum_{j=i+1}^{k} \tau(F_i)\tau(F_j)$
$F \to W_k \uparrow F_1, \ldots, F_k$	$\tau(F) = 1 + \sum_{i=1}^{k} \prod_{j=1}^{i} \tau(F_j)$	$\hat{\tau}(F) = 2\sum_{i=1}^{k-1} \prod_{j=1}^{i} \tau(F_j)$
$F \to P_k \ (k \geq 2)$	$\tau(F) = 1$	$\hat{\tau}(F) = 0$
$F \to D_1$	$\tau(F) = 2$	$\hat{\tau}(F) = 2$
$F \to D_2$	$\tau(F) = 2$	$\hat{\tau}(F) = 0$
$F \to D_3$	$\tau(F) = 1$	$\hat{\tau}(F) = 0$
$F \to D_4$	$\tau(F) = 1$	$\hat{\tau}(F) = 0$
$F \to K_{2,k} \ (k \geq 2)$	$\tau(F) = k$	$\hat{\tau}(F) = k(k-1)$
$F \to W_k \ (k \geq 2)$	$\tau(F) = k+1$	$\hat{\tau}(F) = 2(k-1)$

Acknowledgements This work was supported by EPSRC on the PROMISE project of the DTI/EPSRC Safety-critical Systems research programme. I am grateful to Jennie Rogers for her helpful comments on drafts of this chapter, and to Bill Jackson for discussions about some of the material.

References

1. A. V. Aho, R. Sethi and J. D. Ullman, *Compilers. Principles, Techniques and Tools*, Addison-Wesley, 1986.
2. E. Ashcroft and Z. Manna, Translating program schemas to while-schemas, *SIAM J. Comput.* **4** (1975), 125–146; MR 53#7094.

3. J. R. Bainbridge, Defining testability metrics axiomatically, *Software, Testing, Verification and Reliability* **4** (1994), 63–80.
4. M. Bauderon and B. Courcelle, Graph expressions and graph rewritings, *Math. Systems Theory* **20** (1987), 83–127; *MR* **89f**: 68038.
5. J.-C. Bermond, J. Bond, M. Paoli and C. Peyrat, Graphs and interconnection networks: diameter and vulnerability, *Surveys in Combinatorics* (ed. E. K. Lloyd), Cambridge University Press, 1983, pp. 1–30; *MR* **85h**: 05057.
6. C. Boehm and G. Jacopini, Flow diagrams, Turing machines and languages with only two formation rules, *Comm. Assoc. Comput. Mach.* **9** (1966), 366–371.
7. R. S. Cohen, Star height of certain families of regular events, *J. Comput. System Sci.* **4** (1970), 281–297; *MR* **45**#3134.
8. B. Courcelle, Graph grammars, monadic second-order logic and the theory of graph minors, *Graph Structure Theory* (ed. N. Robertson and P. Seymour), Contemp. Math. **147**, pp. 565–590; *MR* **94m**: 68107.
9. R. Downey, An invitation to structural complexity, *New Zealand J. Math.* **21** (1992), 33–89; *MR* **93i**: 68072.
10. L. C. Eggan, Transition graphs and the star-height of regular events, *Michigan Math. J.* **10** (1963), 385–397; *MR* **28**#1069.
11. S. Eilenberg, *Automata, Languages, and Machines*, Vol. A, Academic Press, 1974; *MR* **58**#26604a.
12. N. E. Fenton, R. W. Whitty and A. A. Kaposi, A generalised mathematical theory of structured programming, *Theoret. Comput. Sci.* **36** (1985), 145–171; *MR* **87e**: 68007.
13. N. E. Fenton, *Software Metrics: A Rigorous Approach*, Chapman and Hall, 1991.
14. F. Harary and E. M. Palmer, *Graphical Enumeration*, Academic Press, 1973; *MR* **50**#9682.
15. K. Hashiguchi, Algorithms for determining relative star height and star height, *Inform. and Comput.* **78**, 2 (1988), 124–169; *MR* **90k**: 68085.
16. M. S. Hecht, *Flow Analysis of Computer Programs*, North-Holland, 1977; *MR* **57**#8120.
17. J. E. Hopcroft and J. D. Ullman, *Formal Languages and their Relation to Automata*, Addison-Wesley, 1969; *MR* **38**#5533.
18. T. Kasai, Translatability of flowcharts into while programs, *J. Comput. System Sci.* **9** (1974), 177–195; *MR* **50**#9035.
19. D. E. Knuth, *The Art of Computer Programming, Vol. 1: Fundamental Algorithms*, 2nd edn, Addison-Wesley, 1973; *MR* **51**#14624.
20. S. R. Kosaraju, Analysis of structured programs, *J. Comput. System Sci.* **9** (1974), 232–255; *MR* **53**#4598.
21. J. C. Lagarias, The $3x + 1$ problem and its generalizations, *Amer. Math. Monthly* **92** (1985), 3–23; *MR* **86i**: 11043.
22. R. C. Linger, H. D. Mills and B. I. Witt, *Structured Programming: Theory and Practice*, Addison-Wesley, 1979.

23. T. Motoki, An application of Cohen's result on star height to the theory of control structures, *J. Comput. System Sci.* **29** (1984), 312–329; *MR* **86g**: 68016.
24. R. E. Prather and S. G. Giulieri, Decomposition of flowchart schemata, *Computer J.* **24** (1981), 258–262.
25. B. K. Rosen, Robust linear algorithms for cutsets, *J. Algorithms* **3** (1982), 205–217; *MR* **83h**: 68046.
26. R. E. Tarjan and J. Valdes, Prime subprogram parsing of a program, *Proc. 7th ACM Conf. Principles of Programming Languages*, ACM Press, 1980, pp. 95–105.
27. R. W. Whitty, Multidimensional software metrics, *Formal Aspects of Measurement* (ed. B. T. Denvir, R. Herman and R. W. Whitty), Springer-Verlag, 1992, pp. 148–169.

16
Artificial Neural Networks

MARTIN ANTHONY

'Artificial neural networks' are machines (or models of computation) based loosely on the ways in which the brain is believed to work. In this chapter, we discuss some links between graph theory and artificial neural networks. We describe how some combinatorial optimization tasks may be approached by using a type of artificial neural network known as a Boltzmann machine. We then discuss 'learning' in neural networks, where we focus on the amount of training data required and the computational complexity.

16.1 Introduction

There has recently been intense, and fast-growing, interest in 'artificial neural networks'. These are machines (or models of computation) based loosely on the ways in which the brain is believed to work. Neurobiologists are interested in using these machines as a means of modelling biological brains, but much of the impetus comes from their applications. For example, engineers wish to create machines that can perform 'cognitive' tasks, such as speech recognition, and economists are interested in financial time series prediction using such machines. Inevitably, there is a certain amount of hype associated with the subject, particularly in relation to neurobiological modelling.

Here, we take what may be called an 'engineering approach' to artificial neural networks. In such an approach, one is not concerned with the issue of whether artificial neural networks are plausible models of real neural networks, but rather, we start from the fact that they exist and are being used extensively. The questions that mathematicians might then ask include the following.

What can artificial neural networks do for mathematics?

What can mathematics say about artificial neural networks that might be of interest or benefit to practitioners?

In this chapter, we discuss some connections between the theory of artificial neural networks and graphs. In one answer to the first of the above questions, we explain how certain types of artificial neural network may be useful for solving combinatorial optimization problems in graph theory. We then turn our attention to a theory of 'learning' in artificial neural networks, where the graph structure of the network and the hardness of graph-colouring tell us something about the complexity of learning.

16.2 Artificial neural networks

It appears that one reason why the human brain is so powerful is the sheer complexity of connections between neurons. In computer science parlance, the brain exhibits huge parallelism, with each neuron connected to many others. This has been reflected in the design of artificial neural networks. One advantage of such parallelism is that the resulting network is *robust*: in a serial computer, a single fault can make computation impossible, whereas in a system with a high degree of parallelism and many computation paths, a small number of faults may be tolerated with little or no upset to the computation.

Two types of neural network are discussed here: *Boltzmann machines* and *feedforward networks*. Loosely speaking, an artificial neural network consists of a directed graph, with *computation units* situated at the vertices and weights on the arcs. Some of the computation vertices may be distinguished as *input nodes*, which receive signals from the outside world, and some as *output nodes*. The vertices have *activations*, and their activations influence those of their neighbours, either stochastically, as in the Boltzmann machine, or deterministically, as in feedforward networks. The degree to which the activation of a vertex influences those of its neighbours is determined by the weights on the arcs. The process of 'learning' is the adjustment of the weights.

16.3 Boltzmann machines

In a Boltzmann machine, a type of *stochastic* artificial neural network, the underlying digraph has symmetric connections: if $ij \in A$, then $ji \in A$.

Furthermore, the weights are always constrained so that $w_{ij} = w_{ji}$. A Boltzmann machine M may therefore be described as an undirected graph (which may have loops) with weighted edges and stochastic computation units at the vertices. More precisely, a *Boltzmann machine* M is a pair (G, Ω), where $G = (V, E)$ is a graph with n vertices and m edges, and $\Omega \subseteq \mathbf{R}^m \times \{0,1\}^n$ is a set of allowable *states*. For each state $\omega = (w_{e_1}, w_{e_2}, \ldots, w_{e_m}, S_1, \ldots, S_n) \in \Omega$, the vector $\mathbf{w} = (w_{e_1}, w_{e_2}, \ldots, w_{e_m})$ describes the weights assigned to the edges $e_1, e_2, \ldots, e_m$ of the graph, and the vector $(S_1, S_2, \ldots, S_n)$ describes which of the vertices (units) $1, 2, \ldots, n$ is 'on', where vertex i is 'on' in state ω if and only if $S_i = 1$. The *consensus function* of the Boltzmann machine M is the function $C : \Omega \to \mathbf{R}$ given by

$$C(\omega) = \sum_{ij \in E} w_{ij} S_i S_j.$$

Computation proceeds in the machine in a stochastic manner in such a way as to increase consensus. Thus, if w_{ij} is positive then there is a tendency for units i and j to be either both on or both off, while if the weight is negative, then there is a tendency for them to have different activations. When a weight is positive we refer to it as *excitatory*, and when negative as *inhibitory*.

We now describe how the state of the network evolves when the weights are fixed. This is, of course, a very restricted analysis since, in general, the weights may also change through 'learning' (see [1, 14]). However, for the applications we have in mind, the weights are determined explicitly by an instance of a combinatorial optimization problem.

If i is a vertex and ω is a state, then $\omega[i \to \bar{i}]$ is the state obtained from ω by changing S_i to $1 - S_i$—that is, by 'flipping' the activation of i. If such a flip is made, then the resulting change in the consensus function is

$$\Delta C_\omega(i) = C(\omega[i \to \bar{i}]) - C(\omega).$$

If this is fairly large, then it is advantageous to flip the activation of i, and if it is negative with large absolute value, then it is disadvantageous to do so. The decision on whether to flip the activation of unit i is made stochastically, based on $\Delta C_\omega(i)$.

The simplest model of computation in a Boltzmann machine is the sequential model. Here, the machine has an internal clock, and on the tth tick of the clock, a vertex i_t is chosen, uniformly at random. The activation of i_t is then flipped, and the state changed to $\omega[i_t \to \bar{i}_t]$, with probability

$$\text{Prob}\left(\omega \to \omega[i_t \to \bar{i}_t]\right) = \frac{1}{1 + \exp(-\beta \Delta C_\omega(i_t))},$$

for some constant $\beta > 0$. A more complicated procedure is that in which the machine's computations are in parallel. Then, at time t, a subset S^t of the vertices is chosen according to some probability distribution on the subsets and, for each $i \in S^t$, a decision is made, as described above, as to whether to flip the activation of that unit. Note that, although this may involve making a large number of such decisions simultaneously, the computations involved in this procedure are local, since $\Delta C_\omega(i)$ depends only on the activations of the units j adjacent to i and on the weights w_{ij}. The result is some state ω' which is obtained from ω by flipping some of the S_i, for $i \in S^t$. It can be proved (see Aarts and Korst [1]) that the sequential mode of computation results in a stationary distribution on the set of all states, in which the probability that the state of the machine is ω is proportional to $\exp(\beta C(\omega))$. The same conclusion holds if the computations are *synchronous, with limited parallelism*. This is the special case of parallel computation, in which only independent sets of vertices are generated, and each independent set is generated with equal probability. In both cases, in the stationary distribution, states of high consensus are more likely.

The mysterious parameter β has an interesting interpretation. Its reciprocal $T = 1/\beta$ is often called *temperature*. There are indications [1] that the process of *simulated annealing*, whereby the temperature is slowly decreased as computation proceeds (so that β is no longer constant, but increases with time), may be helpful. Aarts and Korst [1] have shown that, as $T \to 0$ and as time tends to infinity, the limiting stationary distribution of states is uniform over all states that maximize the consensus function.

16.4 Optimization with Boltzmann machines

A number of combinatorial optimization problems can be realized as the problem of maximizing consensus in Boltzmann machines. The book by Aarts and Korst [1] contains a full discussion of this approach; here, we shall describe the general approach and give two examples.

The general approach Suppose that we have a combinatorial optimization problem. It is often possible to construct a Boltzmann machine with fixed weights determined by the instance of the problem so that maximizing consensus in the machine is equivalent to solving the optimization problem. A general approach, as described in [1], is as follows.

(1) phrase the optimization problem as a $\{0,1\}$-valued linear programming problem;

(2) construct a Boltzmann machine with vertices corresponding to the variables in the linear programming problem;

(3) define the edges and weights of the machine in such a way that the following *correspondence conditions* hold:

local maxima of the resulting consensus function correspond to feasible solutions of the optimization problem;

if S_1 and S_2 are two feasible solutions of the optimization problem, with S_1 better than S_2, then the consensus of the state corresponding to S_1 is higher than that corresponding to S_2.

Note that we may always take the underlying graph of a Boltzmann machine to be complete, since setting a weight to zero is equivalent to omitting an edge. If the procedure just described can be carried out, then, in theory, one way to attempt to solve the optimization problem is to construct the Boltzmann machine with these properties and let it evolve, perhaps decreasing the temperature parameter β^{-1} as time progresses. We have already mentioned that such an approach may be promising, since the machine has certain convergence properties. We remark, however, that even when we know from the theory that the state of the machine will converge to one maximizing consensus, it may not do so feasibly fast.

Maximum cuts To illustrate the Boltzmann machine approach, we start with the problem of finding a maximum cut in a weighted graph. Given a graph $G = (V, E)$ of order n with weighted edges, the problem is to determine a partition of the vertex set into subsets V_1 and V_2 in such a way that the total sum of the weights of the edges that join vertices in the two sets is maximum.

This may easily be phrased as a $\{0,1\}$ linear programming problem: introduce a variable x_i for each vertex $i \in V$, and take $x_i = 1$ to mean that $i \in V_2$ and $x_i = 0$ to mean that $i \in V_1$. Then the problem is to maximize the objective function

$$F = \sum_{i=1}^{n} \sum_{j=i+1}^{n} W_{ij} \left((1 - x_i)x_j + (1 - x_j)x_i \right),$$

where the W_{ij} are the weights on the edges of the graph. Our Boltzmann machine will have as its underlying graph the instance graph, together with a loop on each vertex. We assign weights as follows: for each edge $ij \in E$, set the weight w_{ij} on this edge in the Boltzmann machine to be $-2W_{ij}$, where W_{ij} is the corresponding weight in the original graph, and for each $i \in V$, set the weight w_{ii} on the loop to be $\sum_{j=1}^{n} W_{ij}$. The resulting Boltzmann machine then satisfies the correspondence conditions.

The travelling salesman problem An instance of the travelling salesman problem is a weighted graph $G = (V, E)$, where the vertices represent cities and the weight d_{ij} on the edge ij is the distance from i to j. The aim is to find a Hamiltonian cycle of minimum total length. A number of neural approaches to this problem have been made. We describe here a fairly simple Boltzmann machine implementation, due to Garfinkel [11]. Following the general approach outlined above, we first describe the problem as a linear programming problem. Suppose that the vertex set of the graph is $V = \{1, 2, \ldots, n\}$ and, for i and a between 1 and n, introduce a variable x_{ia}, which, for a particular tour, equals 1 if and only if vertex i is the ath vertex in the tour. Let

$$u_{ijab} = \begin{cases} d_{ij}, & \text{if } a \equiv b+1 \pmod{n}, \\ 0, & \text{otherwise}. \end{cases}$$

The travelling salesman problem is then equivalent to minimizing

$$F = \sum_{i,j,a,b} u_{ijab} x_{ia} x_{jb},$$

subject to the constraints

$$\sum_{i=1}^{n} x_{ia} = 1 \ (1 \leq a \leq n) \quad \text{and} \quad \sum_{a=1}^{n} x_{ia} = 1 \ (1 \leq i \leq n).$$

We now construct a Boltzmann machine whose consensus function satisfies the properties described at the beginning of this section. It has n^2 vertices, one for each pair (i, a), corresponding to x_{ia}. Define the weights as follows:

$$w_{(i,a)(i,a)} = 1 + \max\{d_{ik} + d_{il} : l \neq k\};$$
$$w_{(i,a)(j,b)} = -d_{ij}, \quad \text{for } i \neq j \text{ and } a \equiv b+1 \pmod{n};$$
$$w_{(i,a)(i,b)} = -1 - \min(w_{(i,a)(i,a)}, w_{(i,b)(i,b)}), \quad \text{for } a \neq b;$$
$$w_{(i,a)(j,a)} = -1 - \min(w_{(i,a)(i,a)}, w_{(j,a)(j,a)}), \quad \text{for } i \neq j.$$

Then the correspondence conditions are satisfied, and the Boltzmann machine will 'solve' the travelling salesman problem, although not generally in polynomial time.

16.5 Feedforward networks

In a feedforward network, the underlying directed graph $G = (V, A)$ is *acyclic*—that is, there are no directed cycles. For the networks considered here, there is a specified set I of vertices called *input nodes*, and a single *output node* $z \notin I$. The underlying idea is that all nodes receive and transmit signals: the input nodes receive their signals from the outside world, and the output node transmits a signal to the outside world, while all other nodes receive and transmit along the relevant arcs of the directed graph.

Each arc rs has a weight w_{rs} that represents the strength of the connection between the nodes r and s. A positive weight corresponds to an 'excitatory' connection, and a negative one corresponds to an 'inhibitory' connection. All nodes except the input nodes are 'active' in that they transmit a signal that is a predetermined function of the signals they receive.

There is an *activation function* f_r for each non-input computation node r, and the activity of such a node is specified in two stages. First, the signals arriving at r are aggregated by taking their weighted sum according to the connection strengths on the arcs with terminal node r, and then the function f_r of this value is computed. Thus, the action of the entire network may be described in terms of two functions $p : V \to \mathbf{R}$ and $q : V \to \mathbf{R}$, that represent the received and transmitted signals respectively.

It is convenient to assume that one of the input nodes is a *bias node*, for which the applied signal is always 1. Suppose that the input nodes are labelled $j_0, j_1, j_2, \ldots, j_n$, where j_0 is the bias node. We assume that a vector of real-valued signals $\mathbf{y} = (1, y_1, y_2, \ldots, y_n)$ is applied externally to the input nodes, and $p(j_k) = q(j_k) = y_k$, for $k = 1, 2, \ldots, n$. (Note that the input vector has 1 as first entry; this is the bias signal.) For every other node l, the received and transmitted signals are defined as follows:

$$p(l) = \sum_{\{i : il \in A\}} q(i) w_{il} \quad \text{and} \quad q(l) = f_l(p(l)).$$

The output is the value $q(z)$ transmitted by the output node z. Since the underlying digraph is acyclic, it is possible to partition the nodes into *layers* $l_1, l_2, \ldots, l_k$, in such a way that the nodes in layer l_1 are the input nodes, layer l_k consists solely of the output node, and all arcs go from lower to higher layers. Computation then proceeds upwards through the layers.

Sometimes the activation functions f_r are chosen to be 'sigmoidal': these are smooth, non-decreasing functions mapping into $[0, 1]$. A popular choice is the *standard sigmoid function*

$$\sigma(x) = \frac{1}{1 + \exp(-x)}.$$

Often, however, the *linear threshold function* is taken to be the activation function of each non-input node. This is the function $\mathcal{H} : \mathbf{R} \to \{0, 1\}$ such that $\mathcal{H}(x) = 1$ if $x \geq 0$, and $\mathcal{H}(x) = 0$ if $x < 0$. A feedforward network of this type is known as a *linear threshold network*. A network that consists of a number of input nodes and just one other node, a linear threshold output node, is a special type of *perceptron*, known as the *simple perceptron* (see [20, 18]). This is illustrated in Fig. 16.1. A feedforward linear threshold network is essentially a composition of simple perceptrons. As an example, consider the threshold network illustrated in Fig. 16.2. It is easy to verify that if f is the function computed by this network in the state determined by the weights shown, then $f(101) = f(110) = 1$ and $f(100) = f(111) = 0$. Thus, if we restrict attention to $\{0, 1\}$-valued inputs, this network computes the exclusive-or function of its two non-bias inputs.

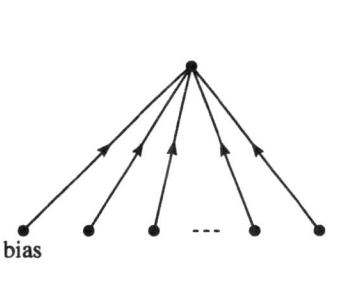

FIG. 16.1

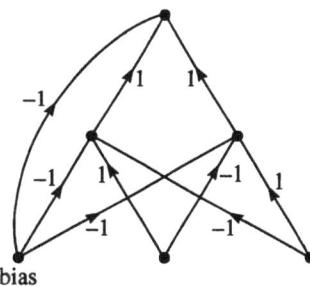

FIG. 16.2

16.6 Supervised learning in feedforward networks

PAC learning One of the main reasons why neural networks have proved so attractive is that they are, in a sense, capable of learning. The use of such anthropomorphic language might be considered controversial, but in a mathematical or engineering approach to neural networks, 'learning' simply means changing the weights of the network in response to some input data. When 'successful' or 'convergent' learning is possible, there is no need to program the network explicitly to perform a particular task; in other words, we need not know in advance how to set the weights. The neural network adjusts its weights according to a *learning algorithm*, in response to some classified training examples, with (roughly speaking) the state of the network converging to the 'correct' one. In this sense, the neural network 'learns from experience'.

In any model of supervised learning, it is assumed that there is some 'target function' t, which is the function to be learned. The target function is to be thought of as the 'correct' function we should like the network to compute. It simplifies matters greatly if we assume that there *is* a correct function t, and that this function can be computed by the network with some set of weight assignments. However, the basic PAC ('probably approximately correct') model has been extended to deal with situations in which neither of these assumptions can be made (see [2], for example).

A *labelled example* for the target function t is a pair $(x, t(x))$, where x is an input pattern to the network; for instance, if the neural network has n input nodes accepting real inputs, then $x \in \mathbf{R}^n$. The network is given a *training sample*, a sequence of such labelled examples: this constitutes its 'experience'. In response to this, its weights are altered by applying a learning algorithm.

More formally, suppose that the set of all possible examples is $X = \mathbf{R}^n$, or $X = \{0,1\}^n$, where n is the number of inputs to the network, and that the target function t can be computed by the neural network in some state. A *training sample for t* of length m is an element **s** of $(X \times \{0,1\})^m$, of the form

$$\mathbf{s} = ((x_1, t(x_1)), (x_2, t(x_2)), \ldots, (x_m, t(x_m))).$$

We denote the set of all training samples of length m for t by $S(m, t)$. A *learning algorithm* accepts as input the training sample **s**, and alters the state of the network in some way in response to the information provided by the sample. We should like the function $L(\mathbf{s})$ that is computed by the network after 'learning' to be an approximation to the target function, or 'closer' to the target function than the function computed before learning.

Many learning algorithms are currently in use. One of the most popular is the *back-propagation algorithm* for feedforward networks with sigmoidal activation functions (see, for example, [14]). We describe here the very simple perceptron learning algorithm, devised by Rosenblatt [20].

For any *learning constant* $\nu > 0$, the *perceptron learning algorithm* L_ν acts on the training sample 'on-line' in the following manner. The algorithm L_ν maintains at each stage a *current state* of the network, $\mathbf{w} = (w_0, w_1, \ldots, w_n)$. Here, w_0 is the weight of the arc from the bias input node to the output node, and $w_1, w_2, \ldots, w_n$ are the weights of the arcs from the other input nodes to the output node (see Fig. 16.1). This current state is updated on the basis of a labelled example $(x, t(x))$. (The initial state is, for example, that in which all of the weights are zero.) Suppose that the current state is $\mathbf{w}$ and that a labelled example $(x, t(x))$ is presented. Denote by $h_\mathbf{w}$ the function computed by the network in its current state. Then the algorithm forms the new state $\mathbf{w}'$, where

$$\mathbf{w}' = \mathbf{w} + \nu \left(t(x) - h_\mathbf{w}(x) \right) x.$$

This learning algorithm, which makes an incremental adjustment of the weights in the event of misclassification, is an example of a 'Hebbian' learning algorithm (see [13]). The *perceptron convergence theorem* (see [20, 18]) asserts that the perceptron learning algorithm aids convergence toward the target function: no matter how many examples are presented, the algorithm makes only a finite number of changes or updates, provided that ν (which can be a function of n), is small enough.

In order to analyse learning in feedforward networks, we first need a mathematical framework in which to define the goal of a successful learning algorithm.

We briefly describe the basic PAC model of learning introduced by Valiant [21], as it applies to feedforward networks in which there is a single output node, giving as output either 0 or 1; a more detailed treatment may be found in [3] or [2]. A fundamental assumption of the PAC model is that the network receives training samples

$$\mathbf{s} = ((x_1, t(x_1)), (x_2, t(x_2)), \ldots, (x_m, t(x_m))),$$

in which $x_1, x_2, \ldots, x_m$ are chosen independently and at random, according to some fixed (but unknown) probability distribution on the set of all examples.

In order to assess how effective a learning algorithm is, we need some measure of how close $L(\mathbf{s})$ is to t. Since there is assumed to be some probability distribution P on the set of all examples, we may define the *error* $\text{er}_P(h, t)$ of a function h (with respect to t) to be the P-probability that a randomly chosen example is classified incorrectly by h—that is,

$$\text{er}_P(h, t) = P(\{x \in X : h(x) \neq t(x)\}).$$

The aim is to ensure that the error of $L(\mathbf{s})$ is 'usually small'. Since each of the m examples in the training sample is drawn randomly and independently, according to P, the sample vector $\mathbf{x}$ is drawn randomly from X^m, according to the product probability distribution P^m. Thus we want it to be true that, with high P^m-probability, the sample $\mathbf{s}$ arising from $\mathbf{x}$ is such that the function $L(\mathbf{s})$ computed after training has small error with respect to t. This is a reasonable goal: some samples will be 'unrepresentative', but such samples will have low probability of being presented. This leads us to the following formal definition of PAC learning. *A learning algorithm L is a PAC learning algorithm for a network if, for any $\delta, \varepsilon > 0$, there is a sample length $m_L(\delta, \varepsilon)$ such that, for all target functions t computable by the network and for all probability distributions P on the set of examples,*

$$P^m (\{\mathbf{s} \in S(m, t) : \text{er}_P(L(\mathbf{s}), t) > \varepsilon\}) < \delta,$$

whenever $m \geq m_L(\delta, \varepsilon)$. In other words, provided that the sample has length at least $m_L(\delta, \varepsilon)$, then it is 'probably' the case that, after training on that sample, the function computed by the network is 'approximately' correct. (We note that the product probability distribution P^m is really defined not on subsets of $S(m, t)$ but on sets of vectors in X^m. However, this abuse of notation is convenient: for a fixed t, there is a clear one-to-one correspondence between vectors $\mathbf{x} \in X^m$ and training samples $\mathbf{s} \in S(m, t)$.)

Note that the probability distribution P occurs twice in the definition: both in the requirement that the P^m-probability of a sample be small, and also through the fact that the error of $L(\mathbf{s})$ is measured with reference to P. The crucial feature of the definition is that we require that the sample length $m_L(\delta, \varepsilon)$ be independent of P and of t. It is not immediately clear that this is possible, but the following informal arguments explain why it may be. If a particular example has not been seen in a large sample s, then the chances are that this example has low probability, and therefore its misclassification contributes little to the error of the function $L(\mathbf{s})$. In other words, the penalty paid for misclassification of a particular example is its probability, and, very loosely speaking, the two occurrences of the probability distribution in the definition can therefore 'balance' each other.

An important property that a learning algorithm might have is *consistency*. We say that the learning algorithm L is *consistent* if, given any training sample

$$\mathbf{s} = ((x_1, t(x_1)), \ldots, (x_m, t(x_m))),$$

the functions $L(\mathbf{s})$ and t agree on x_i, for each $i = 1, 2, \ldots, m$. The perceptron algorithm described above is not generally a consistent algorithm. However, it is easy to construct a consistent learning algorithm L from L_ν: given a sample $\mathbf{s} = ((x_1, t(x_1)), \ldots, (x_m, t(x_m)))$ of examples, L acts on $\mathbf{s}$ by applying L_ν repeatedly, cycling through x_1 to x_m in turn, until no updates are made in a complete cycle.

The VC-dimension and the underlying graph The problem of PAC learning can be addressed by means of a combinatorial parameter known as the *Vapnik-Chervonenkis dimension* (abbreviated to VC-dimension). Suppose that $\mathcal{N}$ is a feedforward neural network that outputs 0 or 1. We say that a set T of examples is *shattered* by $\mathcal{N}$ if, for each of the $2^{|T|}$ possible ways of dividing T into two disjoint sets T_1 and T_0, there is *some* function f computable by $\mathcal{N}$ such that $f(x) = 1$ if $x \in T_1$, and $f(x) = 0$ if $x \in T_0$. The *VC-dimension* of $\mathcal{N}$, denoted by $\dim_{\mathrm{VC}}(\mathcal{N})$, is defined to be the largest size of a set of examples shattered by $\mathcal{N}$. The VC-dimension may be thought of as a measure of the 'expressive power' of the network. Vapnik and Chervonenkis [22] defined this parameter (in a more general context) in studying the uniform convergence of relative frequencies to probabilities.

The VC-dimension characterizes fairly precisely the size of training sample that should be used for effective PAC learning. The following result is due to Blumer *et al.* [8] and Ehrenfeucht *et al.* [10].

Theorem 16.1 *If a feedforward network $\mathcal{N}$ has finite VC-dimension $d \geq 1$, then any consistent learning algorithm L for $\mathcal{N}$ is a PAC learning algorithm. Moreover, there is a constant c_1 such that*

$$\frac{c_1}{\varepsilon}(d \log (1/\varepsilon) + \log (1/\delta))$$

is a sufficient sample length $m_L(\delta, \varepsilon)$ for any such algorithm.

On the other hand, there is another constant c_2 such that, for any PAC learning algorithm L for $\mathcal{N}$, the sufficient sample length $m_L(\delta, \varepsilon)$ must be at least

$$\frac{c_2}{\varepsilon}(d + \log(1/\delta)),$$

for all $\varepsilon \leq 1/8$ and $\delta \leq 1/100$.

We now present a result on the VC-dimensions of feedforward linear threshold networks. The first part is due to Baum and Haussler [6], and the second part is due to Maass [16, 17].

Theorem 16.2 *There is a constant $c_1 > 0$ such that, if $\mathcal{N}$ is any feedforward linear threshold network with one output node and whose underlying digraph has n vertices and m arcs, then $\dim_{VC}(\mathcal{N}) \leq c_1 m \log n$.*

Furthermore, there is another constant $c_2 > 0$ such that some feedforward linear threshold network satisfies $\dim_{VC}(\mathcal{N}) \geq c_2 m \log n$, where the underlying digraph has n vertices and m arcs.

The above result relates the VC-dimension of a neural network to its underlying graph, when the activation functions are all linear threshold functions. It provides an upper bound that is tight to within a constant. It is rather disappointing that this relationship involves only the 'size' of the graph, rather than its structure. In particular, this general result does not involve the number of layers in the network. Tighter bounds have been obtained for networks with very few layers; see, for example, [2, 5]. Results have also been obtained for feedforward networks with sigmoidal activation functions; for results on these and other types of network, see [2, 12].

Learning can be as hard as graph-colouring If the process of PAC learning by an algorithm L is to be of practical value, it should be possible to implement the algorithm 'quickly'. We wish to quantify the behaviour of a learning algorithm for a particular neural network architecture with respect to the size of the network. In particular, we wish to consider how the running time of the algorithm varies with the number n of inputs to the network: for a learning algorithm to be efficient, this running time should increase polynomially with n.

However, there is another important consideration in any discussion of efficiency. Clearly, decreasing ε makes the learning task more difficult, and therefore the time taken to produce a probably approximately correct output should be constrained in some appropriate way as ε decreases; the appropriate condition is that the running time must be polynomial in $1/\varepsilon$. Formally, we say that a learning algorithm L is *efficient with respect to accuracy ε, example size n and sample length m* if its running time is polynomial in m and if there is a value of $m_L(\delta, \varepsilon)$, sufficient for PAC learning, that is polynomial in n and $1/\varepsilon$.

Judd [15] was the first to show that learning in neural networks can be hard, in the formal complexity-theoretical sense. We now describe a simple hardness result from [3, 4], along the lines of one due to Blum and Rivest [7]. Before doing so, we recall that, in complexity theory, two important classes of problems, RP and NP, are defined. The class RP is the class of all problems that can be solved by 'randomized' algorithms in polynomial time, while NP is the class of problems that can be solved by non-deterministic Turing machines in polynomial time (see, for example, [9]). It is conjectured, and widely believed, that RP is a strict subset of NP: this is the RP $\neq$ NP conjecture.

The network $\mathcal{N}_n^k$ that we consider has n inputs and $k+1$ linear threshold nodes, for $k \geq 1$ (see Fig. 16.3). The first k linear threshold nodes are 'in parallel', and each of these is connected to all of the inputs. The last threshold node is the output unit; it is connected by arcs with *fixed* weight 1 to the other linear threshold nodes and to the bias node with weight $-k$. The effect of this arrangement is that the output unit acts as a multiple AND gate for the outputs of the other threshold nodes.

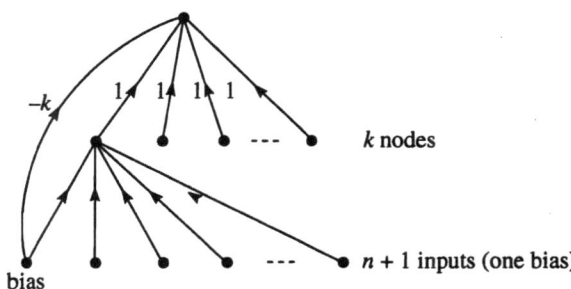

FIG. 16.3

The *consistency problem* for $\mathcal{N}_n^k$, which we refer to as $\mathcal{N}^k$-CONSISTENCY, is as follows.

Instance A sequence $\mathbf{s} = ((x_1, b_1), (x_2, b_2), \ldots, (x_m, b_m))$ of labelled examples, where $x_i \in \mathbf{R}^n$ and $b_i \in \{0, 1\}$, for $1 \leq i \leq m$.

Question Is there a state of the network $\mathcal{N}_n^k$ such that the function t then computed by the network satisfies $t(x_i) = b_i$, for $1 \leq i \leq m$? In other words, is s a training sample for some function computed by the network?

The following result (a special case of one that appears in [19]) relates the consistency problem to the problem of efficient PAC learning.

Theorem 16.3 *If there is a PAC learning algorithm for $\mathcal{N}_n^k$ that is efficient with respect to accuracy, example size and sample length, then there is a randomized polynomial-time algorithm that solves the problem $\mathcal{N}^k$-CONSISTENCY.*

One can prove that $\mathcal{N}^k$-CONSISTENCY is NP-hard for $k \geq 3$, by showing that it is as difficult as graph-colouring. Here, we sketch the reduction; full details can be found in [3]. Let G be a graph with vertex set $V = \{1, 2, \ldots, n\}$ and edge set E. We construct a sequence $\mathbf{s}(G)$ of labelled examples, as follows. For each vertex $i \in V$, we take as a labelled example $(\mathbf{v}_i, 0)$, where $\mathbf{v}_i$ is the vector with 1 in the ith coordinate position, and 0 elsewhere. For each edge $ij \in E$, we take as a labelled example $(\mathbf{v}_i + \mathbf{v}_j, 1)$, and we also take as a labelled example $(\mathbf{0}, 1)$, where $\mathbf{0}$ is the zero vector $00 \ldots 0$. It can be shown that $\mathbf{s}(G)$ is a training sample for some function in $\mathcal{N}_n^k$ if and only if G is k-colourable. It follows that if there is a polynomial-time algorithm for $\mathcal{N}^k$-CONSISTENCY, then there is one for GRAPH k-COLOURING. But GRAPH k-COLOURING is NP-complete for $k \geq 3$, and so the $\mathcal{N}^k$-CONSISTENCY problem is NP-hard if $k \geq 3$. (In fact, the same is true if $k = 2$; this follows from work of Blum and Rivest [7].)

Thus Theorem 16.3 enables us to move from this hardness result for the consistency problem to a hardness result for PAC learning. The theorem tells us that, unless RP = NP, there can be no computationally efficient PAC learning algorithm for this family of neural networks.

References

1. E. Aarts and J. Korst, *Simulated Annealing and Boltzmann Machines: A Stochastic Approach to Combinatorial Optimization and Neural Computing*, Wiley-Interscience, 1989; *MR* **90e**: 90096.
2. M. Anthony, Probabilistic analysis of learning in artificial neural networks: the PAC model and its variants, *The Computational and Learning Complexity of Neural Networks* (ed. I. Parberry), to appear.
3. M. Anthony and N. Biggs, *Computational Learning Theory: An Introduction*, Cambridge Tracts in Theoretical Computer Science **30**, Cambridge University Press, 1992; *MR* **93f**: 68078.
4. M. Anthony and N. Biggs, Computational learning theory for artificial neural networks, *Mathematical Approaches to Neural Networks* (ed. J. G. Taylor), North-Holland Math. Library **51**, Elsevier, 1993, pp. 25–62.
5. P. L. Bartlett, Vapnik–Chervonenkis bounds for 2-layer and 3-layer networks, *Neural Computation* **5** (1993), 371–373.
6. E. Baum and D. Haussler, What size net gives valid generalization?, *Neural Computation* **1** (1989), 151–160.
7. A. Blum and R. L. Rivest, Training a 3-node neural net is NP-complete, *Advances in Neural Information Processing Systems I* (ed. D. S. Touretzky), Morgan Kaufmann, 1989, pp. 494–501; *MR* **92a**: 68004.
8. A. Blumer, A. Ehrenfeucht, D. Haussler and M. K. Warmuth, Learnability and the Vapnik–Chervonenkis dimension, *J. Assoc. Comput. Mach.* **36** (1989), 929–965; *MR* **91f**: 68178.
9. T. Cormen, C. E. Leiserson and R. L. Rivest, *Introduction to Algorithms*, MIT Press, 1990; *MR* **91i**: 68001.
10. A. Ehrenfeucht, D. Haussler, M. Kearns and L. G. Valiant, A general lower bound on the number of examples needed for learning, *Inform. and Comput.* **82** (1989), 247–261; *MR* **91b**: 68048.
11. R. S. Garfinkel, Motivation and modeling, *The Travelling Salesman Problem: A Guided Tour of Combinatorial Optimization* (ed. E. H. Lawler et al.), Wiley, 1985, pp. 17–36.
12. D. Haussler, Decision-theoretic generalizations of the PAC model for neural net and other learning applications, *Inform. and Comput.* **100** (1992), 78–150; *MR* **93i**: 68149.
13. D. O. Hebb, *The Organization of Behaviour*, Wiley, 1949.
14. J. Hertz, A. Krogh and R. G. Palmer, *Introduction to the Theory of Neural Computation*, Addison-Wesley, 1991; *MR* **92f**: 82044.
15. J. S. Judd, Learning in neural networks, *Proc. 1st Workshop on Computational Learning Theory*, Morgan Kaufmann, 1988, pp. 2–8; *MR* **92a**: 68004.

16. W. Maass, Bounds on the computational power and learning complexity of analog neural nets (extended abstract), *Proc. 25th ACM Symp. Theory of Computing*, ACM Press, 1993, pp. 335–344.
17. W. Maass, Neural nets with superlinear VC-dimension, *Neural Computation* **6** (1994), 877–884.
18. M. Minsky and S. Papert, *Perceptrons*, MIT Press, 1988.
19. L. Pitt and L. G. Valiant, Computational limitations on learning from examples, *J. Assoc. Comput. Mach.* **35** (1988), 965–984; *MR* **91f:** 68096.
20. F. Rosenblatt, Two theorems of statistical separability in the perceptron, *Mechanisation of Thought Processes: Proc. Symp. National Physical Laboratory*, Vol. 1, HMSO, 1959; *MR* **22**#13330b.
21. L. G. Valiant, A theory of the learnable, *Comm. Assoc. Comput. Mach.* **27** (1984), 1134–1142.
22. V. N. Vapnik and A. Y. Chervonenkis, On the uniform convergence of relative frequencies of events to their probabilities, *Theory Probab. Appl.* **16** (1971), 264–279; *MR* **44**#6018.

17
International Finance

NORMAN BIGGS

Exchanges among a given set of currencies define a graph, and the rates of exchange satisfy a condition which can be described in terms of a potential function on the graph. Using this approach, the matrix-tree technique is applied to show that the balance of payments determines a unique system of exchange rates.

17.1 Introduction

Let V be a set of currencies and E the set of pairs of currencies that can be exchanged directly. Clearly, there is a corresponding graph $G = (V, E)$. The aim of this chapter is to explain how such a network of exchange rates can be studied in graph-theoretic terms. If every pair of currencies is exchangeable, then the graph is a complete graph, but this will not generally be the case, and the theory is not restricted to complete graphs. However, for our purposes it is reasonable to assume that the graph is connected, which means that any two currencies can be converted indirectly, by a sequence of exchanges involving other currencies.

After some historical motivation, a simple model will be set up. This model is based on the idea of the potential for purchasing a bundle of goods, subsumes classical theories of exchange rate determination, and has the desirable property that it defines a network of exchange rates in which no profit by a cycle of exchanges (known as *cyclic arbitrage*) is possible. The model also describes a crude mechanism, well known in the 18th century, for confining exchange rates within a certain band, although it will appear that, for simple algebraic reasons, this mechanism is not consistent with the cyclic arbitrage condition.

In the opposite direction, it will be shown that any network of exchange rates, satisfying the condition that no profit by cyclic arbitrage is possible, can be represented by a potential function. This result is fairly obvious in the case when the network is complete, but requires some proof in other cases. This idea facilitates the generalization to arbitrary networks of theories that have been much studied in the simplest case, that of two countries in isolation.

The potential-theoretic approach is used to prove that given a network with specified cash flows between the countries, the balance of payments determines a unique system of exchange rates for which no cyclic arbitrage is possible. Furthermore, explicit formulas for the rates can be obtained by the 'matrix-tree' technique. As a digression, it will be shown that the mathematics underlying these results can be interpreted in terms of the problem of ranking the contestants in a tournament.

Finally, the mechanics of an exchange rate network will be considered in the light of the preceding discussion.

17.2 Exchange dealing then and now

We begin with an example from the early 19th century. The following data is taken from Kelly's *The Universal Cambist* [15, Vol. 2, p. 123].

London on	Amsterdam	35	Shillings Flemish per Pound Sterling
	Madrid	38	Pence Sterling per Dollar of Plate
	Paris	24	Livres per Pound Sterling
Amsterdam on	London	34	Shillings Flemish per Pound Sterling
	Paris	53	Grotes per Ecu of 3 Francs
	Madrid	92	Grotes per Ducat of Plate
Paris on	London	$23\frac{1}{2}$	Livres per Pound Sterling
	Madrid	16	Francs per Doubloon of Plate
	Amsterdam	54	Grotes Flemish per Ecu of 3 Francs
Madrid on	London	39	Pence Sterling per Dollar of Plate
	Amsterdam	94	Grotes per Ducat of Plate
	Paris	$16\frac{1}{2}$	Francs per Doubloon of Plate

At this time exchange operations were conducted by means of 'bills of exchange', but for our purposes it is not necessary to understand the mechanics of this procedure. What we do need to know is that Kelly's entry 'X on Y' means that holders of X-units were willing to exchange them for Y-units at the rate given. We shall denote this rate by $\alpha(X, Y)$; precisely, this is the number of Y-units which holders of X-units will take in exchange for one X-unit. For example, if L is London and A is Amsterdam, then $\alpha(L, A) = 35$.

Clearly, some further explanation is needed, because (then as now) there was a mystique surrounding these matters. This manifests itself in conventions that tend to obfuscate rather than illuminate the situation. First we note that the traditional modes of quotation follow an orientation convention different from that contained in our definition of $\alpha(X, Y)$. By our definition, $\alpha(A, L) = 1/34$, whereas the practical convention is to quote the inverse. (There is a valid reason for this, which will appear shortly.) Secondly, although the rates of exchange quoted link the currencies of Amsterdam, London, Paris and Madrid, where the basic money-units were, respectively, the shilling Flemish, the pound sterling, the franc, and the dollar of plate, in practice quotations were made in terms of 'traditional' units that differed from the basic ones. The relevant relationships were as follows.

Amsterdam:	1 shilling = 12 grotes
London:	1 pound = 240 pence
Paris:	3 francs = 1 ecu, 80 francs = 81 livres
Madrid:	4 dollars = 1 doubloon, 375 dollars = 272 ducats

If we express all the rates in terms of the four basic money-units, we can compute the entries in the following table, in which the entry in row X and column Y is $\alpha(X, Y)$.

	Amsterdam	London	Paris	Madrid
Amsterdam	–	1/34	36/53	1125/6256
London	35	–	1920/81	240/38
Paris	54/36	81/1880	–	1/4
Madrid	6392/1125	39/240	33/8	–

This information can be displayed conveniently in graphical form (see Fig. 17.1).

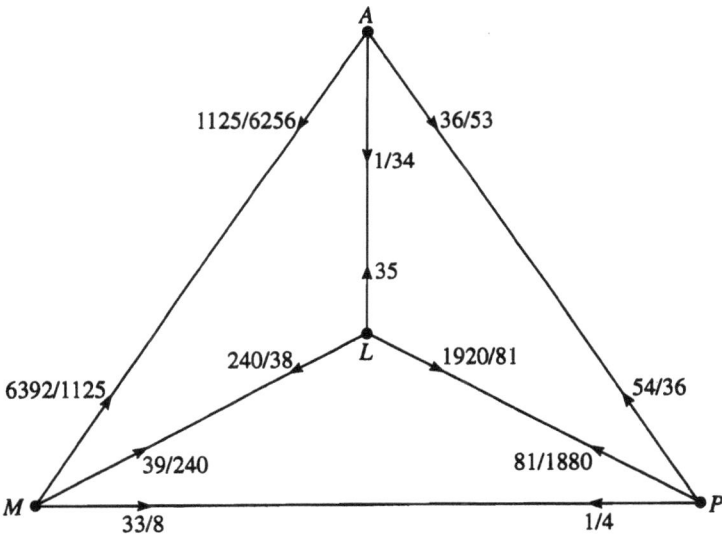

FIG. 17.1

Note that in all cases, $\alpha(X, Y)\alpha(Y, X) > 1$. This can be explained as follows. If I have $\alpha(X, Y)\alpha(Y, X)$ Y-units, then there are (by assumption) people willing to exchange their X-units for my Y-units at the rate $\alpha(X, Y)$, and so I can obtain $\alpha(Y, X)$ X-units. Now I can find someone who is (by assumption) willing to exchange Y-units for my X-units at the rate $\alpha(Y, X)$, and I shall finish up with one Y-unit. Thus the stated condition expresses the fact that it is impossible to make money by switching back and forth between two currencies. Nowadays the word *arbitrage* is used to characterize such procedures and, as will appear, this concept is fundamental to our theory.

In modern terms, the rates $\alpha(X, Y)$ are 'ask' rates. At this point it is important to note that a 'bid' is just an 'ask' viewed from the opposite direction. An 'ask' is an offer by a holder of X to 'sell' one X-unit for $\alpha(X, Y)$ Y-units; this is the same as a 'bid' to 'buy' Y at the rate of $\beta(Y,X)$ X-units per Y-unit, where

$$\beta(Y,X) = 1/\alpha(X,Y).$$

It follows that

$$\beta(X,Y)\beta(Y,X) < 1 \quad \text{and} \quad \alpha(X,Y) > \beta(X,Y).$$

The asymmetry of the exchange markets is thus rather more subtle than the terminological distinction between an ask and a bid. In fact, it depends on the distinction between the agent who offers the deal and the one who accepts. Holders of X who wish to exchange for Y can proceed in one of two ways: they can propose a deal at the $\alpha(X,Y)$ rate, or they can accept a deal proposed by a holder of Y at the $\beta(X,Y) = 1/\alpha(Y,X)$ rate. As one would expect, the rate for the proposer is better than the rate for the accepter, because the accepter is certain of a deal but the proposer is not.

Returning to Kelly's table, we observe that the convention adopted for the quotation of rates on the bilateral $X \leftrightarrow Y$ exchange was to choose one ordering arbitrarily, say (X,Y), and quote the values $\alpha(X,Y)$ and $\beta(X,Y)$; these then determine $\alpha(Y,X)$ and $\beta(Y,X)$. At that time the asymmetry depended in practice upon whether one chose to 'remit' a bill or 'draw' one, and the uncertainty lay in the remit option. Nowadays the same quoting convention is observed, but the uncertainty is much less, and this is reflected in very small 'bid–ask' spreads. For example, at 10.47 GMT on 16 June 1993, the bid rate for Deutschmarks per dollar was 1.6439 and the ask rate was 1.6443. If we allow the term 'transaction cost' to cover the cost of uncertainty, as well as more conventional charges, it is clear that this cost is now very small indeed.

Another form of arbitrage opportunity inherent in this situation was of great importance in Kelly's time, when it was known as Arbitration (Simple and Compound). We shall call it *cyclic arbitrage*. The idea is that by traversing a cycle in the exchange rate graph, one might be able to finish up with a profit. In the 19th century the relatively large bid–ask spreads also played a part in this process, because the advantages of remitting over drawing might easily outweigh the risk and delay involved. The reader is referred to Kelly's book [15, pp. 123–6] for a number of examples of cyclic arbitrage, based on the data set out above. However, it is thought that nowadays the smooth working of the markets has essentially removed such opportunities. Although the precise justification for this claim is unclear, there can be little harm in using it for a mathematical study of the exchange rate networks of today.

17.3 Exchange rate networks

In this section we formalize the ideas developed above. Let $G = (V, E)$ be a graph whose vertex set V is a set of currencies, and whose edge set E is the set of pairs of currencies that are directly exchangeable. As noted earlier, we assume that G is connected, but not necessarily complete. We write vw for the unordered pair of vertices forming an edge, so that for each edge vw there are two *sides*, the ordered pairs (v,w) and (w,v). We denote by S the set of sides of G.

International Finance

Suppose that, for each side (v, w), there is an associated positive real number $\rho(v, w)$ that determines the rate of exchange between v and w. Precisely, we take this to mean that one v-unit can be exchanged for $\rho(v, w)$ w-units. The units here are moneys-of-account, such as the \$ and the DM, which we refer to as *money-units* when we need to emphasize the point. Note that in theoretical economics, our $\rho(\$, \mathrm{DM})$ is usually called the DM/\$ rate—that is, the number of Deutschmarks per dollar. Based on the discussion in the previous section, our first assumption is that for each edge $vw \in E$,

$$\rho(w, v) = \rho(v, w)^{-1}. \tag{17.1}$$

Our second assumption is that there is no possibility of profit by cyclic arbitrage. For example, suppose that uv, vw and uw are edges of the graph. If one u-unit is exchanged for v-units, and the proceeds are subsequently exchanged for w-units, then the resulting number of w-units is $\rho(u, v)\rho(v, w)$. If our assumption is correct this should be same as the result of exchanging one u-unit for w-units directly, so that $\rho(u, v)\rho(v, w) = \rho(u, w)$. Using condition (17.1), we can write this in the form

$$\rho(u, v)\rho(v, w)\rho(w, u) = 1.$$

More generally, the *no-arbitrage condition* is that for each cycle $v_1 v_2 \ldots v_l v_1$ in G, the exchange rates satisfy

$$\rho(v_1, v_2)\rho(v_2, v_3) \ldots \rho(v_l, v_1) = 1. \tag{17.2}$$

We shall take conditions (17.1) and (17.2) as axioms for the contemporary scene. Thus we define an *exchange rate network*, or *ERN*, to be a pair (G, ρ), in which $G = (V, E)$ is a connected graph and ρ is a function defined on the set S of sides of G, taking positive real values and satisfying conditions (17.1) and (17.2). An example is shown in Fig. 17.2, based on the dollar, yen, pound and Deutschmark.

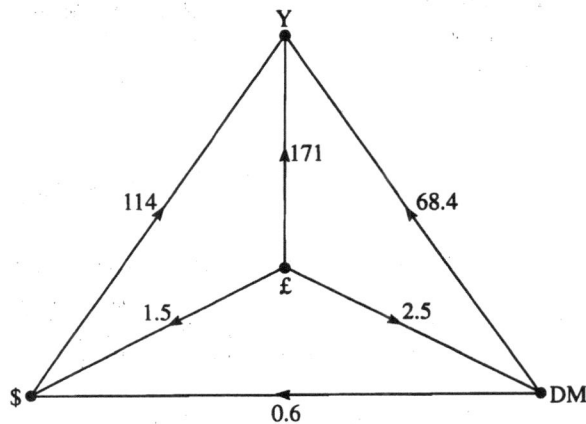

Fig. 17.2

The direction of exchange is determined by the rule that when the arrow points from v to w, the number on the edge is $\rho(v,w)$. The corresponding $\rho(w,v)$ is then determined by condition (17.1), and it can be verified that condition (17.2) holds for all cycles.

17.4 Some classical theory

The early writers on exchange rates, such as J. Harris [12], based their theory on the *Mint Par* between the various currencies. This depended on the legal bullion content of the coinage, a concept that has become vacuous in the present century. An alternative theoretical basis for exchange rates depends on the notion of *Purchasing Power Parity*, or *PPP*, developed by Cassel from about 1916 onwards, although the idea is so fundamental that it is no surprise to find versions of it in much earlier works (see Dornbusch [9]). As we shall see, the Mint Par and PPP theories can both be explained within the same simple framework, and both theories have the desirable feature that they define a network of exchange rates in which no profit by cyclic arbitrage is possible.

We need to postulate a 'country' C_v in which v is the legal currency. If the good j may be bought in country C_v with unit price $p_j^{(v)}$, then the price of a bundle $\mathbf{y} = (y_1, y_2, \ldots, y_b)$ of goods is

$$\langle \mathbf{p}^{(v)}, \mathbf{y} \rangle = p_1^{(v)} y_1 + p_2^{(v)} y_2 + \cdots + p_b^{(v)} y_b.$$

In an ideal situation, the cost of the bundle in another country C_w would be equal to the number of w-units obtained in exchange for its cost in v-units—that is,

$$\langle \mathbf{p}^{(w)}, \mathbf{y} \rangle = \langle \mathbf{p}^{(v)}, \mathbf{y} \rangle \rho(v,w).$$

Thus, ideally, for each $\mathbf{y}$ the quotient $\langle \mathbf{p}^{(w)}, \mathbf{y} \rangle / \langle \mathbf{p}^{(v)}, \mathbf{y} \rangle$ should be equal to the exchange rate between v and w. In practice this is not the case, and it is usual to choose a 'typical' bundle $\mathbf{y}$ and define $\langle \mathbf{p}^{(v)}, \mathbf{y} \rangle$ to be the *price index* for v, based on $\mathbf{y}$. The ratio of the price indices for w and v is the *theoretical exchange rate*, based on purchasing power parity with respect to $\mathbf{y}$:

$$\rho^{\mathbf{y}}(v,w) = \langle \mathbf{p}^{(w)}, \mathbf{y} \rangle / \langle \mathbf{p}^{(v)}, \mathbf{y} \rangle.$$

The following example will be considered throughout the chapter. (Where we leave the example and return to the main text, you will see the symbol ◊.)

Example 17.1 We shall calculate the Mint Par exchange rate between the English and French currencies one hundred years ago, according to the method given by Clare in his *A Money-Market Primer* [6]. We work in terms of the bundle $\mathbf{g} = (1, 0, 0, \ldots, 0)$, where gold is presumed to be the first good. In our notation, we have

$$\langle \mathbf{p}^{(S)}, \mathbf{g} \rangle = p_1^{(S)}, \quad \langle \mathbf{p}^{(F)}, \mathbf{g} \rangle = p_1^{(F)},$$

where $p_1^{(S)}$ and $p_1^{(F)}$ are the prices of a bullion-unit of pure gold in the English and French money-units, the pound sterling and the franc. As we shall see, Clare was able to determine these numbers absolutely, because there was at that time a chain of legal relationships involving money-units, coins, gold and bullion-units.

Under English law, 480 troy ounces of gold, 11/12ths fine, was coined into 1869 coins called sovereigns, each worth one pound sterling.

Under French law, 1000 grams of gold, 9/10ths fine, was coined into 155 coins called Napoleons, each worth 20 francs.

According to the Weights and Measures Act of 1878, 1 ounce troy was equivalent to 31.1035 grams.

Taking the bullion-unit to be 1 gram of pure gold, we have

$$p_1^{(S)} = \frac{1869}{480 \times 31.1035 \times 11/12} = 0.1366,$$

$$p_1^{(F)} = \frac{155 \times 20}{1000 \times 9/10} = 3.4444.$$

Hence the theoretical exchange rate, or Mint Par, was

$$\rho^g(S, F) = p_1^{(F)}/p_1^{(S)} = 25.2215. \quad \diamond$$

Example 17.2   Writing in 1938, Evitt [10, pp. 20–21] described various bundles used to construct price indices at that time. For example, the Ministry of Labour's Index of Retail Prices was, roughly speaking, based on the bundle $(60, 16, 12, 8, 4)$, where the 'goods' are amalgamations representing Food, Rent, Clothing, Fuel and Light, and Miscellaneous Items (including soap, soda, ironmongery, brushes, crockery, tobacco, fares and newspapers). We shall resist the temptation to dwell on the fascinating insights provided by such indices, and pass quickly on to a basic theorem. The proof is by direct verification.

Theorem 17.1   *For any fixed bundle* **y**, (G, ρ^y) *is an ERN; in other words,* ρ^y *satisfies conditions (17.1) and (17.2).*

This theorem shows that the Mint Par and PPP theories have the desirable feature that they define a network of theoretical exchange rates in which no profit by cyclic arbitrage is possible. Historically, the quest for an appropriate bundle **y** has been difficult. The earliest 'Mint Par'-theorists used the bundle consisting of silver alone, but this was abandoned early in the 19th century in favour of the gold standard bundle **g**, as defined above. This in turn was effectively abandoned at the time of the 1914–18 war, as a result of problems that had little to do with exchange rates. The rise of PPP theory could be viewed as an attempt to find an acceptable alternative bundle and, although that aim is overly ambitious, the theory still plays a useful role in modern macroeconomic theory.

Our aim is to show that it is possible to study the foreign exchange market as if it were operating under a kind of PPP theory, even though the theory in its simple form is clearly inapplicable. Before we reach that point, we look at an aspect of the theory that foreshadows some of the difficulties experienced in modern times. In particular, we shall explain how a mechanism for controlling exchange rates can be in conflict with the no-arbitrage conditions.

17.5 The export and import points

Even within a very strictly defined legal context, such as that described in Example 17.1, the actual rate of exchange can deviate from the theoretical one. In this section we look at the mechanism which, in the 18th and 19th centuries, was believed to keep actual exchange rates within a narrow band around the theoretical ones. It will appear that there are inherent problems involving cyclic arbitrage in this situation. In Section 17.9 we consider briefly how the mechanism worked, and how it was able to overcome these problems. Of course, the idea of a band, or target zone, has re-emerged in recent times, and similar problems arise. These difficulties have recently been discussed by Jørgensen and Mikkelsen [13].

We consider the exchange of a bundle $\mathbf{y}$ of goods located in C_v for money-units valid in C_w. There are two possible ways of realizing this exchange. If the bundle is exchanged for money units in C_v and the proceeds are exchanged for w-units, then the amount realized is

$$\langle \mathbf{p}^{(v)}, \mathbf{y} \rangle \, \rho^{act}(v, w),$$

where ρ^{act} is the actual exchange rate in operation. On the other hand, the bundle may be transported to C_w and exchanged there for w-units directly. The transportation will involve some costs, and we suppose that γ^{vw} is the vector describing such costs—precisely, $\gamma^{vw} = (\gamma_1, \gamma_2, \ldots, \gamma_b)$, where γ_i is the cost of sending one unit of the ith good from C_v to C_w, measured in w-units. Thus the cost of sending the bundle $\mathbf{y}$ from C_v to C_w is $\langle \gamma^{vw}, \mathbf{y} \rangle$ w-units, and the net amount realized by the entire process is

$$\langle \mathbf{p}^{(w)}, \mathbf{y} \rangle - \langle \gamma^{vw}, \mathbf{y} \rangle.$$

These calculations show that it is advantageous to 'export' the bundle $\mathbf{y}$ if and only if

$$\langle \mathbf{p}^{(v)}, \mathbf{y} \rangle \, \rho^{act}(v, w) \leq \langle \mathbf{p}^{(w)} - \gamma^{vw}, \mathbf{y} \rangle. \qquad (17.3)$$

It is convenient to express this condition as a relationship between the actual and theoretical exchange rates, with a cost factor that is, as far as possible, independent of the units. Observe that a bundle $\mathbf{y}$ is a representative of a class of 'equivalent' bundles, consisting of all multiples $\alpha \mathbf{y}$ with $\alpha > 0$. One particular bundle equivalent to $\mathbf{y}$ is obtained by taking $\alpha = \langle \mathbf{p}^{(v)}, \mathbf{y} \rangle^{-1}$. The price of this bundle in C_v is one v-unit, and so we shall say that

$$\mathbf{y} / \langle \mathbf{p}^{(v)}, \mathbf{y} \rangle$$

is the *v-unit* of $\mathbf{y}$. Let $\theta^{\mathbf{y}}(v, w)$ be the cost of transporting the v-unit of $\mathbf{y}$ from C_v to C_w, expressed in v-units. Note that, although γ^{wv} and γ^{vw} are defined in terms of different units, the definition of θ^y has been framed so that if the costs are symmetric in real terms, then $\theta^y(v,w) = \theta^y(w,v)$.

Since we are looking at the process from the viewpoint of C_v, we formulate the main result in these terms.

Theorem 17.2 *It is advantageous to export* y *from* C_v *to* C_w *if and only if*

$$\rho^{act}(v,w) \leq (1 - \theta^y(v,w))\,\rho^y(v,w). \qquad (17.4)$$

The proof is a simple algebraic exercise, using the definitions and equation (17.3).

We call the right-hand side of inequality (17.4) the *export point* ρ^y_- associated with the bundle y and the relevant cost—that is,

$$\rho^y_-(v,w) = (1 - \theta^y(v,w))\,\rho^y(v,w).$$

The argument shows that if the actual money-market rate ρ^{act} falls below ρ^y_-, then it is advantageous to settle accounts abroad by exporting y instead of using the money-market.

Example 17.1 (continued). We return to the gold standard bundle g, as defined in Example 17.1. In that case, $\theta^g(v,w)$ is the cost in v-units of transporting one v-unit's worth of gold. In 1893 Clare [6] estimated the cost of shipping gold from London to Paris to be 0.5%. Thus $\theta^g(S,F) = 0.005$, and the export point in 1893 was

$$(1 - 0.005) \times 25.2215 = 25.0954.$$

This was interpreted as meaning that the money-market rate could not fall below 25.0954, because in that case English merchants would prefer to settle their accounts in France by sending gold, rather than by using bills of exchange. ◊

Condition (17.4), with v and w interchanged, determines the export point for the direction (w,v). Looking at this from the perspective of C_v, we obtain the condition that expresses the fact that debtors in C_w will find it advantageous to pay their creditors in C_v by sending y, rather than by exchange in the money-market. This is called the *import point* $\rho^y_+(v,w)$ for the bundle y with cost vector γ^{wv}—that is,

$$\rho^y_+(v,w) = \frac{1}{\rho^y_-(w,v)}.$$

This can be expressed in terms of ρ^y and θ^y as follows:

$$\rho^y_+(v,w) = \frac{1}{(1 - \theta^y(w,v))\rho^y(w,v)} = \rho^y(v,w)/(1 - \theta^y(w,v)).$$

If transportation costs are symmetric in real terms, we have

$$\theta^y(w,v) = \theta^y(v,w),$$

so if θ denotes their common value, then the actual exchange rate lies within a band defined in terms of θ and ρ^y. For example, under the gold standard with symmetric real costs θ, the actual exchange rate should never move outside the band

$$(1-\theta)\rho^g \leq \rho^{act} \leq (1-\theta)^{-1}\rho^g,$$

because if either limit were to be breached, merchants would make alternative arrangements involving the shipment of bullion.

At this point we draw attention to the difficulties of this crude mechanism from the network point of view. The following simple example shows that the bands defined by the import and export points do not fit well with the algebraic constraints of the no-arbitrage condition (17.2).

Example 17.3 Suppose that C_u, C_v and C_w are three countries, geographically equidistant, so that the transportation costs between any two in either direction can be represented by a constant θ. Let a, b, c be the theoretical exchange rates, as indicated in Fig. 17.3, and let α, β, γ be the corresponding actual rates. We assume that the theoretical rates are defined in terms of a fixed bundle, and so they form an ERN (see Theorem 17.1). We should like to ensure that the actual rates form an ERN also, because that would mean that the money-market can still operate efficiently to prevent profit by cyclic arbitrage.

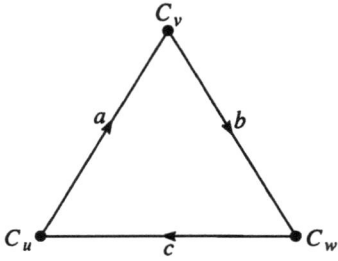

FIG. 17.3

Let us suppose that the rates α and β fall close to, but not outside, the export points a_- and b_-; for definiteness, let us say that

$$\alpha = (1 - 0.9\theta)a > a_- \quad \text{and} \quad \beta = (1 - 0.9\theta)b > b_-.$$

Then, if it were possible for the actual rates to form an ERN, we should have

$$\gamma = \alpha\beta = (1 - 1.8\theta + 0.81\theta^2)ab = (1 - 1.8\theta + 0.81\theta^2)c,$$

since a, b, c form an ERN. But this implies that, for realistically small values of θ, we must have $\gamma < c_- = (1-\theta)c$. For example, if $\theta = 0.1$, we must have $\gamma = 0.8281c$, whereas $c_- = 0.9c$. This means that, even if the rates between C_u and C_v, and between C_v and C_w, fall within the accepted bands, the cyclic arbitrage condition may force the direct rate between C_u and C_w outside the band. If we accept that the rate cannot move outside the band, then the conclusion must be that opportunities for profit by cyclic arbitrage will arise.

17.6 Potential theory

In this section we recast the theory outlined above in a form suggested by links with other network problems, such as the flow of water or electricity. A 'hydraulic model' for a single exchange rate operating under the gold standard was described long ago by Barker [1], and the famous Phillips machine exploits similar ideas. The electrical analogy has also been used in descriptive economics; for example, 'wiring diagrams' are used in work of Copeland [7], but I have not found any mathematical treatment.

Recall that the Mint Par definition of exchange rates is obtained by taking **y** to be a bundle consisting of a single good, such as gold. Explicitly, putting **g** = $(1, 0, 0, \ldots, 0)$, we obtain

$$\rho^{\mathbf{g}}(v, w) = \frac{\langle \mathbf{p}^{(w)}, \mathbf{g} \rangle}{\langle \mathbf{p}^{(v)}, \mathbf{g} \rangle} = \frac{p_1^{(w)}}{p_1^{(v)}} = \frac{1/p_1^{(v)}}{1/p_1^{(w)}}.$$

In other words, the Mint Par (gold standard) exchange rate is the ratio of the amount of gold that can be bought for one v-unit to the amount that can be bought for one w-unit. More generally,

$$\Phi^{\mathbf{y}}(v) = \langle \mathbf{p}^{(v)}, \mathbf{y} \rangle^{-1}$$

is the amount of the bundle **y** that can be bought for one v-unit, and we can think of $\Phi^{\mathbf{y}}(v)$ as the 'potential' for purchasing **y**, in v-units. Thus the PPP theoretical exchange rate

$$\rho^{\mathbf{y}}(v, w) = \Phi^{\mathbf{y}}(v)/\Phi^{\mathbf{y}}(w)$$

is expressed as a 'potential quotient'.

Theorem 17.1 shows that PPP theory (including the Mint Par theory as a special case) has the desirable feature that it defines a network of theoretical exchange rates in which no profit by cyclic arbitrage is possible. Clearly, this result depends on the fact that PPP exchange rates can be expressed as potential quotients, as above. Indeed, if we are given *any* function $\Phi : V \to \mathbf{R}^+$, then the associated function defined by $\rho(v, w) = \Phi(v)/\Phi(w)$ satisfies conditions (17.1) and (17.2).

The significant point is that the converse of the result described in the preceding paragraph is also true—that is, any function defined on the sides of a graph satisfying conditions (17.1) and (17.2) can be written as a potential quotient.

Theorem 17.3 *If S is the set of sides of a connected graph G, and $\rho : S \to \mathbf{R}^+$ satisfies conditions (17.1) and (17.2), then there is a function $\Phi : V \to \mathbf{R}^+$ such that, for each edge vw,*

$$\rho(v, w) = \Phi(v)/\Phi(w).$$

The proof depends on the fact that if T is a spanning tree of G, then there is a unique path in G from any vertex to any other vertex *using only edges in T*. Choose a vertex z in G, and, for any other vertex v, let $vab\ldots yz$ be the path in T from v to z. Set

$$\Phi(z) = 1, \quad \Phi(v) = \rho(v,a)\rho(a,b)\ldots\rho(y,z).$$

Now it can be checked (see [4] for the details) that $\rho(v,w) = \Phi(v)/\Phi(w)$ for each edge vw of G.

When the conclusion of Theorem 17.3 holds, we say that Φ is a *potential* for ρ. Since there are, in general, many different spanning trees, it appears as if there may be many unrelated potentials for a given ρ. However, it turns out that all such potentials are related in a very simple way: if Φ_1 and Φ_2 both satisfy the conclusion of Theorem 17.3, then there is a constant c such that $\Phi_1(x) = c\Phi_2(x)$, for all $x \in V$. This implies that if we pick any vertex z, there is always a potential for ρ which is 1 at z. In a more practical vein, the proof provides an explicit verification of the following useful fact:

> if the ERN conditions hold, then the complete set of exchange rates is determined by the subset of those exchange rates corresponding to any spanning tree.

This fact was noted independently by Jørgensen and Mikkelsen [14]; of course, the basic idea is very familiar in algebraic graph theory [3].

The general significance of the theorem is as follows. If the foreign exchange markets are efficient and adjust to prevent cyclic arbitrage, then, even though they may have no direct connection with fundamentals, they establish for themselves an 'ideal potential'. In Section 17.7 we show how such a potential may be determined by the totality of cash flows within the network.

Example 17.1 (continued). In the network of Example 17.1, we can take $\Phi(Y) = 1$, $\Phi(DM) = 68.4$, $\Phi(\pounds) = 171$ and $\Phi(\$) = 114$. It is easy to check that these values produce the given set of exchange rates; for example,

$$\rho(DM,\$) = \Phi(DM)/\Phi(\$) = 68.4/114 = 0.6. \quad \Diamond$$

This example is transparent to common sense: if we know the potential of each currency relative to the yen, then in the absence of arbitrage, we can work out all the exchange rates. Indeed, this idea yields a simple proof of Theorem 17.3 in the case where the underlying graph is complete. For more complicated networks that are connected but not necessarily complete, the proof that a potential function always exists requires slightly more sophisticated arguments.

Finally, we note that in the traditional applications of potential theory, $\phi = \log \Phi$ would be called the *potential function*. In that situation, the potential difference $\lambda(v,w) = \phi(v) - \phi(w)$ satisfies the additive analogues of conditions (17.1) and (17.2):

$$\lambda(w,v) = -\lambda(v,w); \tag{17.5}$$
$$\lambda(v_0,v_1) + \lambda(v_1,v_2) + \cdots + \lambda(v_{l-1},v_0) = 0. \tag{17.6}$$

These conditions will be familiar to those who have studied the theory of electrical networks.

17.7 Determination of exchange rates by cash flows

The elements of the model thus far can be summarized as follows. We have a graph $G = (V, E)$, whose vertex set V is a set of currencies v and whose edge set E defines the pairs of currencies that can be exchanged. S is the set of *ordered* pairs associated with E, and there is an exchange rate function $\rho : S \to \mathbf{R}^+$ satisfying conditions (17.1) and (17.2), so that, by Theorem 17.3, we can write $\rho(v,w) = \Phi(v)/\Phi(w)$ for some function $\Phi : V \to \mathbf{R}^+$.

We now introduce a new function $c : S \to \mathbf{R}^+$. The suggested interpretation is that, for each (v,w) in S, the value $c(v,w)$ is the amount of cash transferred from country C_w to country C_v in a given time period, measured in w-units. In general, $c(v,w)$ is the sum of two components, the cash received by agents in C_v in return for exports to C_w, and the value of capital investments transferred from C_w to C_v. For our present purposes, there is no reason to separate the two; in both cases, cash that was previously held in w-units has been exchanged for v-units.

Fixing attention on v, we see that the 'balance of payments' for that currency can be calculated as follows. Between C_v and C_w, the amounts transferred (measured in v-units) are $c(v,w)\rho(w,v)$ from C_w to C_v and $c(w,v)$ from C_v to C_w, so the surplus at C_v is

$$c(v,w)\rho(w,v) - c(w,v).$$

Thus the overall balance of payments at C_v is

$$B(v) = \sum_{w \sim v} \{c(v,w)\rho(w,v) - c(w,v)\},$$

where the sum is taken over all vertices w adjacent to v.

We shall show that, given a connected graph G and a cash flow function $c : S \to \mathbf{R}^+$, the balance of payments conditions

$$B(v) = 0, \quad \text{for each } v \in V,$$

determine a unique function $\rho : S \to \mathbf{R}^+$ such that (G, ρ) is an ERN. The balance of payments equations for the case $n = 3$ were written down by Krugman [16], who stressed that they do *not* imply that each bilateral market is in balance.

Since the required (G, ρ) is an ERN we are, in effect, looking for a potential Φ for ρ. Writing $x_v = \Phi(v)$, we note that the quantities x_v must all be positive, and that $\rho(w,v) = x_w/x_v$. So we can substitute in the formula for $B(v)$ and multiply through by x_v, thereby obtaining the following form of the equation $B(v) = 0$:

$$\sum_{w \sim v} c(v,w) x_w - \left(\sum_{w \sim v} c(w,v) \right) x_v = 0. \tag{17.7}$$

Using this form we can express the system of equations $B(v) = 0$ as a matrix equation. Let **x** be the column vector (x_v), and let **C** be the square matrix whose rows and columns are labelled by the elements of V and whose entries are

$$C_{vw} = \begin{cases} c(v,w), & \text{if } vw \in E, \\ 0, & \text{if } vw \notin E, \\ -\sum_{x \neq v} c(x,v), & \text{if } w = v. \end{cases}$$

The sum is taken over all $x \neq v$ only for convenience; it is enough to sum over those x adjacent to v. With these definitions, the balance of payments conditions are clearly equivalent to the matrix equation $\mathbf{Cx} = \mathbf{0}$. We construct an explicit solution of this equation, in which each x_v is positive, by using the *matrix-tree* technique (see Tutte [18] and the references given in Section 17.8).

Given a spanning tree T of G and a vertex v, we have a rooted tree $T[v]$ with root v. We give the edges of $T[v]$ a direction, by directing them away from v so that they are sides of the graph. Specifically, an edge $rs \in T$ corresponds to the two sides (r,s) and (s,r), and we assign one side to $T[v]$ by the following rule: the side (r,s) is in $T[v]$ if the path in T from v to s contains r.

Given the function c (or the associated matrix **C**), for each vertex v we let

$$x_v(T) = \prod_{(r,s) \in T[v]} c_{rs}.$$

Then the required solution is provided by the following theorems, whose proofs may be found in [4].

Theorem 17.4 *A solution of the system $\mathbf{Cx} = \mathbf{0}$ is given by*

$$x_v = \sum_T x_v(T),$$

where the sum is taken over all spanning trees of G.

Theorem 17.5 *Let G be a connected graph and c be a function from S to $\mathbf{R}^+$. Then there is a function $\rho : S \to \mathbf{R}^+$ such that*

(a) *(G, ρ) is an ERN;*
(b) *$B(v) = 0$, for each $v \in V$.*

In general, ρ is uniquely determined by conditions (a) and (b), and is given by $\rho(v,w) = x_v/x_w$, where x_v is defined above.

These theorems establish what might be called the 'universal' form of PPP theory. Recall that the theoretical exchange rate $\rho^y(v,w)$, based on purchasing power parity with respect to a bundle of goods **y**, is given by the ratio of the amount of **y** that can be bought for one v-unit to the amount that can be bought for one w-unit. Theorem 17.5 defines exchange rates by a similar ratio, based on cash flows in the network as a whole, rather than the cost of a specific bundle in the two currencies. Instead of the 'potential' of currency v being measured by how much **y** it purchases, it is measured by a sum of products

$$\sum_T x_v(T) = \sum_T \prod_{(r,s)\in T[v]} c_{rs},$$

involving all cash flows in the network.

Example 17.4 We shall write down explicitly the solution for the case of four countries forming a complete graph, as in Figs 17.1 and 17.2. From the general case we can deduce what happens if the network is not complete, by putting certain terms c_{vw} equal to zero.

The following table lists the terms in each x_v corresponding to the sixteen spanning trees of the complete graph with vertex set $\{1,2,3,4\}$. A spanning tree is denoted by its three edges, with a specific orientation of each edge. The symbol in Table 17.1 indicates which orientations of the edges give the term occurring in the relevant x_v. For example, the symbol $-++$ in the 12.13.14 row and x_2 column indicates that the corresponding term in x_2 is $c_{21}c_{13}c_{14}$.

Table 17.1

	x_1	x_2	x_3	x_4
12.13.14	+ + +	− + +	+ − +	+ + −
12.13.24	+ + +	− + +	+ − +	− + −
12.13.34	+ + +	− + +	+ − +	+ − −
12.14.23	+ + +	− + +	− + −	+ + −
12.14.34	+ + −	− + −	+ − +	+ − −
12.23.24	+ + +	− + +	− − +	− + −
12.23.34	+ + +	− + +	− − +	− − −
12.24.34	+ + −	− + −	− − +	− − −
13.23.24	+ − +	− + +	− − +	− + −
13.23.34	+ − +	− + +	− − +	− − −
13.24.34	+ − +	− + −	− − +	− − −
13.14.23	+ + −	+ − +	− + −	+ − −
13.14.24	+ + −	+ − +	− + +	+ − −
14.23.24	+ + −	− + +	− − +	− + −
14.23.34	+ − −	− + +	− − +	− − −
14.24.34	+ − −	− + −	− − +	− − −

The advantage of the matrix-tree technique is that it leads to elegant formulas containing only positive terms. However, for practical purposes, the number of terms is uncomfortably large in all but the smallest cases. The number of terms in the formula for x_v is equal to the number of spanning trees in G, and for a complete graph with n vertices this is known to be n^{n-2}, by Cayley's Theorem. Note that determinantal methods for solving the system $\mathbf{Cx} = \mathbf{0}$ lead to even larger numbers of terms, with the added complication that there are terms of opposite sign that have to be cancelled. Of course, in a numerical example we would calculate x_v by a method based on elimination, which is computationally feasible for large graphs.

The explicit formulas provided by the matrix-tree technique allow us to study the effect of changes in cash flows on exchange rates. To do this we must assume that the markets operate so that the rates always form an ERN, and that the rates are determined by the balance of payments. Nevertheless, there are many points of contact between the theoretical approach suggested here and current macroeconomic theories of exchange rate determination (see Gartner [11]).

Given $G = (V, E)$, with $|V| = n$ and $|E| = m$, we regard the $2m$ entries c_{rs} of the matrix $\mathbf{C}$ as independent variables. The 'potentials' x_v ($v \in V$) are functions of these variables; specifically, each x_v is a sum of terms $x_v(T)$, and each $x_v(T)$ is the product of $n-1$ variables.

Consider a given pair $(r, s) \in S$. The form of x_v implies that we can collect the terms in which c_{rs} is a factor and write

$$x_v = D^v_{rs} c_{rs} + E^v_{rs},$$

where D^v_{rs} and E^v_{rs} do not depend on c_{rs}. Clearly, D^v_{rs} is the partial derivative $\partial x_v / \partial c_{rs}$, so that D^v_{rs} tells us how x_v is affected by a change in the cash flow c_{rs}.

In fact, D^v_{rs} itself can be expressed as a sum of products, taken over the set of spanning trees T for which (r, s) is in $T[v]$; that is, the spanning trees for which rs is in T and the path in T from v to s uses the side (r, s). For example, in the complete graph on four vertices (Example 17.4), D^1_{23} is the sum of terms corresponding to the spanning trees 12.14.23, 12.23.24, 12.23.34 and 14.23.24—that is,

$$D^1_{23} = c_{12}c_{14} + c_{12}c_{24} + c_{12}c_{34} + c_{14}c_{42}.$$

An application of the general result is the observation that, for any edge vw, $D^v_{wv} = 0$. This is because if the edge wv is in T, it is oriented away from v in $T[v]$, and so (v, w) rather than (w, v) is in $T[v]$. Thus changes in the cash flows out of v do not affect x_v. However, such flows do affect the values of x_u, for vertices u adjacent to v, and hence they affect the exchange rates $\rho(v, u)$.

17.8 Application to a tournament ranking problem

I am indebted to John Moon for pointing out that the mathematical theory developed in the previous section can be applied to a problem about tournaments. In this context, Theorem 17.5 is a generalization of the work of Berman [2], who applied the matrix-tree technique to a problem previously considered by Daniels [8] and Moon and Pullman [17].

Let the vertices of $G = (V, E)$ be a set of teams, and let an edge indicate a win-or-lose game between the corresponding teams. For each side $(v, w) \in S$, let $c(v, w)$ denote the probability that v defeats w. Suppose that if team v loses a game, then it pays an amount x_v to the team that wins. Then the numbers x_v define a fair system of 'handicaps', for which the expected payoff of each team v is zero, provided that

$$\sum_{w \sim v} c(v, w) x_w - \left(\sum_{w \sim v} c(w, v) \right) x_v = 0.$$

This is exactly the same as our balance of payments condition (17.7), and our theory implies that, up to multiplication by a constant, there is a unique fair system of handicaps. In this application we assume that $c(v, w) + c(w, v) = 1$, but that is clearly not necessary.

17.9 Mechanisms linking exchange rates and trade

Modern macroeconomic theory, as described by Gartner [11] for example, provides us with impressively detailed formulations of the interactions between economies and the exchange rate, in so far as they affect two countries in isolation, or the 'domestic-and-foreign' situation. Our programme is to discuss the case of a network of countries, with special reference to the implications of the ERN conditions (17.1) and (17.2). The notion of a potential function is well suited to the analysis of this problem.

It is instructive to compare and contrast two mechanisms that link the level of international trade with the exchange rate. The first applies when rates are flexible and there is some easily recognized standard. In this, the classical case, the mechanism which was believed to restore actual exchange rates to their theoretical level was based on a crude form of the Quantity Theory. In terms of the gold standard, the argument went roughly as follows. If, as the result of imbalance of trade between C_v and C_w, there is a lack of demand for the currency v and the rate $\rho^{act}(v, w)$ falls to the export point ρ^g_-, then the resulting outflow of gold reduces the stock of money in C_v, and prices fall accordingly. This tends to attract more trade, so that goods flow out, the demand for the currency v increases, and the exchange rate returns to its standard level.

This mechanism seems to have operated quite well in practice. Its defects, such as those outlined in Example 17.3, were probably obscured by the substantial costs of exchange transactions, which exerted a damping effect on the system, and to some extent precluded the possibility of profit by cyclic arbitrage.

The second mechanism operates when rates are fixed arbitrarily, as was the case for much of the 20th century. Let us suppose that a *trade* function $z : S \to \mathbf{R}^b$ is given; that is, for each $(v,w) \in S$ there is a b-vector

$$\mathbf{z}^{vw} = (z_1^{vw}, z_2^{vw}, \ldots, z_b^{vw}),$$

such that z_i^{vw} is the amount of the ith good exported from C_v to C_w in a given time period. Then the cash transferred from C_w to C_v in payment is $\langle \mathbf{p}^{(w)}, \mathbf{z}^{vw} \rangle$ in w-units. If we consider only the current account, this is the quantity denoted by $c(v,w)$ in Section 17.7. Theorem 17.5 tells us that given a graph G, a price function p, and a trade function z, there is a unique system of exchange rates ρ such that (G, ρ) is an ERN and every current account is in balance. In other words, there is a function ρ^* such that $\rho^*(p,z) = \rho$ is the ideal exchange rate function for the given p and z.

Now, suppose that p_0 and z_0 are the existing values of the price and trade functions. If the exchange rates are fixed arbitrarily, the actual rate ρ^{act} may differ from $\rho^*(p_0, z_0)$. Over a period of time, this discrepancy will affect trade in several ways; for example, some countries will have a trade imbalance and will take unilateral measures to correct it, possibly by using a strategy that involves alteration of prices. Consequently, a complete set of new values p_1 and z_1 will arise, for which the ideal rate $\rho^*(p_1, z_1)$ is, in theory, equal to the actual one ρ^{act}. However, what we have just described is not a mechanism for stabilizing exchange rates. Rather, it is a mechanism for encouraging trade and prices to conform to an existing fixed exchange rate, and as such it may be considered undesirable. In that light, the argument provides some justification for flexible, as opposed to fixed, exchange rates.

Since the early 1970s, exchange rates have operated under a range of loosely-defined regimes, quite different from either of those described above. There is no recognized standard like the gold standard, nor are the rates fixed arbitrarily. In the same period it has become apparent that the rates are influenced rather less by the current account (the trade in goods and services), and rather more by the capital account. A start has been made on the problem of using the potential-theoretic approach in this context. It involves the consideration of forward rates and the expectations of investors, and leads to some general conclusions about the compatibility of mechanisms with ERN conditions. Details may be found in [5].

References

1. D. Barker, *Cash and Credit*, Cambridge University Press, 1910.
2. K. A. Berman, A graph-theoretical approach to handicap ranking of tournaments and paired comparisons, *SIAM J. Alg. Discrete Math.* **1** (1980), 359–361; *MR* **81i**: 05072.
3. N. L. Biggs, *Algebraic Graph Theory*, 2nd edn, Cambridge University Press, 1993.

4. N. L. Biggs, Exchange rates and the matrix-tree theorem, Theoretical Economics Discussion Paper TE/94/273, London School of Economics, 1994.
5. N. L. Biggs, Exchange rate networks and mechanisms, Mathematics Preprint Series LSE-MPS-64, London School of Economics, 1994.
6. G. Clare, *A Money-Market Primer*, Effingham Wilson, 1893.
7. M. A. Copeland, *A Study of Money-flows in the United States*, NBER Publications **54**, 1952.
8. H. E. Daniels, Round-robin tournament scores, *Biometrika* **56** (1969), 295–300.
9. R. Dornbusch, Purchasing power parity, *New Palgrave Dictionary of Economics*, Vol. 3, MacMillan, 1987.
10. H. E. Evitt, *A Manual of Foreign Exchange*, 3rd edn, Pitman, 1938.
11. M. Gartner, *Macroeconomics under Flexible Exchange Rates*, Harvester Wheatsheaf, 1993.
12. J. Harris, *An Essay upon Money and Coins*, London, 1757.
13. B. N. Jørgensen and H. O. A. Mikkelsen, An arbitrage free trilateral target zone model, Memo 1993-17, Institute of Economics, University of Aarhus, 1993.
14. B. N. Jørgensen and H. O. A. Mikkelsen, Multilateral models in international finance: a graph theoretical approach, Preprint, November 1993.
15. P. Kelly, *The Universal Cambist*, 2nd edn, London, 1835.
16. P. R. Krugman, Vehicle currencies and the structure of international exchange, *J. Money, Credit and Banking* **12** (1980), 513–526.
17. J. W. Moon and N. J. Pullman, On generalized tournament matrices, *SIAM Rev.* **12** (1970), 384–399; *MR* **42**#7525.
18. W. T. Tutte, *Graph Theory*, Addison-Wesley, 1984; *MR* **87c**: 05001.

Notes on Contributors

Martin Anthony has been a Lecturer in Mathematics at the London School of Economics since 1990. He obtained a BSc in Mathematics from Glasgow University in 1988, and a PhD in Mathematics from the University of London in 1991. He has published research papers on discrete mathematics, neural networks and computational learning theory, and co-authored (with Norman Biggs) *Computational Learning Theory: An Introduction*.

Address: Department of Mathematics, London School of Economics, Houghton Street, London WC2A 2AE, UK.
e-mail: anthony@lse.vax.ac.uk

Lowell Beineke is the Schrey Professor of Mathematics at Indiana University–Purdue University at Fort Wayne, where he has been since receiving his PhD from the University of Michigan. His graph theory interests are broad and, in addition to topological graph theory, include line graphs, tournaments, decompositions and vulnerability. With Robin Wilson, he co-edited *Selected Topics in Graph Theory* (3 volumes) and *Applications of Graph Theory*.

Address: Department of Mathematics, Indiana University–Purdue University, Fort Wayne, IN 46805, USA.
e-mail: beineke@cvax.ipfw.indiana.edu

Norman Biggs has held appointments at the University of Southampton and Royal Holloway College, University of London. In 1988 he became a Professor at the London School of Economics, where he is now also Director of the Centre for Discrete and Applicable Mathematics. His books include *Algebraic Graph Theory*, *Graph Theory 1736–1936* (with E. K. Lloyd and R. J. Wilson), *Interaction Models*, *Permutation Groups and Combinatorial Structures* (with A. T. White), *Discrete Mathematics*, *Introduction to Computing and Pascal* and *Computational Learning Theory* (with M. Anthony).

Address: Department of Mathematics, London School of Economics, Houghton Street, London WC2A 2AE, UK.
e-mail: biggs@lse.vax.ac.uk

Notes on Contributors

Graham Brightwell is a Reader in Mathematics at the London School of Economics, where he went after nine years at Cambridge, as an undergraduate, postgraduate and Research Fellow. His main mathematical interests are in partial orders and in random methods in combinatorics, but he has also co-written papers on pure graph theory, learning theory, and even relativity.

Address: Department of Mathematics, London School of Economics, Houghton Street, London WC2A 2AE, UK.
e-mail: brightwell@lse.vax.ac.uk

Peter Cameron is currently Professor of Mathematics at Queen Mary and Westfield College, London. His mathematical interests are in permutation groups and their operands (which may be logical, algebraic or combinatorial, finite or infinite). He has written lecture notes on permutation groups, parallelisms, and projective and polar geometry and (with J. H. van Lint) a book on graphs, codes and designs, and an undergraduate textbook on combinatorics.

Address: Department of Mathematics, Queen Mary and Westfield College, Mile End Road, London E1 4NS, UK.
e-mail: p.j.cameron@qmw.ac.uk

Roger Cook is Professor of Mathematics at the University of Sheffield. He received a PhD from the University of London for results in number theory, and a DSc for published work on number theory and graph theory. His research interests are primarily number theory and combinatorics. He is an Editor of *Mathematical Spectrum*.

Address: Pure Mathematics Section, University of Sheffield, Hicks Building, Hounsfield Road, Sheffield S3 7RH, UK.
e-mail: roger.cook@sheffield.ac.uk

Robert Curtis studied at Sidney Sussex College, Cambridge, receiving his PhD in 1972 under the supervision of John Conway. After a Research Fellowship at Cambridge working on the *ATLAS of Finite Groups*, he spent three years teaching mathematics at Bowdoin College, Maine. Since 1980 he has lectured in mathematics at the University of Birmingham.

Address: Department of Mathematics, University of Birmingham, Edgbaston, Birmingham B15 2TT, UK.
e-mail: r.t.curtis@bham.ac.uk

Colin McDiarmid is a University Lecturer in Operational Research at Oxford University, and Tutorial Fellow in Mathematics at Corpus Christi College, Oxford. His main research interests are in discrete mathematics—in particular, combinatorial algorithms and probability. He has published many papers on discrete mathematics, on the mathematics of operational research, and on theoretical computer science.

Address: Corpus Christi College, Oxford OX1 4JF, UK.
e-mail: cmcd@stats.ox.ac.uk

Tony Morris joined the Royal Navy as a teenager, where his period of service included an eventful voyage to the South Atlantic. He studied for A-levels aboard ship, and eventually left the Navy to study mathematics at the University of Birmingham, receiving his PhD in 1994 under the supervision of Robert Curtis. He now teaches mathematics at Westminster School.
Address: Westminster School, London SW1.

James Oxley is Professor of Mathematics at Louisiana State University. He received his DPhil from Oxford University in 1978 under the supervision of Dominic Welsh. His main research interests are in matroid theory and graph theory. He is the author of *Matroid Theory*.
Address: Mathematics Department, Louisiana State University, Baton Rouge, LA 70803-4918, USA.
e-mail: oxley@mavais.math.lsu.edu

Ronald Read obtained a BA from Cambridge University in 1948, and a PhD from London University in 1958. He was a Lecturer (later Professor) at the University of the West Indies in Jamaica from 1950 to 1970, and then took up a post at the University of Waterloo. His research interests are in graphical enumeration, chromatic polynomials and computer applications in combinatorics. He officially retired in 1990, but continues research at Waterloo as an adjunct professor.
Address: Department of Combinatorics and Optimization, University of Waterloo, Waterloo, Ontario, Canada N2L 3G1.
e-mail: rcread@math.uwaterloo.ca

Peter Rowlinson is Professor of Mathematics at the University of Stirling, where he has taught since 1969. His DPhil thesis at Oxford University was concerned with finite groups, and in 1975-76 he was Visiting Associate Professor of Mathematics at the California Institute of Technology. He has worked in algebraic graph theory since 1980.
Address: Department of Mathematics and Statistics, University of Stirling, Stirling FK9 4LA, UK.
e-mail: p.rowlinson@stirling.ac.uk

Edward Scheinerman majored in Mathematics at Brown University, and received his PhD in Mathematics from Princeton University in 1984. From there, he went to Johns Hopkins University, where he is a Professor of Mathematical Sciences and Computer Science. His primary mathematical interests are geometric representations of combinatorial structures and random graphs. He serves as a Managing Editor for the *Journal of Graph Theory*.
Address: Department of Mathematical Sciences, Johns Hopkins University, Baltimore, MD 21218, USA.
e-mail: edward.scheinerman@jhu.edu

Dominic Welsh is a Professor of Mathematics at the University of Oxford. His main research has been in probability theory, matroid theory, percolation theory and computational complexity. He is the author of *Matroid Theory*, co-author (with G. R. Grimmett) of *Probability, an Introduction, Codes and Cryptography, Complexity: Knots, Colourings and Counting*, editor of *Combinatorial Mathematics and its Applications*, co-editor (with D. R. Woodall) of *Combinatorics* and (with G. R. Grimmett) *Disorder in Physical Systems*.

Address: Mathematical Institute, 24–29 St Giles', Oxford OX1 3LB, UK.
e-mail: dwelsh@oxford.vax.ac.uk

Robin Whitty is Professor of Software Engineering at South Bank University. He has been Vacation Fellow at British Telecom Laboratories, and is a Fellow of the Institute of Mathematics and its Applications and a Visiting Research Fellow at Goldsmiths College, University of London. His research interests include software measurement and graph theory. In his spare time he juggles and plays the piano (but not at the same time).

Address: South Bank University, Borough Road, London SE1 0AA, UK.
e-mail: whittyr@sbu.vax.ac.uk

Peter Wild is Reader in Pure Mathematics at Royal Holloway College, University of London, which he joined after a short period as Research Scientist with CSIRO, Australia. He received his undergraduate degree in Australia, and his PhD at Queen Mary and Westfield College, London. His research interests lie in discrete mathematics and its applications, including cryptography, statistical design and autocorrelation sequences.

Address: Department of Mathematics, Royal Holloway College,
University of London, Egham, Surrey TW20 0EX, UK.
e-mail: uhah018@rhbnc.vax.ac.uk

Robin Wilson is Dean and Director of Studies in the Faculty of Mathematics and Computing at the Open University. He graduated in mathematics from Oxford University, and received his PhD in number theory from the University of Pennsylvania. He has written and edited many books on graph theory and combinatorics and on the history of mathematics, and his research interests include edge-colouring of graphs and mathematics in Victorian Britain.

Address: Faculty of Mathematics and Computing, Open University,
Walton Hall, Milton Keynes MK7 6AA, UK.
e-mail: r.j.wilson@open.ac.uk

Index

abstract group, 130
achiral link, 182
action state, 235
activation, 248
activation function, 252
acyclic graph, 8, 53
adjacency algebra, 89
adjacency matrix, 4, 86
adjacent edges, 3
adjacent vertices, 3
admittance matrix, 86
A-equivalent set, 39
affine group, 136
age of a graph, 79
alkyl-substituted benzene, 22
almost simple group, 136
almost surely, 195
alternating link, 181
ambient isotopic, 178
amphicheiral link, 182
angle matrix, 93
antichain, 55, 113
antipodal graph, 136
A-optimal, 214, 217
arbitrage, 263
arboricity, 8, 65
arc, 2
arc-homogeneous tournament, 44
arity, 71, 130
asteroidal triple, 145
atomic formula, 71
attribute, 229
attributed grammar, 229
automorphism, 3, 44
automorphism group, 3, 128, 130
autonomous set, 59

back-propagation algorithm, 254
balanced incomplete block design, 216
ball, 149
Berge graph, 198
BIBD, 216
bicentral tree, 8
bicentre of a tree, 8
binary Golay code, 120
binary matroid, 103
bipartite graph, 7
biplanar graph, 160
block, 6, 45, 209
block design, 209

Boltzmann machine, 248
bond, 37
bond-closed set, 37
bound occurrence of a variable, 71
boundary, 46
box, 149
boxicity, 150
bracket polynomial, 186
braid, 182
braid index, 183
bridge, 6
Brooks' Theorem, 11
Burnside's Lemma, 31

canonical efficiency factor, 218
canonical star basis, 95
Cartesian product, 6
Catalan number, 40
Cayley colour digraph, 131
Cayley graph, 133
Cayley's Theorem, 8
cellular embedding, 165
central tree, 8
centre of a tree, 8
chain, 54
character, 35
characteristic polynomial, 4
checkerboard, 179
chemical enumeration, 19
Chinese postman problem, 41
Chomsky hierarchy, 229
chordal graph, 144
chromatic index, 195
chromatic number, 11, 168
chromatic polynomial, 11
circuit elimination axiom, 101
circuit of a matroid, 100, 101
clique, 7
clique number, 7
closed walk, 4
closure of a language, 228
cocircuit of a matroid, 105
codeword, 116
coin representation, 149
colour class, 11
commute time, 202
commuting graph, 137
compact 2-manifold, 164
Compactness Theorem, 72
comparability graph, 53

compilation, 227
complement of a digraph, 32
complete bipartite graph, 7
complete block design, 218
complete graph, 7
complete k-partite graph, 7
complete theory, 71
component of a code, 122
components, 4
composite flowgraph, 237
concurrence, 215
configuration, 15, 16
configuration generating function, 16
congruence, 35
connected design, 214
connected digraph, 4
connected graph, 4
connectedness, 196
connectivity, 6
consecutive 1s property, 146
consensus function, 248
consistency problem, 258
consistent learning algorithm, 256
constant symbol, 71
containment order, 53
content, 15, 16
content of configuration, 16
contractible graph, 5, 156
contracting an edge, 5, 103
contraction, 104
contrast, 213
converse of a digraph, 32
coordinate order, 53
correspondence condition, 250
countable graph, 2
countably categorical theory, 80
covering graph, 54
covering pair, 54
crosscap, 169
crosscap number, 169
crossing number, 160, 181
cubic graph, 3
current graph, 167
current state, 254
cut-edge, 6
cut-vertex, 6
cutset, 6
cycle, 4
cycle graph, 7
cycle index, 17, 38
cycle matroid, 100
cycle type, 17
cyclic arbitrage, 261, 264

cyclic neighbourhood-inclusion property, 148

decision state, 235
decomposition expression, 238
decomposition of a flowgraph, 237
degree list, 3
degree of a vertex, 3
deleting an edge, 5, 103
deletion, 104
deletion-contraction formula, 11
dependency graph, 199
dependent set of a matroid, 103
design, 209
deterministic finite automaton, 227, 230, 231
deterministic graph, 201
diagram, 54
diameter, 4
digraph, 2
Dilworth's Theorem, 55
dimension, 55, 62
Diophantine equation, 35
directed comparability graph, 53
Dirichlet convolution, 36
Dirichlet series, 35
disconnected graph, 4
disjoint union, 6
distance, 4
distance matrix, 89
distance-regular graph, 89
distance-transitive graph, 76, 135
D-optimal, 214
double shift graph, 66
doubly-even code, 117
Downward Löwenheim–Skolem Theorem, 72
drawing of a graph, 156
D-structure, 240
dual graph, 10
dual matroid, 59

Earth–Moon map-colouring problem, 159
edge, 1
edge colouring, 195
edge covering by cliques, 143
edge set, 1
edge-connectivity, 6
edge-cut, 6
edge-deleted subgraph, 5
edge-disjoint path, 4
edge-independence number, 3
edge-transitive group, 3
effective resistance, 202

efficiency balanced, 220
efficiency factor, 218
efficient learning algorithm, 257
eigenvalues of a graph, 4, 86
elimination of quantifier, 79
embedding of a graph, 156, 165
empty graph, 7
end-vertex, 3
E-optimal, 214
equivalent configurations, 15, 16, 178, 239
ERN, 265
error of a function, 255
estimable contrast, 213
Eulerian graph, 9, 41
Eulerian trail, 9, 41
Euler's function, 34
Euler's Polyhedron Formula, 10, 157
eutactic star, 93
exchange rate network, 265
excitatory weight, 248
excluded minor, 112
expander, 199
expansion coefficient, 46
experimental design, 209
export point, 269
extended Hamming code, 118
extension property, 78

face, 9, 66
face of a graph, 157
Fáry's Theorem, 163
feedback vertex number, 234
feedforward network, 248
figure, 15
figure generating function, 16
finite automaton, 231
finite intersection property, 235
finite state machine, 227
finitely axiomatizable property, 72
first-order, 71
first-order property, 72
first-order theory, 71
Fisher information matrix, 213
fixed-point-free involution, 121
flipping, 59
flow-difference, 61
flype, 190
forest, 8
formula, 71
Four Colour Theorem, 11
Fraïssé limit, 79
free amalgamation, 82
free occurrence of a variable, 71

function symbol, 70

general graph, 1
generating function, 14
genus of a graph, 167
genus of a surface, 164
girth, 4
glue, 122
Golay code, 118
good class, 82
graph, 1
graph grammar, 227
graphical enumeration, 13
ground set of a matroid, 101

Hadamard matrix, 45
Hadwiger number, 135
Hajós conjecture, 199
Hall's Theorem, 56
Hamiltonian cycle, 9, 196
Hamiltonian graph, 9
Hamiltonian path, 9
Hamming ball, 205
Hamming distance, 116
Hamming graph, 136
Hamming weight, 116
Hamming weight enumerator, 117
height, 55
hereditary class, 142
hitting time, 197
homeomorphic graphs, 5, 156
homogeneous graph, 78, 137
hyperedge, 2
hypergraph, 2
hypergraph-colouring, 200

identity graph, 31
import point, 269
in-degree, 3
incidence matrix, 4, 210
incidence order, 53
incident edge, 3
incomparability graph, 53
independence number, 3, 195
independent set of a matroid, 103
independent set of edges, 3
independent set of vertices, 3
index, 87
induced subgraph, 5
inequivalent configurations, 16
infinite graph, 2
infinite region, 9
inherited attribute, 229
inhibitory weight, 248

input node, 248, 252
intersection array, 90
intersection graph, 141
intersection number, 143
intersection of graphs, 6
interval graph, 144
interval order, 66
i-path, 64
irreflexive, 52
isolated vertex, 3
isomer, 21
isomorphic graphs, 3, 101
isomorphic t-tuples, 136
isomorphism of plane trees, 39
isotopy, 178
isthmus, 6

join, 3, 6
Jones polynomial, 185

Kauffman bracket polynomial, 185
k-chromatic graph, 11
k-colourable graph, 11
k-connected graph, 6
k-connectedness, 107
k-cube, 8
k-dimensional octahedron, 8
k-edge-connected graph, 6
Kirchhoff matrix, 221
Kleene's Theorem, 232
knot, 177
k-regular graph, 3
Kuratowski's Theorem, 10, 156

labelled example, 254
labelled flowgraph, 236
labelled graph, 2
λ-categorical set, 80
language, 228
Laplacian matrix, 86
learning algorithm, 253, 254
Legendre symbol, 35
length of a code, 116
lexical analyser, 230
line graph, 6
linear arboricity, 200
Linear Arboricity Conjecture, 200
linear code, 116
linear crossing number, 163
linear drawing, 163
linear forest, 200
linear order, 52
linear threshold function, 253
linear threshold network, 253

link, 177
locally finite graph, 2
locally H graph, 137
locally uniform graph, 137
loop, 1
Lovász Local Lemma, 199
Löwenheim–Skolem Theorem, 72

MacWilliams identities, 117
magnification coefficient, 46
main angle, 94
map, 10
matching, 3
Matrix-Tree Theorem, 9, 222
matroid, 101
matroid represented by a matrix, 103
maximal element, 54
maximal planar graph, 10
maximal-clique incidence matrix, 146
maximum (weighted) cut problem, 201
maximum clique, 7
maximum cut, 250
m-dimensional octahedron, 223
medial graph, 180
median, 205
Menger's Theorem, 6
minimal element, 54
minimal forbidden family, 171
minimally 3-connected matroid, 108
minimally k-connected graph, 108
minimum cut, 201
minimum Hamming distance, 117
minor, 5, 113, 171
minor-closed, 111
minor-minimal family, 172
Mint Par, 266
mirror image, 182
Möbius function, 34
Möbius Inversion Formula, 36
Molien series, 118
money-units, 265
monosubstituted alkane, 19
(M, S)-optimal, 217
multigraph, 1
multiple edges, 1
multiple-interval graph, 147
multiplicative function, 34

natural inner product, 117
negative Pell equation, 48
neighbourhood, 3
nesting complexity, 241
neural network, 247

$\mathcal{N}^k$-consistency, 258
(n,k,c)-expander, 46
(n,k,c)-magnifier, 46
no-arbitrage condition, 265
non-orientable genus, 169
non-orientable surface, 169
non-separable graph, 6
non-terminal symbol, 228
normal equations, 211
normalized Hadamard matrix, 45
null graph, 7

odd graph, 48
oligomorphic group, 80
1-factor, 3
open path, 4
order of a graph, 1
Ore's Theorem, 9
orientable surface, 164
orientable surface S_h with h handles, 164
oriented knot, 185
orthogonal vectors, 117
Otter's formula, 29
out-degree, 3
outer polynomial, 191
outerplanar graph, 10
output node, 248, 252

PAC learning, 253
PAC model, 253
Paley design, 45
Paley graph, 42
Paley tournament, 44
parse tree, 227
partial order, 52
partially ordered set, 52
path, 4
Pell's equation, 35
perceptron, 253
perceptron convergence theorem, 254
perceptron learning algorithm, 254
perfect elimination order, 147
perfect graph, 11, 55, 198
Perfect Graph Theorem, 56
perfect matching, 3
permutation graph, 150, 205
permutation group, 130
permutation of a braid, 182
Petersen graph, 7
ϕ-bounded graph, 81
planar graph, 9, 156
plane graph, 9
plane tree, 39

planted tree, 39
Platonic graph, 7
plot, 209
Pólya's Enumeration Theorem, 18, 39
polynomial of a link, 185
poset, 52
potential function, 272
power group, 38
Power Group Enumeration Theorem, 31
PPP, 266
predimension of a graph, 82
price index, 266
prime flowgraph, 237
prime knot, 182
prime number, 34
primitive graph, 91
primitive group, 76
probability, 194
product of languages, 228
production rule, 228
program structuredness, 239
program testability, 242
property Q, 62
pseudo-elementary property, 74
Purchasing Power Parity, 266

quadratic residue, 35, 44
quadratic residue graph, 42
quadratic residue tournament, 44
quotient of languages, 228

Rado's graph, 77
Ramanujan graph, 47
Ramsey number, 198
random Cayley graph, 134
random graph, 77, 194
random graph process, 197
random walk, 202
rank 3 graph, 136, 137
rank of a matroid, 105
rank of a set of edges, 187
rank-r whirl, 108
realizer, 55
recognition problem, 179
Reconstruction Theorem, 96
reduced diagram, 188
reduced form, 232
reduced normal equations, 213
region, 9
region of a graph, 157
region of an embedding, 165
regular expression, 228
regular graph, 3

regular graph design, 216
regular language, 228
regular representation, 130
Reidemeister moves, 178
Reidemeister's Theorem, 179
relation symbol, 70
Riemann zeta function, 36
rim, 108
Ringel–Youngs Theorem, 167
Robertson–Seymour Theorem, 171
root of a tree, 28, 39
rooted forest, 28
rooted graph, 2
rooted tree, 28, 39
rotation rule, 166
Rotation Scheme Theorem, 166
rotation-extension, 196
row–column design, 224
RP $\neq$ NP conjecture, 257

s-arc transitive graph, 136
Schnyder labelling, 63
Schnyder's Theorem, 61
Seidel matrix, 86
Seifert graph, 184
Seifert surface, 183
self-complementary graph, 6
self-dual code, 117
self-sufficient graph, 82
sentence, 71
separating set of vertices, 6
shattered set, 256
side, 264
Σ-graph, 142
Σ-representation, 142
signed graph, 2
simple contrast, 214
simple digraph, 2
simple graph, 1
simple matroid, 107
simple perceptron, 253
simplicial vertex, 147
simulated annealing, 249
site, 15, 16
size of a function, 143
span, 189
spanning subgraph, 5
spanning tree, 8
spectral decomposition, 92
sphericity, 149
split graph, 148
spoke, 108
S-structured flowgraph, 239

stability number, 3
stable set of vertices, 3
standard sigmoid function, 252
star basis, 95
star cell, 95
star height, 233
star partition, 95
start state, 231
start symbol, 228
state, 231
states model, 187
statistics, 208
Steinitz' Theorem, 157
strength of a multigraph, 31
string graph, 148
Strong Perfect Graph Conjecture, 198
strongly connected digraph, 4
strongly regular graph, 46
structure, 71
subdividing an edge, 5
subflowgraph, 237
subgraph, 5
Superposition Theorem, 31
supervised learning, 253
switching, 78
synchronous, with limited parallelism, 249
synthesized attribute, 229

Tait colouring, 179
Tait conjectures, 188
Tait's flyping conjecture, 190
tame knot, 178
temperature, 249
term, 71
terminal symbol, 228
termination state, 231
tetrad, 122
theoretical exchange rate, 266
thickness of a graph, 158
3-connected matroid, 107
3-transposition, 137
tolerance graph, 147
total chromatic number, 196
Total Colouring Conjecture, 196
tournament, 7, 41, 44
traceable graph, 9
trail, 4
training sample, 254
transitive closure, 54
transitive group, 3, 52, 144
transitive group action, 134
transitive orientation, 144
travelling salesman problem, 251

treatment, 209
tree, 8, 28
trefoil knot, 177
triad, 109
triangle, 4
triangle problem, 203
triangular graph, 223
triangulation, 10
trivial flowgraph, 237
t-separable flowgraph, 240
t-tuple transitive, 136
t-tuple transitive graph, 136
Turán's brick factory problem, 161
Turán's Theorem, 143
Tutte polynomial, 186
Tutte's Wheels and Whirls Theorem, 108
Tutte's Wheels Theorem, 108
2-cell embedding, 165
2-closed permutation group, 130
2-connected matroid, 106
2-sum, 107
two-way infinite path, 4
type $[n, k]$ of a code, 116
type $[n, k, d]$ of a code, 117

uncountably categorical theory, 80
underlying graph, 3
uniform matroid, 103
union of graphs, 6
union of languages, 228
unit interval graph, 146
universal class, 131
unknot, 177
unknotting problem, 179
unlabelled flowgraph, 237

unlabelled graph, 14
upward drawing, 54
Upward Löwenheim–Skolem Theorem, 72

Vapnik–Chervonenkis dimension, 256
variable, 71
variety, 45, 209
variety concurrence graph, 215
VC-dimension, 256
vector matroid, 103
vertex set, 1
vertex splitting, 240
vertex-colouring, 195
vertex-deleted subgraph, 5
vertex-disjoint path, 4
vertex-homogeneous tournament, 44
vertex-primitive graph, 135
vertex-transitive graph, 134
vertical connectedness, 107
Vizing's Theorem, 195
(v, k, λ) t-design, 45
voltage graph, 167

walk of length r, 4
walk-regular graph, 94
wheel, 7
Whitney connectedness, 107
width, 55
word, 228
wreath product, 132
writhe, 185

yield, 209

zeta function, 37